Understanding Homeland Security

Third Edition

This book is dedicated to Olumide, Sarah, and Ola Jane.

The invariable mark of wisdom is to see the miraculous in the common.

—Ralph Waldo Emerson

Sara Miller McCune founded SAGE Publishing in 1965 to support the dissemination of usable knowledge and educate a global community. SAGE publishes more than 1000 journals and over 600 new books each year, spanning a wide range of subject areas. Our growing selection of library products includes archives, data, case studies and video. SAGE remains majority owned by our founder and after her lifetime will become owned by a charitable trust that secures the company's continued independence.

Los Angeles | London | New Delhi | Singapore | Washington DC | Melbourne

Understanding Homeland Security

Third Edition

Gus Martin

California State University, Dominguez Hills

Los Angeles | London | New Delhi
Singapore | Washington DC | Melbourne

FOR INFORMATION:

SAGE Publications, Inc.
2455 Teller Road
Thousand Oaks, California 91320
E-mail: order@sagepub.com

SAGE Publications Ltd.
1 Oliver's Yard
55 City Road
London EC1Y 1SP
United Kingdom

SAGE Publications India Pvt. Ltd.
B 1/I 1 Mohan Cooperative Industrial Area
Mathura Road, New Delhi 110 044
India

SAGE Publications Asia-Pacific Pte. Ltd.
18 Cross Street #10-10/11/12
China Square Central
Singapore 048423

Acquisitions Editor: Jessica Miller
Editorial Assistant: Sarah Manheim
Content Development Editor: Adeline Grout
Production Editor: Andrew Olson
Copy Editor: Diane Wainwright
Typesetter: C&M Digitals (P) Ltd.
Proofreader: Scott Oney
Indexer: Kathy Paparchontis
Cover Designer: Dally Verghese
Marketing Manager: Jillian Ragusa

Printed in the United States of America

Library of Congress Cataloging-in-Publication Data

Names: Martin, Gus, author.

Title: Understanding homeland security / Gus Martin, California State University, Dominguez Hills.

Description: Third edition. | Los Angeles : SAGE, [2019] | Includes bibliographical references and index.

Identifiers: LCCN 2019007057 | ISBN 9781544355801 (pbk. : alk. paper)

Subjects: LCSH: United States. Department of Homeland Security. | Terrorism—United States—Textbooks. | Civil defense—United States—Textbooks. | National security—United States—Textbooks. | Emergency management—United States—Textbooks.

Classification: LCC HV6432 .M377 2019 | DDC 363.325—dc23
LC record available at https://lccn.loc.gov/2019007057

This book is printed on acid-free paper.

SUSTAINABLE FORESTRY INITIATIVE

Certified Chain of Custody
Promoting Sustainable Forestry
www.sfiprogram.org
SFI-01268

SFI label applies to text stock

19 20 21 22 23 10 9 8 7 6 5 4 3 2 1

BRIEF CONTENTS

DETAILED CONTENTS

PART III • THE TERRORIST THREAT AND HOMELAND SECURITY

PART IV • PREPAREDNESS AND RESILIENCE

ACKNOWLEDGMENTS

I am indebted to many people for their support and encouragement in bringing this venture to completion and would like to express special appreciation for the very professional and expert attention given to this project by the editorial group at SAGE. Without their patient professionalism and constructive criticism, this project would not have incorporated the comprehensiveness and completeness that was its objective from the beginning.

I extend thanks to colleagues who shared their expert advice and suggestions for crafting this volume. Deep appreciation is also given to the panel of peer reviewers assembled by the very able editors and staff of SAGE Publications during several rounds of review. The insightful, constructive comments and critical analysis of the following reviewers were truly invaluable.

Ryan Baggett, Eastern Kentucky University

Thomas J. Carey, Monmouth University

Ron L. Lendvay, University of North Florida

Finally, I thank my wife, children, and family for their constant support, encouragement, and humor during the course of this project.

ABOUT THE AUTHOR

C. Augustus "Gus" Martin is a Professor of Criminal Justice Administration at California State University, Dominguez Hills, where he regularly teaches a course on the subject of terrorism and extremism. He has also served as Associate Vice President for Human Resources Management, Acting Associate Dean of the College of Business Administration and Public Policy, Associate Vice President for Faculty Affairs, and Chair of the Department of Public Administration. He began his academic career as a member of the faculty of the Graduate School of Public and International Affairs, University of Pittsburgh, where he was an Administration of Justice professor. His current research and professional interests are terrorism and extremism, homeland security, the administration of justice, and juvenile justice.

Dr. Martin is the author of several books on the subjects of terrorism and homeland security, including *Terrorism: An International Perspective* (with Fynnwin Prager; SAGE, 2019); *Essentials of Terrorism: Concepts and Controversies* (SAGE, 2019); *Understanding Terrorism: Challenges, Perspectives, and Issues* (SAGE, 2018); *The SAGE Encyclopedia of Terrorism*, Second Edition (SAGE, 2011); *Terrorism and Homeland Security* (SAGE, 2011); and *The New Era of Terrorism: Selected Readings* (SAGE, 2004). He is also the author of *Juvenile Justice: Process and Systems* (SAGE, 2005).

Prior to joining academia, Dr. Martin served as Managing Attorney for the Fair Housing Partnership of Greater Pittsburgh, where he was also director of a program created under a federal consent decree to desegregate public and assisted housing. He was also Special Counsel to the Attorney General of the U.S. Virgin Islands on the island of St. Thomas. As Special Counsel, he occupied a personal and confidential position in the central office of the Department of Justice; sat as hearing officer for disciplinary hearings and departmental grievances; served as chair of the Drug Policy Committee; served as liaison to the intergovernmental Law Enforcement Coordinating Committee as well as to the Narcotics Strike Force; and provided daily legal and policy advice to the Attorney General. Prior to serving as Special Counsel, he was a "floor" Legislative Assistant to Congressman Charles B. Rangel of New York. As Legislative Assistant, he researched, evaluated, and drafted legislation in areas of foreign policy, foreign aid, human rights, housing, education, social services, and poverty; he also drafted House floor statements, *Congressional Record* inserts, press releases, and news articles; and he composed speeches, briefing materials, and legislative correspondence.

INTRODUCTION AND RATIONALE

Welcome to *Understanding Homeland Security, Third Edition*, a comprehensive textbook for students and professionals who wish to explore the phenomenon of modern homeland security. Readers who fully engage themselves in the recommended course of instruction offered in the pages that follow will acquire a solid foundation for understanding the nature of issues addressed by the homeland security enterprise. Readers will also discover that their facility for critically assessing homeland security issues in general—and plausible incidents in particular—will be greatly improved.

At the outset, it is important to understand that the study of homeland security is, first and foremost, an investigation into how to secure society from the threat of violent extremism and other potential disasters. Courses that investigate homeland security must, therefore, review the policies, procedures, and administrative networks that anticipate and respond to plausible threats of political violence. None of these considerations can be discussed in isolation from the others if one wishes to develop facility in critically evaluating the nature of homeland security. Thus, the study of homeland security is also one of the most dynamic subjects in the social sciences.

This book is designed to be a primary resource for university students and professionals who require fundamental expertise in understanding homeland security. The content of *Understanding Homeland Security, Third Edition* is directed to academic and professional courses of instruction whose subject areas include homeland security, terrorism, criminal justice administration, political conflict, armed conflict, and social environments. It can be incorporated into classes and seminars covering security studies, the administration of justice, conflict resolution, political theory, and other instruction in the social sciences. It is intended for undergraduate and master's-level university students as well as professionals who require instruction in understanding terrorism.

No prerequisites are specifically recommended, but grounding within one of the following disciplines would be helpful: political science, government, administration of justice, or public administration.

COURSE OVERVIEW AND PEDAGOGY

Understanding Homeland Security, Third Edition introduces readers to homeland security in the modern era, focusing on the post–September 11, 2001, period as its primary emphasis. It is a review of theories, agency missions, laws, and regulations governing the homeland security enterprise. It is also a review of the many threat scenarios and countermeasures that exist in the post–September 11 era. Very importantly, a serious exploration will be made of the underlying reasons for constructing an extensive homeland security system—for example, threats of extremist violence, potential nonterrorist hazards, and historical episodes of challenges to homeland security.

The pedagogical approach of *Understanding Homeland Security, Third Edition* is designed to stimulate critical thinking. Students, professionals, and instructors will find that each chapter follows a sequence of instruction that builds on previous chapters and, thus, incrementally enhances the reader's knowledge of each topic. Chapters incorporate the following features:

Chapter Learning Objectives. Using Bloom's taxonomy, chapter objectives are summarized at the beginning of each discussion.

Opening Viewpoints. At the beginning of each chapter, Opening Viewpoints present relevant examples of theories and themes discussed in each chapter and serve as "reality checks" for readers.

Chapter Perspectives. Chapters incorporate focused presentations of perspectives that explore people, events, organizations, and movements relevant to the subject matter of each chapter.

Global Perspectives. Selected chapters incorporate presentations of international perspectives that explore global people, events, organizations, and movements relevant to the subject matter of each chapter.

Discussion Boxes. Discussion Boxes present provocative information and pose challenging questions to stimulate critical thinking and further debate.

Chapter Summary. A concluding discussion recapitulates the main themes of each chapter.

Key Terms and Concepts. Important terms and ideas introduced in each chapter are listed for review and discussion. These Key Terms and Concepts are further explored and defined in the Glossary.

Recommended Websites and Web Exercises. Web exercises at the ends of chapters have been designed for students, professionals, and instructors to explore and discuss information found on the Internet.

Recommended Readings. Suggested readings are listed at the end of each chapter for further information or research on each topic.

STUDENT STUDY SITE

SAGE Publications has dedicated an online Student Study Site to *Understanding Homeland Security, Third Edition*. The companion website enables readers to better master the course and book material and provides instructors with additional resources for enriching the quality of their course of instruction. The URL for the Student Study Site is edge.sagepub.com/martinhs3e.

CHAPTER GUIDE

This volume is organized into five thematic units, each consisting of pertinent chapters. A Glossary is included after the substantive chapters.

Part I. Foundations of Homeland Security

Part I comprises chapters that provide historical and definitional background; discuss all-hazards issues, the legal foundations of homeland security, and civil liberties debates; and supply an organizational overview of the system.

Chapter 1. History and Policy: Defining Homeland Security

Chapter 1 presents an introduction to the concept of homeland security. This chapter begins with a review of the historical context of homeland security. This historical perspective serves as the prelude to a conceptual analysis of homeland security in the modern era. The discussion concludes with a review of policy options for promoting domestic security.

Chapter 2. Homeland Security and the All-Hazards Umbrella

The discussion in Chapter 2 investigates the broad conceptualization of homeland security known as the all-hazards umbrella—this conceptualization encompasses both terrorist hazards and nonterrorist hazards. The terrorism nexus is discussed within the context of conventional and unconventional weapons and hazards. The all-hazards nexus is discussed within the context of nonterrorist hazards, such as natural disasters and technological scenarios.

Chapter 3. The Legal Foundations of Homeland Security

In Chapter 3, readers become familiar with central legal concepts underlying the homeland security enterprise. International and historical perspectives and events are explored, as are pertinent laws passed in the pre–9/11 era as well as legislation passed after 9/11. The scope of the USA PATRIOT Act of 2001 is outlined and discussed, including discussion of post–9/11 legislation such as the USA FREEDOM Act of 2015.

Chapter 4. Civil Liberties and Securing the Homeland

Chapter 4 investigates the implications of implementing the homeland security system on civil liberties. The careful balance between achieving security and preserving civil liberties is evaluated. A historical context of challenges to civil liberties is presented to provide an instructive perspective on the modern era.

Part II. Homeland Security Agencies and Missions

Part II discusses the homeland security organizational enterprise and its mission.

Chapter 5. Agencies and Missions: Homeland Security at the Federal Level

Chapter 5 discusses and evaluates the federal level of the homeland security enterprise. The scope of the federal homeland security bureaucracy is discussed, as is the role of the Department of Homeland Security. The discussion includes assessments of the roles of other sector-specific federal agencies. It also explores the mission of the military in supporting the homeland security enterprise.

Chapter 6. Prediction and Prevention: The Role of Intelligence

Chapter 6 discusses and evaluates the mission of the U.S. intelligence community and its presence as a member of the homeland security enterprise. This chapter investigates the configuration and central role of the intelligence community in securing the homeland.

Chapter 7. Agencies and Missions: Homeland Security at the State and Local Levels

Chapter 7 discusses and evaluates the state and local levels of the homeland security enterprise. The purpose of this presentation is to investigate administrative systems and resources available at local levels of governance, since it is from these levels that first responders are deployed when an incident occurs. State systems, local initiatives, and the roles of law enforcement agencies are discussed.

Part III. The Terrorist Threat and Homeland Security

Part III probes terrorist threat environments.

Chapter 8. Sea Change: The New Terrorism and Homeland Security

The nature of terrorism in the modern era is investigated in Chapter 8. This chapter compares and contrasts the "Old Terrorism" and the New Terrorism, explores the role of religion in modern terrorism, and examines new modes of terrorism and warfare. Asymmetrical warfare, netwar, and the destructive use of technologies are discussed. This chapter also discusses policy options for countering extremism and terrorism.

Chapter 9. The Threat at Home: Terrorism in the United States

Chapter 9 presents an overview of terrorism in postwar America. It probes the background of political violence from the left and right and presents a detailed discussion of leftist and rightist terrorism in the United States. The chapter also evaluates international terrorism and prospects for violence emanating from modern religious extremists on the left and right. The phenomenon of lone-wolf terrorism in the United States is explored.

Part IV. Preparedness and Resilience

Part IV discusses resilience, prevention, protection of security nodes, planning, and the role of responders at every level.

Chapter 10. Porous Nodes: Specific Vulnerabilities

Chapter 10 explores sensitive sectors of the homeland security enterprise. The purpose of this chapter is to examine the vulnerability of critical security nodes that may plausibly be targeted by violent extremists. It begins with a discussion of challenges to cybersecurity and continues with examinations of issues related to aviation, border, and port security.

Chapter 11. Always Vigilant: Hardening the Target

Chapter 11 investigates target hardening within the context of several vulnerable sectors. Information security is discussed within the contexts of cyberwar as a counterterrorist option and the use of surveillance technologies. Protecting critical infrastructure, border control, and transportation security are also discussed.

Chapter 12. Critical Resources: Resilience and Planning

Chapter 12 investigates the roles of resilience and proper planning, including the importance of prevention and mitigation planning. Within this context, responses to terrorist deployment of chemical, biological, radiological, and nuclear hazards are examined.

Chapter 13. Critical Outcomes: Response and Recovery

Chapter 13 investigates response and recovery mechanisms, focusing on administrative coordination and planning. Within this framework, the discussion delivers an overview of federal, state, and local response and recovery coordination and planning. The challenge of reactive planning is also presented.

Part V. Homeland Security: An Evolving Concept

Part V discusses the future of homeland security.

Chapter 14. The Future of Homeland Security

In Chapter 14, readers are challenged to critically assess trends and other factors that can be used to project near-future issues involving the homeland security enterprise. In particular, this chapter presents fresh discussions and data. Likely scenarios for homeland security challenges and threat environments of the near future are offered.

NEW TO THIS EDITION

- Reflecting the importance of intelligence within homeland security, a new chapter has been added (Chapter 6. Prediction and Prevention: The Role of Intelligence) that examines the mission of the U.S. intelligence community and its presence as a member of the homeland security enterprise.

- The content of several chapters has been reorganized to better reflect the changing homeland security environment.
  - Coverage of cybersecurity is now included in Chapter 11 on target hardening.
  - Coverage of border control is now discussed in Chapter 10 on specific vulnerabilities.
  - Discussion of CBRN threats has been moved to Chapter 12 on resilience planning.

- Critical topics have been added or expanded in the new edition, including
  - the role of FEMA and preparedness planning;
  - the different types of bombs that can be used in terrorist attacks;
  - the role of civil liberty and countering extremism through reform;
  - the responsibilities of the National Guard in responding to emergencies and restoring civil order;
  - asymmetrical warfare and the contagion effect, particularly in the case of motorized vehicle attacks;
  - resilience and the need for rapid recovery from emergencies;
  - the whole-community approach to local planning and preparedness;
  - the militarization of the police; and
  - electronic surveillance by government agencies.

Recent events, terrorist attacks, and cyberattacks have been included, such as the pipe bomb clusters in Manhattan and New Jersey, the hackings during the 2016 and 2018 elections, the mailing of pipe bombs to prominent politicians and other public individuals in 2018, recent school shootings, and the shooting at the Tree of Life synagogue in Pittsburgh, Pennsylvania.

Hijacked United Airlines Flight 175 from Boston crashes into the south tower of the World Trade Center and explodes at 9:03 a.m. on September 11, 2001, in New York City.

Spencer Platt/Getty Images

FOUNDATIONS OF HOMELAND SECURITY

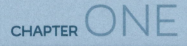
HISTORY AND POLICY
Defining Homeland Security

Chapter Learning Objectives

This chapter will enable readers to do the following:

1. Apply a working definition of homeland security

2. Describe historical perspectives on homeland security in the United States

3. Explain the modern concept of homeland security and its dynamic qualities

4. Analyze policy options and response categories for threats to the homeland

Opening Viewpoint: The Concept of Homeland Security

Events on the morning of September 11, 2001, profoundly impacted how the people of the United States perceived the quality of violence posed by modern terrorism. The United States had certainly experienced domestic terrorism for much of its history but never on the scale of the 9/11 attacks and never with the underlying understanding that Americans themselves were primary targets. In previous generations and recent history, terrorist attacks were primarily the work of domestic extremists, and cross-border violence was perceived as an exception that occurred mostly beyond the borders of the American homeland. For this reason, domestic security initiatives prior to the era of homeland security were conceptually centered on suppressing domestic dissidence rather than responding to threats from abroad.

After the September 11 attacks, a profound and fundamental policy shift occurred in the American approach to domestic security. A new concept, *homeland security*, was adopted to coordinate preparedness and response initiatives at all levels of society. The new homeland security enterprise marshaled the resources of federal, state, local, and private institutions. The intention was to create an ongoing and proactively dynamic nationwide culture of vigilance. This new concept supplanted previously reactive and largely decentralized approaches to extremist violence.

In the current domestic security environment, the new homeland security enterprise is conceptually dynamic in the sense that it evolves and adapts with changing security threats and terrorist environments.

Unlike previous security environments, modern homeland security policies must necessarily be configured to link domestic policies to emerging international events; this is a dynamic and ongoing policymaking process. Depending on national and political necessities, its purview has also been expanded to include hazards other than extremist violence. At the same time, core initiatives and goals drive homeland security so that it has become an integral component of security preparedness and response efforts at all levels of government and society. Thus, the post–9/11 era has become a period of history wherein the concept of homeland security is common to the domestic security culture of the United States.

Homeland security is a relatively new concept that, however defined, exists to safeguard the domestic security of the United States and broadly promote the stability of society when man-made and natural disasters occur. Although originally configured to describe national responses to domestic terrorist incidents in the aftermath of the September 11, 2001, terrorist attacks, homeland security was conceptually expanded after Hurricane Katrina in 2005 to include preparedness and recovery from natural and hazard-related incidents. Nevertheless, it is the domestic security mission of the homeland security enterprise that continues to be its fundamental and underlying tenet in the modern era. An extraordinarily large amount of resources—human and financial—are devoted to strengthening domestic security and coordinating this effort at all levels of government.

In the modern era, the threat of terrorism and other challenges to domestic security have significantly affected the missions of government agencies, nationally and locally. Every level of each domestic security organization, law enforcement agency, and emergency response institution incorporates homeland security contingency planning and training. Homeland security has become endemic to the modern domestic security environment and is arguably the domestic counterpart to international counterterrorist initiatives undertaken by national security and national defense institutions. However, although the concept of homeland security has created a fresh and pervasive domestic security environment in the modern era, similar security environments have existed periodically in the history of the United States. This historical perspective is often misunderstood and commonly forgotten in the current security environment.

This chapter investigates definitional issues in the study of homeland security. Here you will probe the historical and cultural nuances of these issues and develop a critical understanding of why defense of the homeland became a central policy initiative in the United States. Historically, perceived threats to domestic security have resulted in the designation of sometimes controversial security environments. For example, periodic anticommunist Red Scares occurred during the twentieth century in which authoritarian procedures were adopted to preempt perceived threats of sedition. (Full consideration of the Red Scares is provided in Chapter 4.) Within this context, it must be remembered that the development of modern homeland security theory evolved within a practical and real-life framework—in other words, a nontheoretical reality in which actual and verifiable threats to domestic security do exist. Such threats emanate from both foreign and domestic sources. General categories of policy options in response to domestic threats are presented in this chapter to facilitate your understanding of definitional perspectives. These policy options represent examples of the domestic application of homeland security intervention.

The discussion in this chapter will review the following topics:

- The past as prologue: The historical context of homeland security

- Defining an era: What is homeland security?

- Domestic security and threats to the homeland: Policy options

THE PAST AS PROLOGUE: THE HISTORICAL CONTEXT OF HOMELAND SECURITY

In the aftermath of the September 11, 2001, terrorist attacks on the United States, the federal government exercised swift leadership in significantly altering the domestic security culture. It did this by aligning national response mechanisms with the newly emergent threat environment. The post–9/11 threat environment proved to be dynamic in the sense that it posed new challenges for the homeland security enterprise over time—for example, the emergent prominence of the Islamic State of Iraq and the Levant, also known as Islamic State of Iraq and al-Sham (ISIS), in 2014. For this reason, national response mechanisms were likewise required to be nimble in designing responsive policies.

It is important to understand that this modern alignment was not the first time the United States adapted its domestic security culture to perceived or actual threat environments. There are many historical examples that predate the post–9/11 era, and these examples provide historical context to the study of the modern concept of homeland security. Table 1.1 summarizes several historical homeland security environments.

Table 1.1   The Past as Prologue: The Historical Context of Homeland Security

The modern homeland security environment grew from the need to design a systematic approach toward responding to threats to domestic security. Several historical periods predated the modern environment. The following table summarizes these historical periods, plausible threats, and defining events.

Historical Period	Activity Profile	
	Plausible Threats	Defining Events
Early Republic (External Threats)	Frontier conflicts Border security	Native American warfare War of 1812 1916 Mexican Expedition
Early Republic (Domestic Threats)	Early disturbances Regional conflict Labor and ideological conflict Racial terrorism	Whiskey Rebellion Civil War and Reconstruction Haymarket Riot; Homestead Strike; Anarchist terrorism KKK terrorism
Modern Era (Post–WWII)	Cold War Domestic discord International religious terrorism	Civil defense Civil Rights movement; 1960s protests Mass-casualty attacks

From its inception, the United States responded to foreign and domestic crises and threats during periods when the concept of homeland security did not exist in its modern context. Responses to emergencies and threats differed markedly depending on the security environment characterizing each period. Nevertheless, the perceived threats were deemed, at the time, to be significant enough to warrant intensive policy intervention.

External Threats to the Early Republic

During the colonial and early republic periods, most security threats emanated from frontier conflicts between Native Americans and settlers, and the burden of responding to such emergencies initially fell to local and state militias. Border security became paramount in the aftermath of British incursions during the War of 1812, resulting in federal coordination of the construction and garrisoning of forts and coastal defenses. Border defense, frontier expansion, and occasional military campaigns (such as the Mexican Expedition of 1916) were typical security priorities. Nevertheless, until the Second World War, the national budget for centralized security spending in the United States traditionally remained low, except in times of war.

Whiskey Rebellion: One of the early post-independence disturbances was popularly known as the Whiskey Rebellion, an anti-tax uprising in western Pennsylvania (1791–1794).

Domestic Threats to the Early Republic

Aside from early post-independence disturbances, such as the anti-tax **Whiskey Rebellion** in western Pennsylvania (1791–1794), security threats originating from domestic disputes were rare and short-lived. The Civil War and postwar Reconstruction in the American South were, of course, exceptions to this pattern. Federal policies during the Civil War and Reconstruction included what would be labeled civil liberties abrogations in the modern era as well as the use of national institutions (such as the army and federal marshals) to maintain order in the occupied South. As we will discuss in Chapter 4, restrictions on liberty have historically been enacted to address what were, at the time, deemed serious threats to the national security of the United States.

Haymarket Riot of 1886: During Chicago's Haymarket Riot of 1886, an anarchist threw a dynamite bomb at police officers, who then opened fire on protesters. Scores of officers and civilians were wounded.

It was not until the end of the nineteenth century that labor-related and ideological discord garnered national attention. American workers began to organize labor unions during the Civil War era, and thousands of workers were union members by the 1880s. In May 1886, large demonstrations inspired by a strike against the McCormick Harvesting Machine Company occurred in Chicago. On May 1, 1886, a large May Day parade was held at the McCormick plant, and two days later a worker was killed during a demonstration at the plant. On May 4, a large rally at Haymarket Square in Chicago precipitated

the **Haymarket Riot of 1886**, when an anarchist threw a dynamite bomb at police officers who were attempting to disperse the crowd. The police then opened fire on protesters. Seven police officers and three civilians were killed, and scores were wounded. During the **Homestead Steel Strike of 1892** on the Monongahela River near Pittsburgh, a strike by steelworkers resulted in a pitched gun battle between striking workers and hundreds of Pinkerton agents (in which the strikers prevailed). The strike was eventually suppressed

▶ **Photo 1.1** The Haymarket bombing and riot on May 4, 1886. Chicago police fired into the crowd after an anarchist threw a dynamite bomb that killed several officers.

following intervention by the Pennsylvania state militia. Both incidents are examples of serious labor-related discontent. In addition, ideological extremists, such as violent anarchists and communists, were responsible for events such as the 1901 assassination of President William McKinley (by an anarchist), the Wall Street bombing of 1920 (which killed and wounded more than 170 people and was never solved), and numerous other bombings and attempted assassinations. Federal soldiers and state militias were deployed on hundreds of occasions during this period. Racial terrorism, often committed by the Ku Klux Klan, also contributed to the perceived need for nationwide responses to extremist violence. In this environment, laws were passed to suppress activism and extremism. These included the Espionage Act of 1917, the Immigration Act of 1918, and the Sedition Act of 1918. During this period, known as the first Red Scare, federal and state government agents were deployed to disrupt perceived subversive groups and detain suspected extremists.

Modern Precursors to Homeland Security

After the Second World War, the international community entered a prolonged period of competition and conflict between the United States and the Soviet Union and their allies. Known as the Cold War, the period from the late 1940s to the late 1980s was a time of threatened nuclear warfare, actual and extensive warfare in the developing world, and domestic security tension in the United States. The threat of nuclear war spawned an extensive network of civil defense programs in the United States, extending from the national level to the local level. Virtually every community engaged in civil defense drills and contingency planning. Federal civil defense initiatives were subsumed under and coordinated by a succession of agencies. These included the Civil Defense Administration, the Office of Defense Mobilization, the Office of Civil and Defense Mobilization, and the Office of Civil Defense. In 1979, the Federal Emergency Management Agency (FEMA) was established for the overall coordination of disaster relief.

DEA/A. DAGLI ORTI/De Agostini/Getty Images

▶ **Photo 1.2** U.S. president William McKinley is shot on September 6, 1901, by anarchist Leon Czolgosz, who hid his gun in a handkerchief and fired as the president approached to shake his hand. McKinley died eight days later.

During the Cold War, domestic disturbances in the United States led to the initiation of federal, state, and local efforts to monitor activist activity and quell disorder. These disturbances included civil rights marches in the American South, urban riots during the 1960s, student activism on college campuses, rioting at the 1968 Democratic National Convention, and terrorist attacks by ideological and nationalist extremists. Disorders gradually receded with the passage of civil rights laws, the end of the Vietnam War, and the end of the Cold War, brought about by the dissolution of the Soviet Union.

Following the Cold War, significant new threats to domestic security arose from extremists who had no compunction against launching mass-casualty attacks against civilian "soft targets." The 1993 World Trade Center and 1995 Oklahoma City bombings were deliberate attempts to maximize civilian casualties and damage to the intended targets. The September 11, 2001, attacks on the World Trade Center and the Pentagon were the final incident prior to the modern era of homeland security.

DEFINING AN ERA: WHAT IS HOMELAND SECURITY?

The catastrophic terrorist attack on September 11, 2001, was a defining moment for the United States. With nearly 3,000 fatalities, the nation found itself at war against an enemy who was clearly adept at converting modern technology into weapons of mass destruction. Thus, the dawn of the twenty-first century witnessed the birth of the modern era of homeland security. Pervasive domestic security systems became a new norm for the United States, and internationally, the nation embarked on its longest war. Ironically, the death of al-Qaeda leader Osama bin Laden in May 2011 occurred on the eve of the 10th commemoration of the September 11 attack on the U.S. homeland. Significantly, homeland security continued to serve as an essential institution for maintaining vigilance against terrorist threats, as evidenced by emergency response and domestic security procedures following the April 2013 Boston Marathon bombing. Chapter Perspective 1.1 discusses the successful hunt for Osama bin Laden and events leading to his death.

CHAPTER PERSPECTIVE 1.1

The Death of Osama bin Laden

Al-Qaeda founder Osama bin Laden was killed during a raid by U.S. naval special forces on May 2, 2011, in Abbottabad, Pakistan. The successful attack by a unit popularly known as SEAL Team Six ended an intensive manhunt for the most wanted terrorist leader in the world.

The successful hunt for Osama bin Laden originated from fragments of information gleaned during interrogations of prisoners over several years, beginning in 2002. Believing that bin Laden retained couriers to communicate with other operatives, interrogators focused their attention on questioning high-value targets about the existence and identities of these couriers. This focus was adopted with an assumption that bin Laden and other al-Qaeda leaders would rarely communicate using cell phone technology as a precaution against being intercepted by Western intelligence agencies.

Early interrogations produced reports that a personal courier did indeed exist, a man whose given code name was Abu Ahmed al-Kuwaiti. In about 2007, intelligence officers learned al-Kuwaiti's real name, located

him, and eventually followed him to a recently built compound in Abbottabad, Pakistan. U.S. intelligence operatives observed the compound locally from a safe house and concluded that it concealed an important individual. Based on other surveillance and circumstantial intelligence information, officials surmised that Osama bin Laden resided at the compound with his couriers and their families.

Options for assaulting the compound included a surgical strike by special forces, deploying strategic bombers to obliterate the compound, or a joint operation with Pakistani security forces. The latter two options were rejected because of the possibility of killing innocent civilians and distrust of Pakistani security agencies. Approximately two dozen SEAL commandos practiced intensely for the assault and were temporarily detailed to the CIA for the mission. A nighttime helicopter-borne attack was commenced on May 2, 2011. The courier al-Kuwaiti and several others were killed during the assault, and women and children found in the compound were bound and escorted into the open to be found later by Pakistani security forces. Osama bin

(Continued)

Laden was located on an upper floor of the main building and shot dead by SEALs. Four others were killed in addition to bin Laden, whose body was taken away by the assault team. He was subsequently buried at sea.

Al-Qaeda threatened retribution for the attack and named Ayman al-Zawahiri as bin Laden's successor in June 2011. Subsequent to bin Laden's death, al-Qaeda's leadership brand faced competition from a new Islamist movement calling itself the Islamic State of Iraq and the Levant (ISIS).

Discussion Questions

1. What effect did the successful hunt for Osama bin Laden have on domestic homeland security?

2. Which options are most desirable when conducting global manhunts for terrorist suspects?

3. How can homeland security agencies and assets best be coordinated internationally?

Department of Homeland Security. (Publishers thank the Department of Homeland Security for its cooperation and assistance. The Department of Homeland Security's cooperation and assistance does not reflect an endorsement of the contents of the textbook.)

▶ **Photo 1.3** Official seal of the U.S. Department of Homeland Security.

The term *homeland security* was, at first, considered to be a rather vague and imprecise descriptor. It nevertheless became a conceptually integral element in designing policies to protect the United States from violent extremists. This section will discuss this concept by exploring homeland security within the following contexts:

- The modern era of homeland security

- Conceptual foundation: Central attributes of homeland security

- The homeland security environment: A dynamic construct

- A new focus: The *Quadrennial Homeland Security Review Report*

The Modern Era of Homeland Security

The modern era of homeland security began with the rapid implementation of a series of policy initiatives in the immediate aftermath of the September 11, 2001, terrorist attacks. These initiatives heralded the establishment of a new security culture in the United States, one that significantly affected the work of government and the everyday lives of residents. The new homeland security environment unfolded very quickly in the following sequence:

- On September 20, 2001, President George W. Bush announced that a new Office of Homeland Security would be created as a unit in the White House.

- On September 24, 2001, President Bush stated that he would propose the passage of new homeland security legislation titled the "Uniting and Strengthening America by Providing Appropriate Tools Required to Intercept and Obstruct Terrorism Act," popularly known as the **USA PATRIOT Act of 2001**.

- On October 8, 2001, President Bush issued **Executive Order 13228**. This executive order was titled "Establishing the Office of Homeland Security and the Homeland Security Council" and stated that "the functions of the Office [of Homeland Security] shall be to coordinate the executive branch's efforts to detect, prepare for, prevent, protect against, respond to, and recover from terrorist attacks within the

USA PATRIOT Act of 2001: On October 26, 2001, President George W. Bush signed the "Uniting and Strengthening America by Providing Appropriate Tools Required to Intercept and Obstruct Terrorism Act of 2001," commonly known as the USA PATRIOT Act, into law. It was an omnibus law whose stated purpose was, in part, to "deter and punish terrorist acts in the United States and around the world" by expanding the investigative and surveillance authority of law enforcement agencies.

United States."[1] This statement of purpose by the United States was the first to result from the September 11 crisis and continues to guide the implementation of the concept of homeland security in relation to counterterrorist policies.

- Also on October 8, 2001, Executive Order 13228 established a Homeland Security Council, charging it "to develop and coordinate the implementation of a comprehensive national strategy to secure the United States from terrorist threats or attacks."

- On October 26, 2001, the USA PATRIOT Act of 2001 was signed into law. Its stated purpose was, in part, to "deter and punish terrorist acts in the United States and around the world" by expanding the investigative and surveillance authority of law enforcement agencies.

- On October 29, 2001, the first **Homeland Security Presidential Directive (HSPD)** was issued by President Bush. Chapter Perspective 1.2 summarizes the first reported compilation of HSPDs as released by the Committee on Homeland Security of the U.S. House of Representatives.

- On November 25, 2002, the cabinet-level Department of Homeland Security was established when President Bush signed the Homeland Security Act of 2002 into law.

Executive Order 13228: On October 8, 2001, President Bush issued Executive Order 13228, titled "Establishing the Office of Homeland Security and the Homeland Security Council."

Homeland Security Presidential Directives (HSPDs): On October 29, 2001, President Bush released the first of many HSPDs, which implement policies and procedures constituting the homeland security enterprise.

CHAPTER PERSPECTIVE 1.2

Homeland Security Presidential Directives

In the aftermath of the September 11, 2001, terrorist attack, President George W. Bush issued a series of Homeland Security Presidential Directives (HSPDs). The House of Representatives' Homeland Security Committee published the first compilation of HSPDs in January 2008.[a] The following list summarizes the committee's first compilation. Classified HSPDs are noted as they occurred in the initial compilation, but they have since been declassified.

    HSPD-1. Organization and Operation of the Homeland Security Council

    HSPD-2. Combating Terrorism Through Immigration Policies

    HSPD-3. Homeland Security Advisory System

    HSPD-4. National Strategy to Combat Weapons of Mass Destruction

    HSPD-5. Management of Domestic Incidents

HSPD-6. Integration and Use of Screening Information to Protect Against Terrorism

HSPD-7. Critical Infrastructure Identification, Prioritization, and Protection

HSPD-8. National Preparedness

HSPD-9. Defense of United States Agriculture and Food

HSPD-10. Biodefense for the 21st Century

HSPD-11. Comprehensive Terrorist-Related Screening Procedures

HSPD-12. Policy for a Common Identification Standard for Federal Employees and Contractors

HSPD-13. Maritime Security Policy

HSPD-14. Domestic Nuclear Detection

HSPD-15. (Classified—Not Available)

HSPD-16. National Strategy for Aviation Security

(Continued)

(Continued)

HSPD-17. (Classified—Not Available)

HSPD-18. Medical Countermeasures Against Weapons of Mass Destruction

HSPD-19. Combating Terrorism Use of Explosives in the United States

HSPD-20. National Continuity Policy

HSPD-21. Public Health and Medical Preparedness

Discussion Questions

1. Are HSPDs a valuable tool in framing homeland security policy?

2. How practical are HSPDs for implementing specific strategies?

3. Are alternative sources of leadership, other than the executive branch, viable centers for framing homeland security policy?

Note

a. Committee on Homeland Security of the House of Representatives, *Compilation of Homeland Security Presidential Directives (Updated Through December 31, 2007)* (Washington, DC: U.S. Government Printing Office, 2008).

An interesting international corollary is that, in the post–9/11 era, homeland security has been adapted conceptually to the unique domestic environments of a number of Western democracies. In the European context, what is now considered homeland security was historically framed under the concept of security and (recently) *interoperability* among partners in the European Union. This approach reflected Europe's long experience with combating domestic terrorism conducted by ideological and nationalist extremists. Regardless of the preferred phraseology among Western nations, the homeland security concept expanded considerably during the post–9/11 era.

Conceptual Foundation: Central Attributes of Homeland Security

Because homeland security is a dynamic and evolving concept, it is instructive to identify its central attributes, that is, key features that influence modern approaches to applying homeland security initiatives to domestic threats. These central attributes are distinguishing features and concepts that define the current homeland security environment, and they include the following:

Maj/Maj/Time Life Pictures/Getty Images

▶ **Photo 1.4** Osama bin Laden. Bin Laden was killed during a raid by a U.S. naval special forces unit in Abbottabad, Pakistan, on May 2, 2011.

- The terrorist threat
- The federal bureaucracy
- State and local agencies
- Collaboration on conceptual foundations for comprehensive homeland security

The Terrorist Threat

The modern homeland security environment was created as a direct result of the terrorist attack on the American homeland on September 11, 2001. Plausible threat scenarios include strikes by international terrorists, such as Islamists influenced by the al-Qaeda network. Possible scenarios also include attacks by homegrown ideological extremists as well as domestic sympathizers of religious extremism.

Subsequent attempts by violent extremists to launch domestic strikes have necessitated an unending effort to design and apply innovative domestic security policies and initiatives. As a result, verified conspiracies from international and domestic extremists have been detected and thwarted by law enforcement and intelligence agencies. A considerable number of domestic terrorist plots have been neutralized, and successful prosecutions of suspects have resulted in guilty verdicts and incarceration of conspirators.

The Federal Bureaucracy

The cabinet-level Department of Homeland Security (DHS) encompasses a large number of formerly independent agencies and casts an exceptionally wide, mission-focused net. Many DHS agencies have significant arrest and investigative authority, thus creating a massive (and potentially intrusive) regulatory bureaucratic enterprise. Other federal agencies not subsumed under the DHS are also tasked with engaging in domestic security missions.

Some agency roles overlap and are not clearly defined, but the federal bureaucracy is nevertheless responsible for framing general and specific homeland security policies as well as national responses. In effect, the federal bureaucracy provides overall leadership for the nation's homeland security enterprise and disburses resources and assistance to guide state and local authorities.

State and Local Agencies

Similar to the federal bureaucracy, states have created homeland security bureaus and agencies as a matter of necessity. Many state and local initiatives are undertaken using federal financial resources, which were widely disbursed following the September 11 attack. The result has been the permeation of homeland security offices and initiatives at all levels of government.

Homeland security training is a critical necessity, and a significant number of local law enforcement agencies regularly train personnel on how to respond to domestic security events. Non-law enforcement agencies, such as fire departments and emergency medical response agencies, similarly engage in homeland security training.

Collaboration on Conceptual Foundations for Comprehensive Homeland Security

Although achieving agreement on the conceptual foundations of homeland security appears to be a fruitless endeavor, one conceptualization is embodied in *The 2014 Quadrennial Homeland Security Review* (QHSR), discussed further in this chapter. The 2014 QHSR identifies "five basic homeland security missions . . . : Prevent Terrorism and Enhance Security . . . Secure and Manage Our Borders . . . Enforce and Administer Our Immigration Laws . . . Safeguard and Secure Cyberspace . . . Strengthen National Preparedness and Resilience."[2]

These missions are the conceptual foundation for a comprehensive approach to homeland security that includes effective integration of all homeland security operations. Operational integration ideally includes emergency preparedness, managing incident responses, and recovery efforts. However, as a practical matter, it is often difficult to seamlessly integrate these components. This is because the selection and implementation of preferred homeland security operations is part of an evolving and sometimes vigorous policy debate. Nevertheless, planning and responding agencies generally attempt to collaborate on designing response options. There is general consensus that several fundamental response components are necessary and that these essential response operations require administrative integration at all levels of government. Thus, collaboration on the comprehensive conceptual framework presented in the QHSR is a desired goal in theory, if not always in fact.

The Homeland Security Environment: A Dynamic Construct

An important step with respect to defining homeland security is the need to develop an understanding of its relevance to the synonymic concept of *domestic security*. Both embody response options to threat environments existing within the borders of the United States. Within this context, although homeland security can certainly be significantly affected by threats originating from international sources (such as al-Qaeda), the concept of *defending the homeland* inside its borders is at the heart of homeland security. In comparison, the international dimension of waging the war on terrorism extends outside the borders of the United States and resides under the authority of diplomatic missions, intelligence agencies, and the defense establishment. Defining homeland security is largely an exercise in addressing the question of how to protect the nation within its borders from threats domestic and foreign.

In the modern era, homeland security is a dynamic concept that constantly evolves with the emergence of new terrorist threats and political considerations. This evolution is necessary because domestic counterterrorist policies and priorities must adapt to ever-changing political environments and emergent threat scenarios. Factors that influence the conceptualization and implementation of homeland security include changes in political leadership, demands from the public, and the discovery of serious terrorist plots (both successful and thwarted). Keeping this in mind, the following statement by the U.S. Department of Homeland Security exemplifies the conceptual framework for homeland security in the United States (with emphasis added):

> *Protecting the American people from terrorist threats is the reason the Department of Homeland Security was created, and remains our highest priority*. Our vision is a secure and resilient nation that effectively prevents terrorism in ways that preserve our freedom and prosperity. . . . Terrorist tactics continue to evolve, and we must keep pace. Terrorists seek sophisticated means of attack, including chemical, biological, radiological, nuclear and explosive weapons, and cyber attacks. Threats may come from abroad or be homegrown. We must be vigilant against new types of terrorist recruitment as well, by engaging communities at risk [of] being targeted by terrorist recruiters. . . . The Department's efforts to prevent terrorism are centered on a risk-based, layered approach to security in our passenger and cargo transportation systems and at our borders and ports of entry. It includes new technologies to:
>
> - Detect explosives and other weapons
> - Help [protect] critical infrastructure and cyber networks from attack
> - Build information-sharing partnerships
>
> We do this work cooperatively with other federal, state, local, tribal and territorial law enforcement as well as international partners.[3]

Thus, domestic security and protecting the homeland from terrorist threats must be considered core concepts when defining homeland security. These core definitional concepts embody the central mission of the homeland security community at all levels of government.

Although the evolution and expansion of the homeland security umbrella will, from time to time, incorporate additional missions (depending on contemporary political demands), the central focus on protection from violent extremism is an enduring and basic definitional component.

A New Focus: The Quadrennial Homeland Security Review Report

In February 2010, the U.S. Department of Homeland Security published a document intending to consolidate the definition of homeland security by presenting the concept as encompassing a broader and more comprehensive mission than previously envisioned. The document was titled *Quadrennial Homeland Security Review Report: A Strategic Framework for a Secure Homeland* (QHSR), and it was the first of what were projected to be regular quadrennial assessments of homeland security.

The intended purpose of the 2010 QHSR was to "outline the strategic framework to guide the activities of participants in homeland security toward a common end."[4] In this report, Homeland Security secretary Janet Napolitano explained that the core concept for this strategic framework is a new policy-related comprehensiveness, which she termed the **homeland security enterprise.** Napolitano stated,

> The QHSR identifies the importance of what we refer to as the homeland security enterprise—that is, the Federal, State, local, tribal, territorial, nongovernmental, and private-sector entities, as well as individuals, families, and communities who share a common national interest in the safety and security of America and the American population. The Department of Homeland Security (DHS) is one among many components of this national enterprise. In some areas, like securing our borders or managing our immigration system, the Department possesses unique capabilities and, hence, responsibilities. In other areas, such as critical infrastructure protection or emergency management, the Department's role is largely one of leadership and stewardship on behalf of those who have the capabilities to get the job done. In still other areas, such as counterterrorism, defense, and diplomacy, other Federal departments and agencies have critical roles and responsibilities, including the Departments of Justice, Defense, and State, the Federal Bureau of Investigation, and the National Counterterrorism Center. Homeland security will only be optimized when we fully leverage the distributed and decentralized nature of the entire enterprise in the pursuit of our common goals.[5]

The second QHSR assessment, titled *The 2014 Quadrennial Homeland Security Review*, was published in June 2014. The purpose of the 2014 QHSR was summarized as follows:

> More than 12 years after the attacks of September 11, 2001, the United States is poised to begin a new era in homeland security. Long-term changes in the security environment and critical advances in homeland security capabilities require us to rethink the work DHS does with our partners—the work of building a safe, secure, and resilient Nation.[6]

QHSR assessments are deemed necessary because homeland security is an evolutionary concept, and documentary reports such as the QHSR acknowledge the critical need to formally review and assess the homeland security mission.

The foregoing approach broadens the definition of homeland security and clearly reflects the dynamic evolution of the concept in the modern era. As noted previously, the

Quadrennial Homeland Security Review Report: A Strategic Framework for a Secure Homeland (QHSR): A document published by the Department of Homeland Security intending to consolidate the definition of homeland security by presenting the concept as encompassing a broader and more comprehensive mission than previously envisioned. The QHSR is a documentary acknowledgment that homeland security is evolving conceptually.

homeland security enterprise: Homeland Security secretary Janet Napolitano explained that a new comprehensiveness, termed the *homeland security enterprise*, serves as the core concept for the *Quadrennial Homeland Security Review Report's* strategic framework.

2014 QHSR identifies five homeland security missions, each comprising two or more goals. These missions and goals are summarized as follows:

Mission 1: Prevent Terrorism and Enhance Security

- Goal 1.1: Prevent Terrorist Attacks

- Goal 1.2: Prevent and Protect Against the Unauthorized Acquisition or Use of Chemical, Biological, Radiological, and Nuclear Materials and Capabilities

- Goal 1.3: Reduce Risk to the Nation's Critical Infrastructure, Key Leadership, and Events

Mission 2: Securing and Managing Our Borders

- Goal 2.1: Secure U.S. Air, Land, and Sea Borders and Approaches

- Goal 2.2: Safeguard and Expedite Lawful Trade and Travel

- Goal 2.3: Disrupt and Dismantle Transnational Criminal Organizations and Other Illicit Actors

Mission 3: Enforce and Administer Our Immigration Laws

- Goal 3.1: Strengthen and Effectively Administer the Immigration System

- Goal 3.2: Prevent Unlawful Immigration

Mission 4: Safeguard and Secure Cyberspace

- Goal 4.1: Strengthen the Security and Resilience of Critical Infrastructure

- Goal 4.2: Secure the Federal Civilian Government Information Technology Enterprise

- Goal 4.3: Advance Law Enforcement, Incident Response, and Reporting Capabilities

- Goal 4.4: Strengthen the Ecosystem

Mission 5: Strengthen National Preparedness and Resilience

- Goal 5.1: Enhance National Preparedness

- Goal 5.2: Mitigate Hazards and Vulnerabilities

- Goal 5.3: Ensure Effective Emergency Response

- Goal 5.4: Enable Rapid Recovery[7]

The foregoing missions and goals represent an all-hazards approach to the homeland security enterprise; full discussion of the all-hazards umbrella is provided in Chapter 2. Figure 1.1 summarizes the QHSR's representation of the homeland security enterprise.

Figure 1.1 *The 2014 Quadrennial Homeland Security Review: Homeland Security Enterprise Mission*

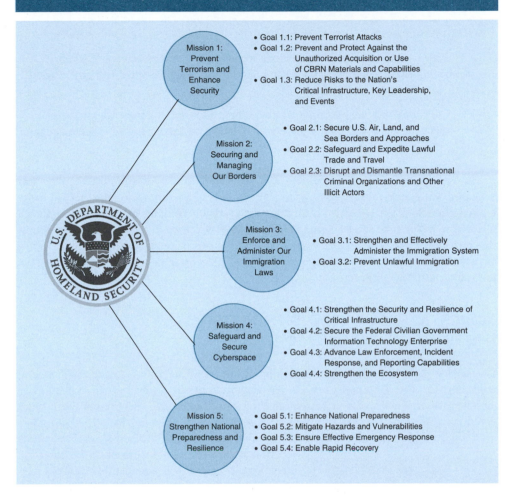

Mission 1: Prevent Terrorism and Enhance Security
- Goal 1.1: Prevent Terrorist Attacks
- Goal 1.2: Prevent and Protect Against the Unauthorized Acquisition or Use of CBRN Materials and Capabilities
- Goal 1.3: Reduce Risks to the Nation's Critical Infrastructure, Key Leadership, and Events

Mission 2: Securing and Managing Our Borders
- Goal 2.1: Secure U.S. Air, Land, and Sea Borders and Approaches
- Goal 2.2: Safeguard and Expedite Lawful Trade and Travel
- Goal 2.3: Disrupt and Dismantle Transnational Criminal Organizations and Other Illicit Actors

Mission 3: Enforce and Administer Our Immigration Laws
- Goal 3.1: Strengthen and Effectively Administer the Immigration System
- Goal 3.2: Prevent Unlawful Immigration

Mission 4: Safeguard and Secure Cyberspace
- Goal 4.1: Strengthen the Security and Resilience of Critical Infrastructure
- Goal 4.2: Secure the Federal Civilian Government Information Technology Enterprise
- Goal 4.3: Advance Law Enforcement, Incident Response, and Reporting Capabilities
- Goal 4.4: Strengthen the Ecosystem

Mission 5: Strengthen National Preparedness and Resilience
- Goal 5.1: Enhance National Preparedness
- Goal 5.2: Mitigate Hazards and Vulnerabilities
- Goal 5.3: Ensure Effective Emergency Response
- Goal 5.4: Enable Rapid Recovery

Source: Quadrennial Homeland Security Review Report. U.S. Department of Homeland Security.

At the same time, the QHSR reiterates the centrality of domestic security and protecting the homeland against violent extremists. As explained in the 2014 QHSR, there are six "prevailing challenges that pose the most strategically significant risk" to the security of the United States:[8]

The threats, hazards, trends, and other dynamics reflected in the drivers of change suggest several prevailing strategic challenges that will drive risk over the next five years:

- The terrorist threat is evolving and, while changing in shape, remains significant as attack planning and operations become more decentralized. The United States and its interests, particularly in the transportation sector, remain persistent targets.

- Growing cyber threats are significantly increasing risk to critical infrastructure and to the greater U.S. economy.

- Biological concerns as a whole, including bioterrorism, pandemics, foreign animal diseases, and other agricultural concerns, endure as a top homeland security risk because of both potential likelihood and impacts.

- Nuclear terrorism through the introduction and use of an improvised nuclear device, while unlikely, remains an enduring risk because of its potential consequences.

- Transnational criminal organizations are increasing in strength and capability, driving risk in counterfeit goods, human trafficking, illicit drugs, and other illegal flows of people and goods.

- Natural hazards are becoming more costly to address, with increasingly variable consequences due in part to drivers such as climate change and interdependent and aging infrastructure.[9]

In essence, then, the dynamic nature of homeland security in the post–9/11 era has trended toward comprehensive integration at all levels of government and society in order to strengthen domestic security. The primary focus of modern homeland security originated in response to terrorist threats against the homeland, which continue to provide its central mission, but the newly articulated homeland security enterprise embodies the trend toward encompassing other domestic emergencies. The QHSR represents a systematic review of the homeland security enterprise.

DOMESTIC SECURITY AND THREATS TO THE HOMELAND: POLICY OPTIONS

A necessary element for framing a definition of homeland security is to acquire a general understanding of available policy options that promote domestic security. When challenged by genuine threats of violence from extremists, the United States may select responsive courses of action from a range of policy initiatives. Domestically, constitutional and legal considerations constrain homeland security policy options. Internationally, counterterrorist options permit aggressive and often extreme measures to be employed. Regardless of whether domestic or international considerations predominate in the selection of response options, the underlying consideration is that targets are selected by terrorists because of their meaningful symbolic value. This is factored into the homeland security calculation when alternative options are selected to protect the homeland. Chapter Perspective 1.3 discusses the symbolism of attacks directed against American interests.

The following discussion outlines available options to counter threats of extremist violence. Detailed discussion of these and other domestic policy options will be presented in later chapters.

▶ **Photo 1.5** The Pentagon on the morning of September 11, 2001.

Cpl. Jason Ingersoll, USMC/U.S. Department of Defense

Domestic Policy Options

Policy options for domestic security may be classified within several broad categories. These include enhanced intelligence, security, legal options, and conciliatory options.

CHAPTER PERSPECTIVE 1.3

The Symbolism of Targets

Terrorist Attacks Against the United States

Many targets are selected because they symbolize the interests of a perceived enemy. This selection process requires that these interests be redefined by extremists as representations of the forces against whom they are waging war. This redefinition process, if properly communicated to the terrorists' target audience and constituency, can be used effectively as propaganda on behalf of the cause.

The following attacks were launched against American interests.

Embassies and Diplomatic Missions

- June 1987: A car bombing and mortar attack were launched against the U.S. embassy in Rome, most likely by the Japanese Red Army.

- February 1996: A rocket attack was launched on the American embassy compound in Greece.

- August 1998: The U.S. embassies were bombed in Dar es Salaam, Tanzania, and Nairobi, Kenya. More than 200 people were killed.

- September 2012: Islamist insurgents attacked a U.S. diplomatic compound and an annex in Benghazi, Libya. The U.S. ambassador and a foreign service officer were killed at the compound. Two CIA contractors were killed at the annex.

International Symbols

- April 1988: A USO club in Naples, Italy, was bombed, most likely by the Japanese Red Army. Five people were killed.

- November 1995: Seven people were killed when anti-Saudi dissidents bombed an American military training facility in Riyadh, Saudi Arabia.

- November 2015: Four people, including two American trainers, were shot and killed by a Jordanian police captain at a police training facility near Amman, Jordan.

Symbolic Buildings and Events

- January 1993: Two were killed and three injured when a Pakistani terrorist fired at employees outside the Central Intelligence Agency headquarters in Langley, Virginia.

- February 1993: The World Trade Center in New York City was bombed, killing six and injuring more than 1,000.

- September 2001: Attacks in the United States against the World Trade Center and the Pentagon killed approximately 3,000 people.

- January 2011: A viable antipersonnel pipe bomb was found in Spokane, Washington, along the planned route of a memorial march commemorating the Reverend Dr. Martin Luther King Jr.

Symbolic People

- May 2001: The Filipino Islamic revolutionary movement Abu Sayyaf took three American citizens hostage. One of them was beheaded by members of the group in June 2001.

- January 2002: An American journalist working for the *Wall Street Journal* was kidnapped in Pakistan by Islamic extremists. His murder was later videotaped by the group.

- August and September 2014: ISIS broadcast the beheadings of two captive American journalists.

Passenger Carrier Attacks

- August 1982: A bomb exploded aboard Pan Am Flight 830 over Hawaii. The Palestinian group 15 May committed the attack. The plane was able to land.

- April 1986: A bomb exploded aboard TWA Flight 840. Four were killed and nine injured, including a

(Continued)

mother and her infant daughter, who fell to their deaths when they were sucked out of the plane. Flight 840 landed safely.

- December 2001: An explosive device malfunctioned aboard American Airlines Flight 63 as it flew from Paris to Miami. Plastic explosives had been embedded in the shoe of passenger Richard Reid.

- December 2009: An explosive device malfunctioned aboard Northwest Airlines Flight 253 as it approached Detroit, Michigan. Plastic explosives had been embedded in the underwear of passenger Umar Farouk Abdulmutallab.

Discussion Questions

1. If you were a strategist for an extremist organization, which symbolic targets would you prioritize?

2. Compare and contrast considerations that would make targets high value versus low value in terms of their symbolism.

3. What kind of target would it be a mistake to attack?

Intelligence

Intelligence refers to the collection of data. Its purpose is to create an informational database about extremist movements and to predict their behavior. As applied to homeland security threats, this process is not unlike that of criminal justice investigators who work to resolve criminal cases. In-depth discussion on the role of intelligence and the configuration of the intelligence community is provided in Chapter 6.

Enhanced Security

Target hardening enhances security for buildings, sensitive installations, transportation nodes, and other infrastructure that are potential targets. The purpose of such enhanced security is to deter or prevent terrorist attacks. Typical enhanced security measures include

Table 1.2   Domestic Policy Options: Intelligence and Enhanced Security

The following table summarizes basic elements of intelligence and enhanced security as domestic security policy options.

Domestic Security Option	Activity Profile		
	Rationale	Practical Objectives	Typical Resources Used
Intelligence	Prediction	Calculating the activity profiles of terrorists	Technology Covert operatives
Enhanced security	Deterrence	Hardening of targets	Security personnel Security barriers Security technology

observable security barriers and checkpoints as well as discreet surveillance technologies. Target hardening also involves enhanced technological security, such as innovative computer firewalls designed to thwart sophisticated cyberattacks.

Enhanced security measures are critical components of homeland security planning and preparation and can be applied to critical infrastructure, border security, port security, aviation security, and information nodes. Further discussion of target hardening is provided in greater detail in Chapter 11.

Table 1.2 summarizes the rationale for, practical objectives of, and typical resources used to implement intelligence and enhanced security policy options.

Legal Options

The United States has developed legal protocols to employ in dealing with terrorism. Some of these protocols were implemented to promote international cooperation, and others were adopted as matters of domestic policy. The overall objective of legal responses is to promote the rule of law and regular legal proceedings. Thus, legal options provide a lawful foundation for the homeland security enterprise. The following are examples of these responses:

- *Law enforcement* refers to the use of law enforcement agencies and criminal investigative techniques in the prosecution of suspected terrorists. This adds an element of *rule of law* to counterterrorism and homeland security. Counterterrorist laws attempt to criminalize terrorist behavior. This can be done by, for example, declaring certain behaviors to be criminal terrorism or enhancing current laws, such as those that punish murder.

- *International law* relies on cooperation among states. Those who are parties to international agreements attempt to combat terrorism by permitting terrorists no refuge or sanctuary for their behavior. For example, extradition treaties permit suspects to be taken into custody and transported to other signatory governments. In some cases, suspects may be brought before international tribunals.

Legal considerations are discussed in greater detail in Chapters 3 and 4, and further discussion of the homeland security role of law enforcement is provided in Chapter 7. The role of intelligence will be evaluated in Chapter 6.

Table 1.3 summarizes the rationale, practical objectives, and typical resources used for implementing legal options.

Conciliatory Options

Conciliation refers to communicating with extremists with the goal of identifying subjects of mutual interest. It is arguably a soft-line approach that allows policymakers to develop a range of options that do not involve confrontation, the use of force, or other suppressive methods. The objectives of conciliation depend on the characteristics of the terrorist threat. In some circumstances, conciliatory options may theoretically reduce or end a terrorist environment. Examples of these responses include the following:

- *Negotiation* refers to engaging with terrorists to agree on an acceptable resolution to a conflict. Negotiated solutions can be incident specific, or they can involve sweeping conditions that may completely resolve the conflict.

- *Concessionary options* can be generalized concessions, in which broad demands are accommodated, or incident specific, in which immediate demands are met.

- *Social reform* is an attempt to address the grievances of the terrorists and their championed group. Its purpose is to resolve the underlying problems that caused the terrorist environment to develop.

Table 1.3 Domestic Policy Options: Legalistic Responses

The purpose of legalistic responses is to provide protection to the general public, protect the interests of the state, and criminalize the behavior of the terrorists. The following table summarizes basic elements of legalistic responses.

Domestic Security Option	Activity Profile		
	Rationale	Practical Objectives	Typical Resources Used
Law enforcement	Enhancement of security apparatus Demilitarization of counterterrorist campaign	Day-to-day counterterrorist operations Bringing terrorists into the criminal justice system	Police personnel Specialized personnel
Domestic laws	Criminalization of terrorist behavior	Enhancement of criminal penalties for terrorist behavior Bringing terrorists into the criminal justice system	Criminal justice system Legislative involvement
International law	International consensus and cooperation	Coalitional response to terrorism	International organizations State resources

CHAPTER SUMMARY

This chapter presented an introduction to the concept of homeland security and discussed the definitional issues arising from the dynamic nature of homeland security. Several fundamental concepts were identified that continue to influence the ongoing evolution of homeland security in the modern era.

It is important to understand the elements that help define homeland security. Central components frame the definitional discussion. These include the terrorist threat, the federal bureaucracy, state and local agencies, the integration of homeland security interventions, and agreement on conceptual foundations for comprehensive homeland security. The *Quadrennial Homeland Security Review* is an essential document for framing a definitional conceptualization of homeland security. A general overview of policy options was presented to provide a perspective for understanding responses to perceived and genuine security threats from violent extremists. Historically, periodic challenges to domestic security led to sweeping policy measures, which, at the time, were deemed to be an appropriate means of securing the domestic environment.

DISCUSSION BOX

This chapter's Discussion Box is intended to stimulate critical debate about the aftermath of another catastrophic terrorist attack on the American homeland.

After the Next 9/11

The September 11, 2001, attack on the U.S. homeland produced the most sweeping reorganization of the American domestic security culture in history. The fear that arose following the attacks was matched by concerns that the United States was ill prepared to prevent or adequately respond to determined terrorists. The new concept of homeland security became part of everyday life and culture because of 9/11.

Assuming some degree of terrorist violence is likely to occur domestically, the possibility of another catastrophic attack leaves open the question of what impact such an event would have on society and the conceptualization of homeland security.

In Chapter 2, readers will evaluate the all-hazards umbrella as a conceptual component of the homeland security enterprise.

Although the likelihood of an incident on the scale of the 9/11 attacks may not be high, it is very plausible that domestic attacks could occur on the scale of the March 2004 Madrid train bombings, the July 2005 London transportation bombings, and the November 2015 Paris ISIS attack.

Discussion Questions

1. How serious is the threat of catastrophic terrorism?

2. Can catastrophic attacks be prevented?

3. How would a catastrophic terrorist attack affect the American homeland security culture?

4. How will society in general be affected by a catastrophic attack?

5. What is the likelihood that homeland security authority will be expanded in the future?

KEY TERMS AND CONCEPTS

The following topics were discussed in this chapter and can be found in the glossary.

Executive Order 13228 8
Haymarket Riot of 1886 5
homeland security enterprise 13
Homeland Security Presidential
    Directives (HSPDs) 9

Homestead Steel Strike of 1892 5
*Quadrennial Homeland Security Review
    Report: A Strategic Framework for
    a Secure Homeland* (QHSR) 13

USA PATRIOT Act of 2001 8
Whiskey Rebellion 5

ON YOUR OWN

Get the tools you need to sharpen your study skills. SAGE edge offers a robust online environment featuring an impressive array of free tools and resources.

Access practice quizzes, eFlashcards, video, and multimedia at **edge.sagepub.com/martinhs3e**

RECOMMENDED WEBSITES

The following websites provide information about defining homeland security and the mission of the modern homeland security enterprise:

    Congressional Research Service: www.fas.org/sgp/
    crs/homesec/R42462.pdf

Homeland Security News Wire: www.homeland securitynewswire.com

U.S. Department of Homeland Security: www.dhs.gov

WEB EXERCISE

Using this chapter's recommended websites, conduct an online investigation of the fundamental characteristics of homeland security.

1. What organizational or procedural commonalities can you find at the federal and state levels?

2. Is there anything that strikes you as being particularly controversial in approaches to securing the homeland?

3. Do you have recommendations on how to proceed with strengthening homeland security in the future?

To conduct an online search on approaches to defining homeland security, activate the search engine on your Web browser and enter the following keywords:

"Definitions of homeland security"

"Domestic security"

RECOMMENDED READINGS

The following publications are good analyses of the concept of homeland security and the homeland security bureaucracy:

Aronowitz, Stanley and Heather Gautney, eds. 2003. *Implicating Empire: Globalization and Resistance in the 21st Century World Order*. New York: Basic Books.

Bergen, Peter I. 2011. *A Very Long War: The History of the War on Terror and the Battles With Al Qaeda Since 9/11*. New York: Free Press.

Booth, Ken and Tim Dunne, eds. 2002. *Worlds in Collision: Terror and the Future of Global Order*. New York: Palgrave Macmillan.

Carr, Matthew. 2007. *The Infernal Machine: A History of Terrorism*. New York: New Press.

Coen, Bob and Eric Nadler. 2009. *Dead Silence: Fear and Terror on the Anthrax Trail*. Berkeley, CA: Counterpoint.

Cronin, Isaac, ed. 2002. *Confronting Fear: A History of Terrorism*. New York: Thunder's Mouth Press.

Gage, Beverly. 2009. *The Day Wall Street Exploded: A Story of America in Its First Age of Terror*. Oxford, UK: Oxford University Press.

Kamien, David G. 2006. *The McGraw-Hill Homeland Security Handbook*. New York: McGraw-Hill.

National Commission on Terrorist Attacks Upon the United States. 2004. *The 9/11 Commission Report*. New York: Norton.

Purpura, Philip P. 2007. *Terrorism and Homeland Security: An Introduction With Applications*. New York: Butterworth-Heinemann.

Sage Publications. 2010. *Issues and Homeland Security: Selections From CQ Researcher*. Thousand Oaks, CA: Sage.

Sauter, Mark A. and James Jay Carafano. 2005. *Homeland Security: A Complete Guide to Understanding, Preventing, and Surviving Terrorism*. New York: McGraw-Hill.

CHAPTER TWO

HOMELAND SECURITY AND THE ALL-HAZARDS UMBRELLA

Opening Viewpoint: Understanding the All-Hazards Umbrella

The all-hazards umbrella refers to preparation for all potential disasters, including natural and human-created disasters. In the field of risk management and emergency management, the adoption of an all-hazards approach is not a new concept. Prior to the post–9/11 era of homeland security, FEMA and similar agencies were tasked with preparing for and responding to all hazards. In the post–9/11 era, the dynamic and evolving conceptualization of homeland security has, when necessary, incorporated features of the all-hazards concept.

As a matter of necessity, the United States is obligated to anticipate the likelihood of the occurrence of natural and human-created disasters—in essence, to prepare for the worst and hope for the best. Within the all-hazards umbrella, preparations include preparing for all potential disasters, including industrial accidents, earthquakes, hurricanes, floods, tornadoes, and, of course, terrorist incidents.

Considering the extensive scope of an all-hazards approach, the need for coordination extends to all levels of society. Because of this and the daunting task of preparing for *all* contingencies, the concept of *all hazards* has generated a debate about how to allocate resources (such as training and equipment) and, in fact, whether it is feasible to prepare for every contingency. Nevertheless, the need for consensus on a common approach to disaster preparation is a fundamental necessity for establishing a viable all-hazards umbrella.

Chapter Learning Objectives

This chapter will enable readers to do the following:

1. Define and discuss the all-hazards umbrella

2. Analyze nonterrorist emergency scenarios

3. Differentiate natural hazards from those caused by human activities

4. Explain the association between terrorism and the all-hazards approach

The establishment of emergency response systems is a critical national necessity. They are needed at every level of society in preparation for the unfortunate (but inevitable) eventualities of natural disasters and incidents arising from human conduct. Historically, such response systems significantly predated the modern era of homeland security. The first national consolidation of emergency mitigation, preparedness, and response occurred on June 19, 1978, when President Jimmy Carter delivered Reorganization Plan Number 3 to Congress. Reorganization Plan Number 3 proposed the consolidation of emergency preparedness and response into a single federal agency: the Federal Emergency Management Agency (FEMA). This necessitated the transferal of several agency functions to the new FEMA bureau. These agencies included the Federal Preparedness Agency, Federal Disaster Assistance Administration, National Fire Prevention and Control Administration, Defense Civil Preparedness Agency, Federal Broadcast System, and Federal Insurance Administration. FEMA was officially established pursuant to Executive Order 12127 on March 31, 1979, after congressional review.

The concept of a FEMA-managed emergency response system was expanded over time, so that in the post–9/11 era, considerable functional consolidation occurred as catastrophic natural and nonterrorist human-related emergencies arose, and many of these functions were incorporated into the homeland security enterprise. This conflation of natural and nonterrorist emergencies with homeland security happens periodically—for example, it occurred in the aftermath of Hurricane Katrina on the U.S. Gulf Coast in 2005. It is a logical progression in the ongoing conceptualization of homeland security, particularly in light of the post–9/11 effort, to coordinate nationwide response mechanisms to mitigate threats from violent extremists. The central premise is that the incorporation of additional nonterrorist hazard and threat scenarios within this framework benefits the nation at large.

The concept of an **all-hazards umbrella** has been adapted to the homeland security enterprise as an available mechanism to mobilize a broad array of resources when disasters occur. The underlying conceptualization of homeland security continues to be that of a domestic security response to violent extremism. Nevertheless, the all-hazards umbrella is applied, when needed, as a practical necessity for emergency response to nonterrorist events. As explained in the 2010 *Quadrennial Homeland Security Review Report*,

> Homeland security describes the intersection of evolving threats and hazards with traditional governmental and civic responsibilities for civil defense, emergency response, law enforcement, customs, border control, and immigration. In combining these responsibilities under one overarching concept, homeland security breaks down longstanding stovepipes of activity that have been and could still be exploited by those seeking to harm America. Homeland security also creates a greater emphasis on the need for joint actions and efforts across previously discrete elements of government and society.[1]

Three levels of potential danger must be considered: hazards, emergency events, and disasters. A condition posing potential risks is referred to as a **hazard**, and depending on the nature of the hazard, it can result in either an emergency event or disaster. When a hazard does in fact result in a condition of risk, an **emergency event** occurs. Emergency events necessitate intervention by institutions trained for emergency response operations, such as law enforcement, medical personnel, or firefighting agencies. A **disaster** occurs when emergency response institutions cannot contain the emergency event or stabilize critical services, such as fire control, order maintenance and restoration, providing immediate medical services, or providing shelter. Declarations of disaster are officially made by the president of the United States after a request is received from a governor; such declarations activate protocols for the provision of federal assistance to an affected region. Full consideration of presidential disaster declarations will be provided in Chapter 13.

all-hazards umbrella: All-hazards preparation entails preparation for a wide range of natural and human-made disasters.

hazard: A condition posing potential risks that can result in either an emergency event or disaster.

emergency event: An emergency event occurs when a hazard does, in fact, result in a condition of risk, necessitating intervention by emergency response institutions, such as law enforcement, medical personnel, or firefighting agencies.

disaster: Conditions rise to the level of a disaster when emergency response institutions are unable to contain or resolve one or more critical services, such as fire management, the restoration of order, attending to medical needs, or providing shelter. Official declarations of disaster are made by the president of the United States after he receives a request from a governor.

Figure 2.1 A Summary of One View of the All-Hazards Umbrella, as Presented by the Centers for Disease Control and Prevention

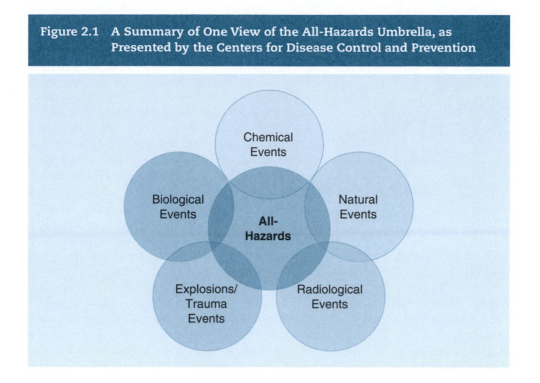

This chapter will discuss the all-hazards umbrella and its applicability in the current homeland security environment. The discussion in this chapter will review the following:

- The terrorism nexus: Conventional and unconventional threats

- The all-hazards nexus: Nonterrorist hazards and threats

- Natural hazards

- Technological scenarios

THE TERRORISM NEXUS: CONVENTIONAL AND UNCONVENTIONAL THREATS

In the modern era, terrorism-related threats can be classified along a sliding scale of technological sophistication and threat potential from weapons deployed by violent extremists. This scale includes conventional weapons and the use of weapons of mass destruction (WMD).

Conventional Weapons

Conventional weapons are military-grade and civilian weapons that are not used as WMD. Many conventional weapons do have the potential to cause mass casualties, but they are accepted as legal weapons of war or appropriate for civilian purposes. Standard firearms and explosives are conventional weapons.

David Stewart-Smith/Contributor/Hulton Archive/Getty Images

▶ **Photo 2.1** Afghan mujahideen (holy warriors) during their jihad against occupying Soviet troops. Osama bin Laden formed the al-Qaeda network during the war.

Firearms

Small arms and other handheld weapons have been and continue to be the most common types of weapons employed by terrorists. These are light- and heavy-infantry weapons and include pistols, rifles, submachine guns, assault rifles, machine guns, rocket-propelled grenades, mortars, and precision-guided munitions. Typical firearms found in the hands of terrorists include submachine guns, assault rifles, rocket-propelled grenades, and precision-guided munitions.

Submachine Guns. Originally developed for military use, submachine guns are now mostly used by police and paramilitary services. Although new models have been designed, such as the famous Israeli Uzi and the American Ingram, World War II–era models are still on the market and have been used by terrorists.

Assault Rifles. Usually capable of both automatic (repeating) and semiautomatic (single-shot) fire, assault rifles are military-grade weapons that are used extensively by terrorists and other irregular forces. The AK-47, invented by Mikhail Kalashnikov for the Soviet Army, is the most successful assault rifle in terms of production numbers and its widespread adoption by standing armies, guerrillas, and terrorists. The American-made M16 has likewise been produced in large numbers and has been adopted by a range of conventional and irregular forces.

Rocket-Propelled Grenades (RPGs). Light, self-propelled munitions are common features of modern infantry units. The RPG-7 has been used extensively by dissident forces throughout the world, particularly in Latin America, the Middle East, and Asia. The weapon was manufactured in large quantities by the Soviets, Chinese, and other communist nations. It is an uncomplicated and powerful weapon that is useful against armor and fixed emplacements, such as bunkers or buildings.

Precision-Guided Munitions. Less commonly found among terrorists, though extremely effective when used, are weapons that can be guided to their targets by using infrared or other tracking technologies. The American-made Stinger is a shoulder-fired surface-to-air missile that uses an infrared targeting system. It was delivered to the Afghan mujahideen (holy warriors) during their anti-Soviet jihad and was used very effectively against Soviet helicopters and other aircraft. The Soviet-made SA-7, also known as the Grail, is also an infrared-targeted surface-to-air missile. Both the Stinger and the Grail pose a significant threat to commercial airliners and other aircraft.

Common Explosives

Terrorists regularly use explosives to attack symbolic targets. Along with firearms, explosives are staples of the terrorist arsenal. The vast majority of terrorists' bombs are self-constructed, improvised weapons rather than premanufactured military-grade bombs. The one significant exception to this rule is the heavy use of military-grade mines by the world's combatants. These are buried in the soil or rigged to be detonated as booby traps. Antipersonnel mines are designed to kill people, and antitank mines are designed to destroy vehicles. Many millions of mines have been manufactured and are available on the international market.

Some improvised bombs are constructed from commercially available explosives such as dynamite and TNT, whereas others are manufactured from military-grade compounds. Examples of compounds found in terrorist bombs include plastic explosives and ammonium nitrate and fuel oil (ANFO) explosives.

Plastic Explosives. Plastic explosives are puttylike explosive compounds that can be easily molded. The central component of most plastic explosives is a compound known as RDX. Nations that manufacture plastic explosives often use chemical markers to "tag" each batch that is made. The tagged explosives can be traced back to their source if used by terrorists. Richard C. Reid, the "shoe bomber" aboard American Airlines Flight 63, attempted to detonate a bomb crafted from plastic explosives molded into his shoe in December 2001. The case of Richard Reid is discussed further in Chapter 9.

Semtex. Semtex is a very potent plastic explosive of Czech origin. During the Cold War, Semtex appeared on the international market, and a large quantity was obtained by Libya. It is popular among terrorists. For example, the Irish Republican Army (IRA) has used Semtex-based bombs in Northern Ireland and England.

Composite-4. Invented in the United States, Composite-4 (C-4) is a high-grade and powerful plastic explosive. It is more expensive and more difficult to obtain than Semtex. The availability of C-4 for use by terrorists became apparent when a renegade CIA agent was convicted of shipping 21 tons of the compound to Libya during the 1970s. About 600 pounds of C-4 was used in the October 2000 attack against the American destroyer USS *Cole* in Yemen, and it was evidently used to bomb the American facility at Khobar Towers in Dhahran, Saudi Arabia, in June 1996.

ANFO Explosives. Ammonium nitrate and fuel oil explosives are manufactured from common ammonium nitrate fertilizer that has been soaked in fuel oil. When ammonium nitrate is used as a base for the bomb, additional compounds and explosives can be added to intensify the explosion. These devices require hundreds of pounds of ammonium nitrate, so they are generally constructed as car or truck bombs. ANFO explosives were used by the IRA in London in 1996; Timothy McVeigh used a two-ton device in Oklahoma City in 1995.

Types of Bombs

Gasoline Bombs. The most easily manufactured (and common) explosive weapon used by dissidents is nothing more than a gasoline-filled bottle with a flaming rag for its trigger. It is thrown at targets after the rag is stuffed into the mouth of the bottle and ignited. Tar, Styrofoam, or other ingredients can be added to create a gelling effect for the bomb, which causes the combustible ingredient to stick to surfaces. These weapons are commonly called "Molotov cocktails," named for Vyacheslav Molotov, the Soviet Union's foreign minister during World War II.

Jeff Fusco/Getty Image

▶ **Photo 2.2** Replica of an explosive suicide vest. Suicide attacks became increasingly common during insurgencies in the post–September 11 era.

The name was invented during the 1939–1940 Winter War by Finnish soldiers, who used the weapon effectively against Soviet troops.

Pipe Bombs. These devices are easily constructed from common pipes, which are filled with explosives (usually gunpowder) and then capped on both ends. Nuts, bolts, screws, nails, and other shrapnel are usually taped or otherwise attached to pipe bombs. Many hundreds of pipe bombs have been used by terrorists. In the United States, pipe bombs were used in several bombings of abortion clinics and at the 1996 Summer Olympics in Atlanta.

Vehicular Bombs. Ground vehicles that have been wired with explosives are a frequent weapon in the terrorist arsenal. Vehicular bombs can include car bombs and truck bombs; they are mobile, are covert in the sense that they are not readily identifiable, are able to transport large amounts of explosives, and are rather easily constructed. They have been used on scores of occasions throughout the world.

Improvised Explosive Devices. These bombs are constructed by nonstate actors and outside of military or regulatory controls by terrorist and insurgent groups. The individuals producing the bombs have expertise in bomb making yet have limited access to equipment or materials. As such, the bomb can vary significantly in power and effectiveness.

Chemical, Biological, Radiological, Nuclear, and Explosive Hazards[2]

CBRNE: Acronym for chemical agents, biological agents, radiological agents, nuclear weapons, and explosives.

Plausible terrorist threat scenarios include the deliberate use of biological and chemical agents against perceived enemies. These threats are a serious concern within the context of an all-hazards approach to homeland security. There are five types of possible weapons, classified as: chemical agents, biological agents, radiological agents, nuclear weapons, and explosives. These are known as **CBRNE** weapons. Explosive weapons were discussed in the previous section. Chemical, biological, radiological, and nuclear weapons and hazards are explored in detail in Chapter 12.

THE ALL-HAZARDS NEXUS: NONTERRORIST HAZARDS AND THREATS

The all-hazards nexus encompasses preparation for responding to natural and human-made disasters and threats. The perceived need for such preparations reflects genuine concern over how to ensure effective emergency management regardless of the cause of the emergency. This concern is based in large part on problems and challenges encountered during actual implementation of preparedness, response, and recovery efforts for recent disasters. An additional concern is the unfortunate fact that recurrent human-initiated crises and incidents have not abated in the present era. For this reason, FEMA and other agencies subsumed under the U.S. Department of Homeland Security (DHS) are required to be prepared to provide leadership for addressing all-hazard scenarios.

Background: Recent Difficulties in Disaster Relief

Prior to the September 11, 2001, terrorist attacks, FEMA was tasked with providing primary leadership in coordinating preparedness, response, and recovery efforts for domestic emergencies. Domestic emergencies were broadly defined, and although the agency had been called upon to provide leadership in recovering from the 1993 World Trade Center and 1995

Oklahoma City terrorist attacks, FEMA's primary mission focus during the pre–9/11 era was the emergency management of natural and industrial disasters. With the establishment of DHS, FEMA was absorbed into DHS and tasked with responding to all hazards.

Unfortunately, two natural disasters occurred that highlighted significant shortcomings in the American emergency management system: one prior to September 11, 2001, and the other after the 9/11 attack. **Hurricane Andrew** struck the United States in 1992, causing approximately $26 billion in damage. Widespread criticism of the federal response argued that federal intervention was unnecessarily slow and uncoordinated. **Hurricane Katrina** struck the Gulf Coast of the United States in 2005, causing billions of dollars in destruction and resulting in nearly 2,000 deaths. Again, strong criticism of the federal response argued that federal intervention was unacceptably inadequate and ponderous. State emergency response authorities in Louisiana were also widely criticized as images of thousands of refugees were broadcast globally.

Experts argued that comparable policy decisions in both cases led to the allegedly poor coordination of federal responses to the disasters. Regarding Hurricane Andrew, the federal government was arguably less prepared than required because of the emphasis during the 1980s on preparing for nuclear attack, which effectively diverted resources and planning from the emergency management of natural disasters. Hurricane Katrina similarly occurred when the United States was devoting enormous resources to mitigating the new terrorist threat, thus diverting resources from preparing for other emergencies, such as natural disasters. Both cases illustrated the need for designing and implementing an all-hazards component to emergency management preparations and response.

All-Hazards Emergency Management: Core Concepts

Managing emergencies under an all-hazards umbrella can be quite complex, and it requires the efficient coordination of several central sequential elements. Such coordination is necessary for the orderly implementation of response efforts when emergencies arise. The sequential elements of the all-hazards umbrella include mitigation of risk, preparedness planning, emergency response operations, and recovery systems. These elements are implemented within the context of a whole-community approach. These elements are frequently referred to as a *phase of disaster model* and the life cycle of emergency management. Figure 2.2 illustrates the application of this concept.

Mitigation of Risk

As defined by FEMA, **mitigation of risk** is the effort to reduce loss of life and property by lessening the impact of disasters. Mitigation is taking action *now*—before the next disaster—to reduce human and financial consequences later (analyzing risk, reducing risk, insuring against risk). Effective mitigation requires that we *all* understand local risks, address the hard choices, and invest in long-term community well-being. Without mitigation actions, we jeopardize our safety, financial security, and self-reliance.[3]

Thus, pre-emergency initiatives are put into place by mitigation mechanisms that are theoretically intended to reduce the potential costs and destructiveness of disasters when they occur. Such initiatives require partnerships between government and private agencies.

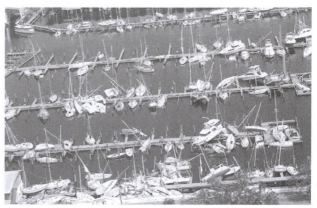

▶ **Photo 2.3** A marina in Louisiana devastated during Hurricane Katrina. The Katrina disaster had a profound effect on the development of the all-hazards approach to homeland security.

Hurricane Andrew: In 1992, Hurricane Andrew struck the United States, causing an estimated $26 billion in damage. The federal response was widely criticized as unnecessarily slow and uncoordinated.

Hurricane Katrina: In 2005, Hurricane Katrina struck the Gulf Coast of the United States, causing nearly 2,000 deaths and billions of dollars in destruction. The federal response was strongly criticized as unacceptably inadequate and ponderous.

mitigation of risk: "The effort to reduce loss of life and property by lessening the impact of disasters." Mitigation mechanisms are pre-emergency initiatives that theoretically reduce the potential costs and destructiveness of disasters when they occur.

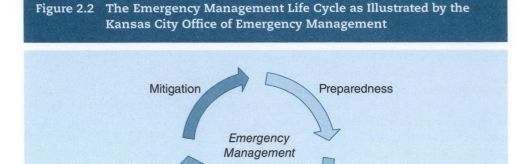

Source: Kansas City Office of Emergency Management.

For example, private insurance for floods, earthquakes, and other emergencies mitigates the financial losses of businesses and private individuals. Other mitigations may include flood-control systems, the posting of evacuation route information, new construction codes for buildings in areas of risk, and the dissemination of information about known hazards.

Preparedness Planning

preparedness planning: The design and adoption of emergency management contingencies prior to the occurrence of an emergency situation. Preparations are made to increase the likelihood that an initial response effort and the subsequent recovery period will ultimately be efficient and successful.

The adoption of emergency management contingencies prior to the occurrence of an emergency situation is a critical component of all-hazards emergency management. Such preparations are termed **preparedness planning**. Ideally, preparedness planning should be designed with the ultimate goal of assuring the efficiency and success of the initial response effort and the subsequent recovery period. Preparedness planning cannot anticipate every contingency or emergency scenario, especially if it is a terrorist threat scenario. However, flexibility and adaptation can mitigate the consequences of unanticipated hazard and threat contingencies when they occur.

Emergency Response Operations

emergency response operations: Coordinated interventions undertaken when emergencies occur.

When emergencies occur, coordinated interventions known as **emergency response operations** are undertaken. Unfortunately, disasters and other emergencies are rarely, if ever, predictable. Hence, emergency responders are usually called upon to reactively implement contingency plans. Significant resources must be efficiently mobilized during emergency response operations, requiring the complex deployment of personnel and equipment to disaster sites. Examples of services rendered during emergency response operations include caring for victims, evacuating endangered populations, maintaining order and security, containing hazardous conditions such as fires, and disseminating critical information.

Recovery Systems

recovery systems: Recovery systems are implemented in the aftermath of emergency events and attempt to return affected regions to predisaster baselines. This requires, at a minimum, rebuilding damaged infrastructure and restoring affected populations to their pre-emergency norms of living. Recovery operations are often quite expensive and long term.

When emergency events occur, the implementation of **recovery systems** is critical to emergency response operations. Virtually synonymous with restoration, recovery operations attempt to return affected regions to predisaster baselines. At a minimum, this requires coordinated intervention to effectively restore affected populations to their pre-emergency norms of living and rebuild damaged infrastructure. Preparations must anticipate the possibility that recovery operations may be quite expensive and long term. Rebuilding infrastructure includes restoring communication, power, and water services. The restoration of transportation requires repair of damaged roadways and overpasses as well as reestablishment of transportation systems. Providing basic services to affected populations includes constructing shelters, providing climate-related assistance, and clothing and feeding those who are forced from their homes. Medical and financial recovery assistance is often essential when significant destruction and dislocation occur.

The Whole-Community Approach to Emergency Management

FEMA has incorporated a policy-planning concept that attempts to meld government and private sector constituencies into a whole-community approach to the homeland security enterprise. FEMA acknowledges that "a government-centric approach to emergency management is not enough to meet the challenges posed by a catastrophic incident"[4] and that "FEMA is only one part of our nation's emergency management team."[5] The whole-community approach envisions collaboration from all segments of society at all levels on the design and implementation of an efficient emergency management system. In essence, "this larger collective emergency management team includes, not only FEMA and its partners at the federal level, but also local, tribal, state and territorial partners; nongovernmental organizations like faith-based and non-profit groups and private sector industry; [and] individuals, families and communities, who continue to be the nation's most important assets as first responders during a disaster."[6] Thus, all segments of society that may potentially be affected by a catastrophic event are ideally to be subsumed under an all-hazards umbrella, thereby promoting an efficient and effective homeland security enterprise.

Case in Point: Nonterrorist Mass Shootings

Nonterrorist disaster relief systems must be prepared to address a wider spectrum of potential crises than natural or technological disasters. (Both are discussed later in this chapter.) Examples of potential crises include intentional human-initiated crises such as nonterrorist mass shootings. When discussing this phenomenon, it is important to first consider that individual possession of firearms in the United States is a fundamental right guaranteed under the Second Amendment to the U.S. Constitution. Firearms are commonly purchased for personal defense, sporting activities, and other responsible purposes.

The United States periodically experiences incidents of nonterrorist mass homicides perpetrated by individuals who typically enter a facility and begin to randomly shoot victims, often using high-powered firearms such as assault rifles and high-caliber handguns. Perpetrators of mass firearm killings rarely justify their actions by citing political motivations, such as ideology, race, or religion, and thus do not fit the modern profile of terrorist operatives or political lone-wolf actors. Rather, those who commit crimes of mass homicide are driven by the same antisocial motivations typically cited by other criminals. The distinctive difference is that they act out their antisocial rationales by engaging in mass firearm killings.

Nonterrorist mass shootings are not common among the world's prosperous democracies. The frequency of these incidents and the overall rate of firearm-related homicides are much higher in the United States than in similar high-income nations. For example, research of 23 populous high-income countries on firearm homicides reported that the US homicide rates were 6.9 times higher than rates in the other high-income countries, driven by firearm homicide rates that were 19.5 times higher. For 15-year olds to 24-year olds, firearm homicide rates in the United States were 42.7 times higher than in the other countries. For US males, firearm homicide rates were 22.0 times higher, and for US females, firearm homicide rates were 11.4 times higher. . . . Among these 23 countries, 80% of all firearm deaths occurred in the United States, 86% of women killed by firearms were US women, and 87% of all children aged 0 to 14 killed by firearms were US children.[7]

The April 20, 1999, mass shooting by two teenagers at Columbine High School in Littleton, Colorado, which resulted in 13 dead and 21 wounded, is arguably a tipping point for studying the profile of mass shootings in the modern era. The two perpetrators, Dylan Klebold and Eric Harris, were heavily armed and methodically shot students and teachers at Columbine High School, eventually committing suicide. Since the Columbine incident, dozens of nonterrorist mass shootings have occurred with a similar sequence of events, in which a shooter methodically shot victims before either committing suicide or being killed or captured by the police. After Columbine, many of the most lethal mass shootings occurred at educational institutions. For example, on December 14, 2012, in Newtown, Connecticut, 20 children and six adults were killed by gunman Adam Lanza at Sandy Hook Elementary School. Similarly, on February 14, 2018, in Parkland, Florida, 17 students were killed and 17 wounded by lone gunman Nikolas Cruz at Marjory Stoneman Douglas High School.

Chapter Perspective 2.1 summarizes incidents occurring after the Columbine High School shootings. This is not an exhaustive list and serves as an instructive "snapshot" of an ongoing societal pattern in the United States.

CHAPTER PERSPECTIVE 2.1

Mass Shootings in the United States After the Columbine[a] Incident

The following timeline summarizes mass-shooting incidents in the United States from the time of the Columbine High School incident on April 20, 1999.

April 20, 1999. Two gunmen killed 13 people and wounded 21 during the Columbine High School mass shooting in Littleton, Colorado.

July 29, 1999. Mark Orrin Barton, 44, murdered his wife and two children with a hammer before shooting up two Atlanta day trading firms. Barton, a day trader, was believed to be motivated by huge monetary losses. He killed 12 including his family and injured 13 before killing himself.

September 15, 1999. Larry Gene Ashbrook opened fire on a Christian rock concert and teen prayer rally at Wedgewood Baptist Church in Fort Worth, TX. He killed

7 people and wounded 7 others, almost all teenagers. Ashbrook committed suicide.

December 26, 2000. Edgewater Technology employee Michael "Mucko" McDermott shot and killed seven of his coworkers at the office in Wakefield, MA. McDermott claimed he had "traveled back in time and killed Hitler and the last 6 Nazis." He was sentenced to 7 consecutive life sentences.

July 8, 2003. Doug Williams, a Lockheed Martin employee, shot up his plant in Meridian, MS, in a racially motivated rampage. He shot 14 people, most of them African American, and killed 7 before killing himself.

March 12, 2005. Persons attending a Living Church of God meeting were gunned down by 44-year-old church member Terry Michael Ratzmann at a Sheraton hotel in Brookfield, WI. Ratzmann was thought to have had religious motivations, and killed himself after executing the pastor, the pastor's 16-year-old son, and 7 others. Four were wounded.

March 21, 2005. Teenager Jeffrey Weise killed his grandfather and his grandfather's girlfriend before opening fire on Red Lake Senior High School, killing 9 people on campus and injuring 5. Weise killed himself.

March 25, 2006. Seven died and 2 were injured by 28-year-old Kyle Aaron Huff in a shooting spree through Capitol Hill in Seattle, WA. The massacre was the worst killing in Seattle since 1983.

October 2, 2006. Children at an Amish schoolhouse in Lancaster, PA, were gunned down by 32-year-old Charles Carl Roberts. Roberts separated the boys from the girls, binding and shooting the girls. Five young girls died, while 6 were injured. Roberts committed suicide afterward.

February 12, 2007. In Salt Lake City's Trolley Square Mall, 5 people were shot to death and 4 others were wounded by 18-year-old gunman Sulejman Talović. One of the victims was a 16-year-old boy.

April 16, 2007. Virginia Tech became the site of the deadliest school shooting in U.S. history when a student, Seung-Hui Choi, gunned down 56 people. Thirty-two people died in the massacre.

December 5, 2007. A 19-year-old boy, Robert Hawkins, shot up a department store in the Westroads Mall in Omaha, NE. Hawkins killed 9 people and wounded 4 before killing himself. The semi-automatic rifle he used was stolen from his stepfather's house.

February 7, 2008. Six people died and two were injured in a shooting spree at the City Hall in Kirkwood, Missouri. The gunman, Charles Lee Thornton, opened fire during a public meeting after being denied construction contracts he believed he deserved. Thornton was killed by police.

February 14, 2008. Steven Kazmierczak, 27, opened fire in a lecture hall at Northern Illinois University, killing 6 and wounding 21. The gunman shot and killed himself before police arrived. It was the fifth-deadliest university shooting in U.S. history.

March 29, 2009. Eight people died in a shooting at the Pinelake Health and Rehab nursing home in Carthage, NC. The gunman, 45-year-old Robert Stewart, was targeting his estranged wife who worked at the home and survived. Stewart was sentenced to life in prison.

April 3, 2009. Jiverly Wong, 41, opened fire at an immigration center in Binghamton, New York, before committing suicide. He killed 13 people and wounded 4.

November 5, 2009. Major Nidal Hasan, a psychiatrist in the U.S. Army, shot to death 13 people at the Fort Hood Army base in Texas. More than 30 others were wounded.

August 3, 2010. Omar S. Thornton, 34, opened fire at Hartford Beer Distributor in Manchester, CT, after getting caught stealing beer. Nine were killed, including Thornton, and two were injured.

January 8, 2011. Former Rep. Gabby Giffords (D-AZ) was shot in the head when 22-year-old Jared Loughner opened fire on an event she was holding at a Safeway market in Tucson, AZ. Six people died, including Arizona District Court Chief Judge John Roll, one of Giffords' staffers, and a 9-year-old girl. Nineteen total were shot. Loughner was sentenced to seven life terms plus 140 years, without parole.

September 6, 2011. Eduardo Sencion, 32, entered an IHOP restaurant in Carson City, NV, and shot 12 people. Five died, including three National Guard members.

October 14, 2011. Eight people died in a shooting at Salon Meritage hair salon in Seal Beach, CA. The gunman, 41-year-old Scott Evans Dekraai, left six women and two men dead, while just one woman survived. It was Orange County's deadliest mass killing.

April 2, 2012. A former student, 43-year-old One L. Goh, killed 7 people at Oikos University, a Korean Christian college in Oakland, CA. The shooting was the sixth-deadliest school massacre in the United States and the deadliest attack on a school since the 2007 Virginia Tech massacre.

April 6, 2012. Jake England, 19, and Alvin Watts, 32, shot 5 black men in Tulsa, Oklahoma, in a racially motivated shooting spree. Three died.

(Continued)

(Continued)

May 29, 2012. Ian Stawicki opened fire on Cafe Racer Espresso in Seattle, WA, killing 5 and himself after a citywide manhunt.

July 20, 2012. During the midnight premiere of *The Dark Knight Rises* in Aurora, CO, 24-year-old James Holmes killed 12 people and wounded 58. Holmes was arrested outside the theater.

August 5, 2012. Six Sikh temple members were killed when 40-year-old U.S. Army veteran Wade Michael Page opened fire in a gurdwara in Oak Creek, Wisconsin. Four others were injured, and Page killed himself.

September 27, 2012. Five were shot to death by 36-year-old Andrew Engeldinger at Accent Signage Systems in Minneapolis, MN. Three others were wounded. Engeldinger went on a rampage after losing his job, ultimately killing himself.

December 14, 2012. Twenty children and six adults were killed during the Sandy Hook Elementary School mass shooting in Newtown, Connecticut.

February 3–12, 2013. Los Angeles police officer Christopher Dorner crossed three counties in southern California during a nine-day shooting spree, killing four people and wounding nine.

September 16, 2013. Lone gunman Aaron Alexis killed 12 people and wounded eight at the headquarters of the Naval Sea Systems Command at the Washington Navy Yard in Washington, D.C.

May 23, 2014. In Isla Vista, California, near the campus of University of California, Santa Barbara, Elliot Rodger killed six people and wounded 13 during a stabbing and shooting spree. His attack was apparently motivated by his hatred of women.

October 24, 2014. Freshman student Jaylen Fryberg shot five students, killing four, at Marysville Pilchuck High School in Marysville, Washington.

February 26, 2015. Lone gunman Joseph Jesse Aldridge shot eight people in four homes, killing seven, in Tyrone, Texas. Four of the fatalities were family members.

October 1, 2015. Student Christopher Harper-Mercer killed nine people and wounded eight on the campus of Umpqua Community College near Roseburg, Oregon. The nine fatalities were a professor and classmates, all shot inside a classroom.

March 9, 2016. Two gunmen in Wilkinsburg, Pennsylvania, killed six people and wounded three during a party. One of the victims was an unborn child.

September 23, 2016. A lone gunman killed five people at the Cascade Mall in Burlington, Washington.

January 6, 2017. A gunman at the Fort Lauderdale–Hollywood International Airport in Broward County, Florida, killed five people and wounded six near a baggage claim area. Thirty-six other people were injured during the ensuing panic.

October 1, 2017. Lone gunman Stephen Paddock shot at concertgoers from a high floor of the Mandalay Bay Casino Hotel in Las Vegas, Nevada. He killed 59 people and wounded 422. A total of 851 people were injured during the ensuing panic.

February 14, 2018. Seventeen people were killed and 17 wounded during the Marjory Stoneman Douglas High School mass shooting in Parkland, Florida.

May 18, 2018. Ten people—eight students and two teachers—were killed and 14 others wounded by gunman Dimitrios Pagourtis at Santa Fe High School in Santa Fe, Texas. Pagourtis was a student at the school. He also constructed explosive devices that were not detonated during the attack.

June 28, 2018. Lone gunman Jarrod Ramos killed five people and wounded two in the offices of the *Capitol Gazette* newspaper in Annapolis, Maryland.

November 7, 2018. Lone gunman David Long entered the Borderline Bar and Grill in Thousand Oaks, California, during a student line-dancing event. He killed 12 people and wounded more than 12 others.

February 15, 2019. A lone gunman killed five people and wounded six at a Henry Pratt Company factory in Aurora, Illinois. He was a recently terminated employee.

Discussion Questions

1. Do similarities exist in mass-shooting incidents?

2. Should such incidents be redefined as acts of terrorism?

3. Can procedures be designed to reduce the frequency or scale of such incidents?

Note

a. Some incident data are derived from Aviva Shen, "A Timeline of Mass Shootings in the US Since Columbine," Think Progress, December 14, 2012, http://thinkprogress.org/justice/2012/12/14/1337221/a-timeline-of-massshootings-in-the-us-since-columbine.

Because of these incidents, adaptive law enforcement tactics have been designed to end mass shootings as quickly as possible. Rather than engaging in prolonged negotiations or otherwise awaiting word from perpetrators, police units are likely to enter affected areas aggressively to end the incident and apprehend suspects. In such scenarios, the primary mission of the first units gaining entry is to hunt and neutralize suspects. Subsequent units evacuate and render aid to potential victims. Concurrently, many business facilities and educational campuses have adopted active-shooter protocols. These protocols, sometimes referred to as *run, hide, fight protocols*, typically parallel recommendations published by the DHS. DHS's publication "Active Shooter: How to Respond" presents a plan for responding to active-shooter scenarios, including the following summarized measures:[8]

1. Evacuate

   If there is an accessible escape path, attempt to evacuate the premises.

2. Hide out

   If evacuation is not possible, find a place to hide where the active shooter is less likely to find you.

3. Take action against the active shooter

   As a last resort, and only when your life is in imminent danger, attempt to disrupt and/or incapacitate the active shooter.

NATURAL HAZARDS

Emergency incidents arising from nonhuman causes are termed **natural hazards**. Phenomena arising from natural environmental conditions cause natural hazard incidents. Examples of natural hazard events include hurricanes, floods, tornadoes, and earthquakes; when they occur, they are capable of eventually being classified as natural disasters. Although incident scenarios for natural hazards can be projected and prepared for, such events are usually unanticipated and their consequences can be unpredictable. The natural hazards discussed in this section represent common and potentially disastrous natural events. Table 2.1 summarizes the characteristics, causes, and destructive potential of natural hazards.

Tropical Cyclonic Storms

A **tropical cyclone** is a "rotating, organized system of clouds and thunderstorms that originates over tropical or subtropical waters and has a closed low-level circulation."[9] The rotating pattern surrounds a calm center, popularly referred to as the *eye* of the storm. Tropical cyclonic

natural hazards: Natural hazards and disasters are emergency incidents arising from nonhuman causes. Such incidents are the consequence of phenomena arising from natural environmental conditions. Events such as earthquakes, hurricanes, tornadoes, and floods are examples of natural hazards.

tropical cyclone: A "rotating, organized system of clouds and thunderstorms that originates over tropical or subtropical waters and has a closed low-level circulation."

Table 2.1 Natural Hazards

Natural hazards are naturally occurring phenomena that are annually responsible for significant human and environmental destruction.

Natural Hazard	Event Profile		
	Primary Characteristics	Causes of the Event	Destructive Potential
Tropical cyclonic storms	Large storm with rotating high winds and thunderstorms	Formation over tropical water	Extensive with prolonged event duration
Earthquakes	Sudden shaking or bouncing	Sudden slip in an Earth fault	Extensive with short event duration
Tornadoes	Violent storm with funnel cloud formation	Strong thunderstorms	Extensive with surprising speed
Floods	Surging water and frequent occurrences	Rain storms and spring thaws	Normally limited because of predictability
Wildland fires	Fire outbreaks in woodland areas	Lightning and human negligence	Extensive natural destruction

tropical depression: A tropical cyclone with maximum sustained winds of up to 38 miles per hour.

tropical storm: A tropical cyclone with maximum sustained winds of 39 to 73 miles per hour.

hurricane: A tropical cyclone with maximum sustained winds of 74 miles per hour or higher.

typhoon: A hurricane in the western North Pacific.

cyclone: A hurricane in the Indian Ocean or South Pacific.

storm surge: A wall of water pushed forward by an approaching hurricane.

Hurricane Sandy: At its peak, Sandy was a Category 3 hurricane, but its eventual merger with a frontal weather system made it an unusually large storm, eventually becoming the geographically largest Atlantic hurricane ever recorded. For this reason, Hurricane Sandy was given the popular nickname "Superstorm Sandy."

storms begin over tropical water and can reach landfall with such intensity that destruction can be quite extensive along coastal regions. Depending on the size of cyclonic storms, they are capable of proceeding far inland and create an extensive swath of destruction. Tropical cyclonic storms are classified by the sheer force of their sustained winds as follows:

- **Tropical depression.** A type of tropical cyclone with maximum sustained winds of 38 mph (33 knots) or less.

- **Tropical storm.** A type of tropical cyclone with maximum sustained winds of 39 to 73 mph (34 to 63 knots).

- **Hurricane.** A type of tropical cyclone with maximum sustained winds of 74 mph (64 knots) or higher. In the western North Pacific, hurricanes are called **typhoons**; similar storms in the Indian Ocean and South Pacific are called **cyclones**.[10]

Hurricanes may last for weeks as they travel over tropical waters. Coastal areas in the path of hurricanes are usually struck by **storm surges**, which are walls of water pushed forward by approaching hurricanes. Chapter Perspective 2.2 discusses the measurement of hurricane intensities.

Case in Point: Superstorm Sandy's Swath of Destruction

Emergency relief systems and procedures were severely challenged in late 2012 when **Hurricane Sandy** struck the United States' eastern seaboard. The hurricane came ashore in the densely populated urban northeastern region of the United States, causing widespread damage. Hurricane Sandy had formed on October 22, 2012, and dissipated by

October 31, 2012. Hurricane Sandy was a Category 3 hurricane at its peak, but it was transformed into an unusually large storm following its merger with a frontal weather system and thereafter became the geographically largest Atlantic hurricane ever recorded. Hurricane Sandy was given the popular nickname "Superstorm Sandy" for this reason.

Superstorm Sandy's path impacted seven nations in the Caribbean, Atlantic, and North America. At least 286 people were killed, and the estimated $65 billion in damage in the United States made it the second-costliest hurricane in U.S. history, behind Hurricane Katrina.[11] Extensive damage was recorded along the entire Atlantic seaboard from Florida to Maine, with particularly severe damage occurring when Sandy struck the New Jersey shore. A total of 24 states were affected, including the Midwestern states of Wisconsin and Michigan.[12]

Emergency response efforts were affected by some political controversy in the United States, primarily because the storm occurred during national election season and because of perceptions that some political leaders delayed adequate relief funding. Significantly, the hurricane was so destructive that the U.S. National Hurricane Center and World Meteorological Organization reported that the name *Sandy* has been permanently retired from named tropical cyclonic storms in the Atlantic and Caribbean region.[13]

Earthquakes

The U.S. Geological Survey defines *earthquake* as "a term used to describe both sudden slip on a fault, and the resulting ground shaking and radiated seismic energy caused by the slip, or by volcanic or magmatic activity, or other sudden stress changes in the earth."[14] When an earthquake occurs, sudden shaking or bouncing of the earth's surface can cause extensive damage to the natural environment and built infrastructure. Although the intensity of earthquakes is measured using accepted measurement scales, it is impossible to predict when or where an earthquake will occur.

The two most accepted measurement scales for earthquakes are the **Modified Mercalli Intensity (MMI) Scale** and the **Richter Scale**. Arguably better known than the MMI, the Richter Scale measures the magnitude of earthquakes using a logarithmic mathematical model in whole numbers and decimal fractions. For example, a magnitude 5.3 might be computed for a moderate earthquake, and a strong earthquake might be rated as magnitude 6.3. Because of the logarithmic basis of the scale, each whole-number increase in magnitude represents a tenfold increase in measured amplitude; as an estimate of energy, each whole-number step in the magnitude scale corresponds to the release of about 31 times more energy than the amount associated with the preceding whole-number value.[15]

The MMI measures the intensity of earthquakes without using the mathematical models employed by the Richter Scale. The MMI is composed of 12 increasing levels of intensity, designated by Roman numerals, that range from imperceptible shaking to catastrophic destruction. Because it does not have a mathematical basis, it is an arbitrary ranking based on observed effects. The MMI value assigned to a specific site after an earthquake gives a more meaningful measure of severity to the nonscientist than does magnitude because intensity refers to the effects actually experienced at that place.[16]

Earthquakes that occur beneath the ocean floor can result in seismic sea waves. Earthquakes or other disturbances on the ocean floor can result in a series of massive waves known as **tsunamis**. They can travel extremely long distances at hundreds of miles per hour and strike land as high as 100 feet or more. Loss of life and damage to populated areas can be massive.

Chapter Perspective 2.3 discusses how the MMI reports the intensity of earthquakes.

CHAPTER PERSPECTIVE 2.3

Reporting Earthquake Intensity: The Modified Mercalli Intensity Scale

The Modified Mercalli Intensity (MMI) Scale reports the intensity of earthquakes by ranking the effects of an earthquake based on observed facts. Measured intensities are reported as follows:

"The following is an abbreviated description of the 12 levels of Modified Mercalli intensity.

I. Not felt except by a very few under especially favorable conditions.

II. Felt only by a few persons at rest, especially on upper floors of buildings.

III. Felt quite noticeably by persons indoors, especially on upper floors of buildings. Many people do not recognize it as an earthquake. Standing motor cars may rock slightly. Vibrations similar to the passing of a truck. Duration estimated.

IV. Felt indoors by many, outdoors by few during the day. At night, some awakened. Dishes, windows, doors disturbed; walls make cracking sound. Sensation like heavy truck striking building. Standing motor cars rocked noticeably.

V. Felt by nearly everyone; many awakened. Some dishes, windows broken. Unstable objects overturned. Pendulum clocks may stop.

VI. Felt by all, many frightened. Some heavy furniture moved; a few instances of fallen plaster. Damage slight.

VII. Damage negligible in buildings of good design and construction; slight to moderate in well-built ordinary structures; considerable damage in poorly built or badly designed structures; some chimneys broken.

VIII. Damage slight in specially designed structures; considerable damage in ordinary substantial buildings with partial collapse.

Tornadoes

Tornadoes are powerful and violent storms originating from strong thunderstorms. They are capable of striking populated areas with exceptional and surprising speed, potentially killing residents and destroying neighborhoods in a few seconds. Tornadoes are characterized by revolving funnel clouds emanating from thunderstorms high in the atmosphere and extending from the storm to the ground.

Although not as large as tropical cyclonic storms, the wind force of tornadoes can potentially far exceed that of Category 5 hurricanes; tornadoes have been measured at up to 300 miles per hour. Tornado intensity is reported in accordance with the **Enhanced Fujita–Pearson Scale** (Enhanced F-Scale) developed in 2007 by the National Oceanic and Atmospheric Administration (NOAA). NOAA describes the Enhanced F-Scale as follows:

> *The Enhanced F-scale . . . is a set of wind estimates (not measurements). . . . [It] uses three-second gusts estimated at the point of damage. . . . These estimates vary with height and exposure. Important: The 3 second gust is not the same wind as in standard surface observations. Standard measurements are taken by weather stations in open exposures, using a directly measured, "one minute mile" speed.*[17]

Table 2.2 reports the Enhanced F-Scale.

Tornadoes typically carve swaths of destruction along the course of their passage, often dozens of miles in length and potentially more than one mile in width. Not every tornado is preceded by signature funnel clouds, so that affected settlements or towns often have no notice—or extremely short notice—of approaching tornadoes. Particularly destructive results have occurred when clusters of tornadoes have struck in concentrated geographic areas. For example, between April 25 and April 28, 2011, more than 350 tornadoes struck the United States, the largest recorded tornado outbreak in history. During the outbreak, on April 27, an EF4 tornado struck the Tuscaloosa–Birmingham, Alabama, area. Its path of destruction stretched more than 80 miles, and more than 60 people were killed. Similarly, in May 2013, several tornadoes struck Oklahoma over several days. On May 20, an EF5 tornado from that storm system, with wind velocities exceeding 200 miles per hour, touched

Enhanced Fujita–Pearson Scale: Reporting of tornado intensity is done in accordance with the Enhanced Fujita–Pearson Scale (Enhanced F-Scale) developed in 2007 by the National Oceanic and Atmospheric Administration (NOAA).

Table 2.2 Reporting Tornado Intensity: The Enhanced Fujita–Pearson Scale						
FUJITA SCALE			**DERIVED EF SCALE**		**OPERATIONAL EF SCALE**	
F Number	**Fastest Quarter Mile (mph)**	**Three-Second Gust (mph)**	**EF Number**	**Three-Second Gust (mph)**	**EF Number**	**Three-Second Gust (mph)**
0	40–72	45–78	0	65–85	**0**	**65–85**
1	73–112	79–117	1	86–109	**1**	**86–110**
2	113–157	118–161	2	110–137	**2**	**111–135**
3	158–207	162–209	3	138–167	**3**	**136–165**
4	208–260	210–261	4	168–199	**4**	**166–200**
5	261–318	262–317	5	200–234	**5**	**Over 200**

Source: U.S. National Oceanic and Atmospheric Administration.

flood zones: National land areas differentiated by the Federal Emergency Management Agency (FEMA) according to their risk of flooding.

U.S. Federal Emergency Management Agency

▶ **Photo 2.4** Destruction in the aftermath of intense tornadoes that struck Moore, Oklahoma, in May 2013.

Special Flood Hazard Areas: Areas that "are at high risk for flooding" and have "a 26 percent chance of suffering flood damage during the term of a 30-year mortgage."

Non-Special Flood Hazard Areas: Areas that have a moderate to low risk of flooding.

down near Moore, Oklahoma. Hundreds of residents were injured, and 23 were killed.

Floods

Floods occur in all 50 states and in all U.S. territories, affecting more people and occurring more frequently than other types of disasters. Fortunately, the geographic location of floods is often predictable, and because of this predictability, FEMA has been able to demarcate national **flood zones** by identifying the risk of flooding in each area. Everyone lives in a flood zone; it's just a question of whether you live in a low-, moderate-, or high-risk area.[18]

Floods originate from a variety of causes, and damage occurs primarily as a result of human proximity to and development on floodplains. Spring thaws and rainstorms are common causes of floods in populated areas. To identify and standardize (label) the likelihood of flooding in specific locations, FEMA classifies flood-prone areas as follows:

- **Special Flood Hazard Areas.** Areas that "are at high risk for flooding" and have "a 26 percent chance of suffering flood damage during the term of a 30-year mortgage."[19]

- **Non-Special Flood Hazard Areas.** Areas with a moderate to low likelihood of flooding.

Because of the frequency and scope of floods, FEMA initiated the **National Flood Insurance Program** (NFIP) in 1978. The program regularly disburses funds for insurance claims, which have totaled nearly $40 billion since its inception. The NFIP maintains and regularly updates **Flood Insurance Rate Maps (FIRMs)**, which are "created . . . for flood-plain management and insurance purposes. . . . A FIRM will generally show a community's base flood elevations, flood zones, and floodplain boundaries."[20]

Wildland Fires

Wildland fires occur in woodland areas and are a perennial problem in the United States. These fires number in the tens of thousands annually, burning millions of acres each year.[21] Wildland fires are often ignited by lightning or human negligence and mistakes. The risk of wildland fires increases during dry, hot conditions, as occurred during California's highly destructive "burn seasons" in 2017 and 2018.

The potential for extensive infrastructure damage and injuries is significant, especially as populations continue to increase near potential fire zones. Property damage and environmental destruction are regular consequences of wildland fires. Environmental damage includes an increased likelihood of flooding and the destruction of natural habitats.

TECHNOLOGICAL SCENARIOS

Unlike natural disasters, which occur as a consequence of environmental, ecological, or geological events, technological scenarios originate in human conduct. Technological scenarios arise when human-created infrastructures and apparatuses malfunction, fail, or are destroyed. New technologies are constantly being created and applied in new and innovative ways. These include new software, machinery, structures, and materials. Examples of potential hazards caused by human-created technologies include grid infrastructure malfunctions, hazardous material accidents, and non-wildland fires. Table 2.3 summarizes technological hazards.

Grid Infrastructure Malfunctions

Communication and power grids are essential to domestic commerce and quality of life. Vital services such as medical and workplace efficiency are dependent on stable communication and power infrastructures. When malfunctions or failures occur, financial consequences can be quite expensive. Power blackouts, computer crashes, and other grid malfunctions are inconvenient and costly. Losses to businesses, government, and private individuals are calculated annually in billions of dollars.

Accidental grid infrastructure malfunctions can be caused by power line surges, lightning, human negligence, or other events. Intentional grid malfunctions may be caused by terrorist events or military attacks. For example, nations possessing nuclear weapons have known from the early years of the nuclear era that detonations of nuclear devices cause electromagnetic pulses (EMPs), which, at intensive levels, can damage or destroy electrical equipment. Weaponized non-nuclear EMP devices exist that can produce the same effect. As reported by the National Research Council and other organizations,[22] deliberate attacks and sabotage against grid infrastructure are very plausible scenarios.

U.S. Federal Emergency Management Agency

▶ **Photo 2.5** A FEMA relief agent interviews a civilian seeking assistance.

Table 2.3 Technological Hazards

Technological hazards are human-originating phenomena that are potentially capable of significant harm.

Technological Hazard	Event Profile		
	Primary Characteristics	Causes of the Event	Destructive Potential
Grid infrastructure malfunctions	Communication and power grid failures	Equipment malfunctions, human negligence, and deliberate sabotage	Uncommon but potential for wide disruption
Hazardous material accidents	Chemical, biological, and radiological accidents	Equipment malfunctions and human negligence	Uncommon but potential for extensive and prolonged event duration
Non-wildland fires	Fire outbreaks on human-constructed infrastructure	Natural phenomena, negligence, intent, and accidents	Extensive destruction

Hazardous Material Accidents

Hazardous materials include chemical, biological, and radioactive materials. These materials are essential for industry, agriculture, and product manufacturing. However, they are also highly regulated and potentially quite dangerous and toxic if mishandled or when accidents occur. Accidents may occur at points of production in facilities where hazardous materials are manufactured. They also occur during transportation of these materials via railways, seaways, and road transport. Some accidents can be quite catastrophic to nearby population centers and the natural environment.

Non-wildland Fires

non-wildland fires: Fires that occur in populated areas and affect human-constructed infrastructure.

Non-wildland fires affect human-constructed infrastructure and occur in populated areas. Many such fires have been structurally devastating to urbanized areas, as reported since the dawn of recorded history. Some fires are ignited by natural hazards, such as earthquakes, lightning, and volcanoes. Other fires may be ignited by arsonists, negligent residents, or industrial accidents. After ignition, the extent of devastation can be exacerbated considerably by flammable construction (such as wooden structures), facilities storing flammable chemicals and other materials, and high winds. Fires in densely populated areas are capable of spreading rapidly.

Statistically, non-wildland fires are a serious hazard. In 2011, according to the National Fire Protection Association, 49 percent of fires were classified as outside or other fires, 35 percent were structure fires, and 16 percent were vehicle fires.[23] That year, there were 1,389,500 fires, resulting in 3,005 civilian fatalities, 17,500 injuries, and damages totaling $11,659,000,000. Home fires caused 84 percent of civilian deaths.

This chapter's Global Perspective reports the disaster that occurred in 2011 when an extraordinarily strong tsunami struck Japan and caused the destruction of a nuclear power installation.

GLOBAL PERSPECTIVE

DISASTER IN JAPAN: TSUNAMI AND THE FUKUSHIMA NUCLEAR DISASTER

On March 11, 2011, an earthquake known as the Tohoku or Great Eastern Japan earthquake occurred approximately 40 miles from the coast of Japan. It was an extremely powerful earthquake, the fourth strongest since measurements began to be recorded in 1900. The earthquake resulted in massive tsunami waves up to approximately 130 feet high that struck as far as six miles inland. More than 18,000 people were killed by the tsunami, including approximately 2,500 missing and presumed dead.

The tsunami wave struck the Fukushima Daiichi nuclear plant, causing severe malfunctions in plant equipment. Three nuclear reactors at Fukushima eventually began to melt down and subsequently released radioactive material. Residents in the surrounding region were quickly evacuated and forced to abandon homes and belongings because of potential radiation contamination. Estimated deaths from plant workers' exposure and the evacuation are approximately 600. It is unknown how many more may become sickened by cancer and other diseases as a result of the released radioactive toxins.

Japanese and international health organizations regularly collaborate to monitor and study the site and the effect of the disaster on potential victims.

CHAPTER SUMMARY

This chapter introduced the all-hazards umbrella of the homeland security enterprise. This approach casts the broadest net in terms of homeland security policy over potential hazards, emergencies, and disasters. Although the founding concept for the homeland security enterprise was that its primary mission is to prepare for and respond to threats from violent extremists, the all-hazards approach has been adapted as needed during emergency events. This is a logical adaptation because of the enormous resources available to the homeland security network.

Terrorist-related hazards include the use of conventional weapons and explosives as well as unconventional chemical agents, biological agents, radiological agents, and nuclear weapons.

Natural hazards pose plausible scenarios that emanate from human proximity to the natural environment. They are largely unpredictable and potentially devastating in scope. Technological scenarios consider potential risks to built infrastructure. They involve hazards and threats to populated centers from emergencies that are often exacerbated by or originate from human activities.

In Chapter 3, the legal foundations of homeland security will be analyzed from the perspectives of global origins and domestic legislative and executive mandates.

DISCUSSION BOX

This chapter's Discussion Box is intended to stimulate critical debate about the difficulty of managing a coordinated homeland security enterprise.

Challenges of Integrating Homeland Security Intervention

Managing an efficient and coordinated homeland security response system is a complicated and challenging process.

The delivery of effective homeland security preparedness, incident response, and disaster recovery are inherently daunting administrative necessities. Nevertheless, coordinating these elements is absolutely vital to assuring an adequate response to all hazards that may arise.

Discussion Questions

1. Can the nation effectively prepare, respond, and recover regardless of what kind of hazard is anticipated?

2. If so, what is the best strategy for preparing for the eventual necessity of dealing with these hazards?

3. If not, should potential events be prioritized? Which ones?

4. Is the all-hazards approach to homeland security a reasonable use of limited resources?

5. Should an all-hazards approach be pursued regardless of cost?

KEY TERMS AND CONCEPTS

The following topics were discussed in this chapter and can be found in the glossary:

ON YOUR OWN

Get the tools you need to sharpen your study skills. SAGE edge offers a robust online environment featuring an impressive array of free tools and resources.

Access practice quizzes, eFlashcards, video, and multimedia at **edge.sagepub.com/martinhs3e**

RECOMMENDED WEBSITES

The following websites provide information about preparedness, response, and recovery efforts under the all-hazards umbrella:

Federal Emergency Management Agency: www.fema.gov

National Fire Protection Association: www.nfpa.org

National Oceanic and Atmospheric Administration: www.noaa.gov

National Weather Service: www.weather.gov

U.S. Geological Survey: www.usgs.gov

WEB EXERCISE

Using this chapter's recommended websites, conduct online research of organizations that monitor the incidence of hazards. Compare and contrast their findings.

1. What are the primary missions of these organizations?

2. How would you describe the differences among research, government, and local organizations?

3. In your opinion, are any of these organizations more comprehensive than other organizations? Less comprehensive?

To conduct an online search on research and monitoring organizations, activate the search engine on your Web browser and enter the following keywords:

"Disaster research"

"Emergency response organizations"

RECOMMENDED READINGS

The following publications provide introductions to the all-hazards designation of homeland security:

Beach, Michael. 2010. *Disaster Preparedness and Management*. Philadelphia: F. A. Davis.

Coppola, Damon P. 2011. *Introduction to International Disaster Management*. Burlington, MA: Butterworth Heinemann.

Griset, Pamela L. and Sue Mahan. 2013. *Terrorism in Perspective*, 3rd ed. Thousand Oaks, CA: Sage.

Hannigan, John. 2012. *Disasters Without Borders: The International Politics of Natural Disasters*. Boston: Polity Press.

Huder, Roger C. 2012. *Disaster Operations and Decision Making*. Hoboken, NJ: Wiley.

Phillips, Brenda D., David M. Neal, and Gary Webb. 2017. *Introduction to Emergency Management*. Boca Raton, FL: CRC Press.

Rubin, Claire B. 2012. *Emergency Management: The American Experience, 1900–2010*. Boca Raton, FL: CRC Press.

CHAPTER THREE

THE LEGAL FOUNDATIONS OF HOMELAND SECURITY

Chapter Learning Objectives

This chapter will enable readers to do the following:

1. Evaluate legal issues related to homeland security

2. Analyze the impact of terrorist environments on the types of legal responses enacted

3. Explain the impact of international counterterrorist legal precedent

4. Discuss the statutory foundations of modern homeland security

Opening Viewpoint: Homeland Security and Applying the Rule of Law

Legalistic responses are law enforcement and law-related approaches for promoting homeland security and managing domestic terrorist environments. These responses apply policies designed to use norms of criminal justice and legal procedures to promote homeland security, investigate alleged threats, and punish those who commit acts of political violence. Legislation, criminal prosecutions, and incarceration are typical policy measures. Challenges arise in the application of law enforcement principles, domestic laws, and international agreements.

Law Enforcement Principles

Because acts of terrorism are considered inherently criminal behaviors under the laws of the United States, state and federal law enforcement agencies often play a major role in homeland security operations. The organizational profiles of these agencies vary from state to state and at different levels of government, with some agencies having large police establishments with extensive homeland security contingencies and others relying more on other police agencies.

Domestic Legal Challenges

An important challenge for lawmakers in democracies such as the United States is balancing the need for counterterrorist legislation against the protection of constitutional rights. In severely strained terrorist environments, it is not uncommon for nations—including democracies—to pass authoritarian laws that promote social order at the expense of civil liberties. Policymakers usually justify these measures by using a balancing argument in which the greater good is held to outweigh the suspension of civil liberties. Severe threats to the state are sometimes counteracted by severe laws.

International Agreements

International law is based on tradition, custom, and formal agreements between nations. It is essentially a cooperative concept because there is no international enforcement mechanism that is comparable to domestic courts, law enforcement agencies, or criminal codes. All of these institutions exist in some form at the international level, but it should be remembered that nations voluntarily recognize their authority. They do this through formal agreements. Bilateral (two-party) and multilateral (multiparty) agreements are used to create an environment that is conducive to legalistic order maintenance.

Prior to adoption of the term *homeland security* in the modern era, a number of terrorism-related laws and treaties were enacted by the United States and the international community. A significant proportion of these laws and treaties were passed to implement cooperative international counterterrorist conventions. In the modern era, the emphasis and focus of terrorism-related laws and treaties consistently evolved to counter new permutations in the global terrorist environment. Thus, during the decades preceding legislation specifically related to homeland security, domestic laws and international conventions were enacted in response to unprecedented international terrorist activity and state sponsorship of violent extremists.

After the September 11, 2001, terrorist attacks, the United States passed new legislation to formally institutionalize homeland security as a new policy concept. In the aftermath of September 11, the United States essentially maintained its previous legislative initiatives and upgraded existing antiterrorist legal approaches. The nation also enacted new homeland security legislation. In this regard, the legal foundations of homeland security are rooted in efforts among nations and within the United States in recent decades to counter international terrorism and state sponsorship of violent extremists.

The discussion in this chapter will review the following:

- Background: Legal precedent and homeland security

- Homeland security: Domestic statutory authority in the United States

- In perspective: Homeland security, counterterrorism, and the law

BACKGROUND: LEGAL PRECEDENT AND HOMELAND SECURITY

During the early 1970s, members of the international community began to enact domestic laws and cooperative treaties to provide adequate legal underpinnings for countering the manner of terrorist violence prevalent during this historical period. Enactment occurred in roughly two phases: During the first phase, international laws and treaties were written to address threats from terrorists on the world stage and growing state sponsorship of these terrorists. During the second phase, new criminal and administrative laws were written that changed the domestic security environment. New criminal laws were enacted that imposed enhanced criminal penalties when already-recognized crimes were committed during terrorist incidents. The new administrative laws and regulations created—especially in the case of the United States after September 11, 2001—a new homeland security environment.

Domestic and International Law: Policy Challenges

Domestic and international counterterrorist laws inherently generate policy questions and political challenges for governments and the international community, which necessarily are the only institutions capable of enforcing legal mandates.

In the United States, domestic policies are affected by international and domestic counterterrorist laws when they are enacted in response to terrorist events. In this regard, homeland security policies adapt to emerging legal environments, which, in turn, must adapt to emerging security environments. Thus, homeland security initiatives reflect legal parameters established under law as well as international and domestic security environments.

Domestic Law and Policy Challenges

As discussed in this chapter's Opening Viewpoint, there exists a delicate challenge of balancing counterterrorist legislation and protecting constitutional and human rights. Even in democracies, serious threats to state authority and stability may result in legalized authoritarian options. Further analysis of civil liberty issues is provided in Chapter 4.

Scott Peterson/Liaison

▶ **Photo 3.1** A typical propaganda portrait commissioned by Iraqi dictator Saddam Hussein. Although Hussein had minimal (if any) ties to Islamist terrorist groups such as al-Qaeda, his regime did provide support and safe haven to a number of wanted nationalist terrorists such as Abu Nidal.

International Law and Policy Challenges

International law is based on tradition, custom, and formal agreements between nations. It is essentially a cooperative concept because there is no international enforcement mechanism comparable to domestic courts, law enforcement agencies, or crimes codes. Although all of these institutions exist in some form at the international level, it should be remembered that nations voluntarily recognize their authority because they are not beholden to any national governing body or constitution.

Nations recognize international legal authority through formal international agreements. Bilateral (two-party) and multilateral (multiparty) agreements are used to create an environment conducive to international order maintenance under the authority of international law. These formal agreements include treaties.

The Historical Context: The Global Threat From International Terrorism

The modern era of international terrorism, in which terrorists were often supported by rival governments, emerged as a considerable security challenge during the 1960s and 1970s. This brand of terrorism quickly became a predominant threat to the international community and societies. The primary reason was uncomplicated: International terrorism emerged as a strategic option during a period in history when violent extremists experimented with ways to publicize their otherwise local grievances.

Reasons for International Terrorism

International terrorism is terrorism that "spills over" onto the world's stage, usually to focus world attention on an otherwise domestic conflict. Targets are selected because of their value as symbols of international interests and the impact that attacks against these targets will have on a global audience.

Terrorism in the international arena first became a common feature of political violence during the late 1960s, when political extremists began to appreciate the value of allowing their revolutionary struggles to be fought in a global arena. By doing so, relatively low-cost incidents reaped significant propaganda benefits that were impossible when radicals limited their revolutions to specific regions or national boundaries. Skillful use of social networking media and the Internet by modern movements such as al-Qaeda and the Islamic State of Iraq and al-Sham (ISIS) not only promulgated their message but also served as effective recruiting tools.

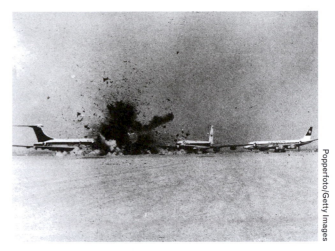

Poppefoto/Getty Images

▶ **Photo 3.2** Aircraft explode at Dawson's Field in Jordan during Black September, 1970. The hijackings and intense fighting afterward marked the beginning of a period when international airline hijackings and Palestinian attacks became common events.

In the modern era of immediate media attention, small and relatively weak movements concluded that worldwide exposure could be achieved by committing acts of political violence against international symbols. These groups discovered that politically motivated hijackings, bombings, assassinations, kidnappings, extortion, and other criminal acts can be quite effective when conducted under an international spotlight. Thus, the international realm guaranteed some degree of attention and afforded greater opportunities for manipulating the world's political, popular, and media sentiments.

Reasons for State Sponsorship of International Terrorism

In the latter half of the twentieth century—and especially in the latter quarter of the century—many governments used terrorism as an instrument of foreign policy. As a policy option, state-sponsored terrorism was a logical option because states cannot always deploy conventional armed forces to achieve strategic objectives. As a practical matter for many governments, it is often logistically, politically, or militarily infeasible to directly confront an adversary. Sponsorship of terrorism is a relatively acceptable alternative for states pursuing an aggressive foreign policy.

Terrorists who sought resources that would allow their cause to spill over into the international arena for maximum media exposure often found willing state sponsors to facilitate them. With the birth of this new security environment, the international community was challenged to design a legal framework for responding to rogue-state sponsors and terrorists who used the world as a global battleground. The approach of the United States was to officially designate some governments as state sponsors of terrorism and some extremist movements as foreign terrorist organizations.

Legitimizing Antiterrorist Legal Authority: Classifying State Sponsors and Foreign Terrorist Organizations

Country Reports on Terrorism is regularly published by the U.S. Department of State. The document identifies and lists **state sponsors of terrorism** and **foreign terrorist organizations (FTOs)**. *State sponsors of terrorism* refers to governments that support terrorist

Country Reports on Terrorism: The U.S. Department of State regularly publishes a document titled *Country Reports on Terrorism*, which identifies and lists state sponsors of terrorism and foreign terrorist organizations (FTOs).

state sponsors of terrorism: Governments that support terrorist activity.

foreign terrorist organizations (FTOs): *Country Reports on Terrorism* reports an updated list of organizations designated by the State Department as FTOs. This list is regularly revised in compliance with Section 219 of the Immigration and Nationality Act.

activity, and the term *FTOs* refers to insurgent terrorist groups. These lists are important for framing the foreign and domestic policies of the United States as well as for creating a legal foundation for responding to terrorist incidents and allegations that perpetrators received sponsorship from a government.

Legal Sanctions Against State Sponsors of Terrorism

Country Reports on Terrorism 2017 designated the following countries as state sponsors of terrorism: Democratic People's Republic of Korea (North Korea), Iran, Sudan, and Syria. The document summarizes the potential legal implications and sanctions of this designation as follows: "In order to designate a country as a State Sponsor of Terrorism, the Secretary of State must determine that the government of such country has repeatedly provided support for acts of international terrorism. Once a country has been designated, it continues to be a State Sponsor of Terrorism until the designation is rescinded in accordance with statutory criteria."

A wide range of sanctions are imposed as a result of a *state sponsor of terrorism* designation, including

- A ban on arms-related exports and sales.

- Controls over exports of dual-use items, requiring 30-day Congressional notification for goods or services that could significantly enhance the terrorist-list country's military capability or ability to support terrorism.

- Prohibitions on economic assistance.

- Imposition of miscellaneous financial and other restrictions.[1]

Interestingly, the State Department's list has historically included countries that significantly reduced their involvement in terrorism, such as Sudan. Also, the list of designated state sponsors of terrorism is dynamic and the designation is sometimes removed, as indicated by the examples of Cuba, Libya, and Iraq, countries which, at one time, were perennially included on the list. Libya's renunciation of support for dissident groups and its cooperation with the world community during the 2000s led to its removal from the list in 2006; the fall of the regime of Muammar Qaddafi ended the government that had sponsored terrorist groups in the past. Iraq, which had also been perennially designated on the list of state sponsors, was removed in October 2004 in the aftermath of the U.S.-led invasion and overthrow of the regime of Saddam Hussein. The North Korean government officially renounced its sponsorship of terrorism and was removed from the list in 2008 but was returned to the list in 2017. Cuba had been a perennial designee but was removed from the list in May 2015 during the process of restoring full diplomatic relations between the United States and Cuba.

Foreign Terrorist Organizations

Country Reports on Terrorism 2017 also reported an updated list of organizations designated by the State Department as foreign terrorist organizations. This list is regularly revised in compliance with Section 219 of the Immigration and Nationality Act (INA) in the following manner:

1. It must be a foreign organization.

2. The organization must engage in terrorist activity, as defined in section 212(a)(3)(B) of the INA (8 U.S.C. § 1182(a)(3)(B)), or terrorism, as defined in section

140(d)(2) of the Foreign Relations Authorization Act, Fiscal Years 1988 and 1989 (22 U.S.C. § 2656f(d)(2)), or retain the capability and intent to engage in terrorist activity or terrorism.

3. The organization's terrorist activity or terrorism must threaten the security of U.S. nationals or the national security (national defense, foreign relations, or the economic interests) of the United States.[2]

Country Reports on Terrorism states that this list is a crucial component of counterterrorist efforts because "FTO designations play a critical role in the fight against terrorism and are an effective means of curtailing support for terrorist activities."[3]

Chapter Perspective 3.1 reports designated FTOs from the 2017 *Country Reports on Terrorism*.

CHAPTER PERSPECTIVE 3.1

Foreign Terrorist Organizations

The Department of State regularly publishes lists of organizations and movements deemed to be foreign terrorist organizations (FTOs). *Country Reports on Terrorism 2017* designated the following organizations to be FTOs:

Abdallah Azzam Brigades (AAB)

Abu Sayyaf Group (ASG)

Al-Aqsa Martyrs Brigade (AAMB)

Ansar al-Dine (AAD)

Ansar al-Islam (AAI)

Ansar al-Shari'a in Benghazi (AAS-B)

Ansar al-Shari'a in Darnah (AAS-D)

Ansar al-Shari'a in Tunisia (AAS-T)

Army of Islam (AOI)

Asbat al-Ansar (AAA)

Aum Shinrikyo (AUM)

Basque Fatherland and Liberty (ETA)

Boko Haram (BH)

Communist Party of Philippines/New People's Army (CPP/NPA)

Continuity Irish Republican Army (CIRA)

Gama'a al-Islamiyya (IG)

Hamas Haqqani Network (HQN)

Harakat ul-Jihad-i-Islami (HUJI)

Harakat ul-Jihad-i-Islami/Bangladesh (HUJI-B)

Harakat ul-Mujahideen (HUM)

Hizballah Hizbul Mujahedeen (HM)

Indian Mujahedeen (IM)

Islamic Jihad Union (IJU)

Islamic Movement of Uzbekistan (IMU)

Islamic State of Iraq and Syria (ISIS)

Islamic State's Khorasan Province (ISIS-K)

ISIL-Libya ISIS Sinai Province (ISIS-SP)

Jama'atu Ansarul Muslimina Fi Biladis-Sudan (Ansaru)

Jaish-e-Mohammed (JeM)

Jaysh Rijal Al-Tariq Al-Naqshabandi (JRTN)

Jemaah Ansharut Tauhid (JAT)

Jemaah Islamiya (JI)

Jundallah Kahane Chai Kata'ib Hizballah (KH)

(Continued)

(Continued)

Kurdistan Workers' Party (PKK)

Lashkar e-Tayyiba (LeT)

Lashkar i Jhangvi (LJ)

Liberation Tigers of Tamil Eelam (LTTE)

Mujahidin Shura Council in the Environs of Jerusalem (MSC)

Al-Mulathamun Battalion (AMB)

National Liberation Army (ELN)

Al-Nusrah Front (ANF)

Palestine Islamic Jihad (PIJ)

Palestine Liberation Front – Abu Abbas Faction (PLF)

Popular Front for the Liberation of Palestine (PFLP)

Popular Front for the Liberation of Palestine-General Command (PFLP-GC)

Al-Qa'ida (AQ)

Al-Qa'ida in the Arabian Peninsula (AQAP)

Al-Qa'ida in the Indian Subcontinent (AQIS)

Al-Qa'ida in the Islamic Maghreb (AQIM)

Real IRA (RIRA)

Revolutionary Armed Forces of Colombia (FARC)

Revolutionary People's Liberation Party/Front (DHKP/C)

Revolutionary Struggle (RS)

Al-Shabaab (AS)

Shining Path (SL)

Tehrik-e Taliban Pakistan (TTP)

Discussion Questions

1. How political is the decision to place a group on the FTO list?

2. What factors would you use to place groups on the FTO list?

3. Should *all* politically violent groups be placed on the FTO list?

Source: https://www.state.gov/j/ct/rls/crt/2017/282850.htm.

The Global Response: International Law and Counterterrorist Cooperation by the World Community

Multinational agreements between partner countries are known as **international conventions**. International conventions state that specified protocols and procedures will be respected among signatories. The quality of terrorist threats existing during specific historical periods is reflected in the underlying purposes of international conventions enacted at the time. The first modern international treaties (agreements) were written during the 1970s to counter recurrent international terrorist incidents, such as hostage taking and aircraft hijacking. During this period, three international conventions prohibited the taking of hostages, and four international conventions were enacted to protect civilian aircraft. During the 1980s, the hijacking of the cruise ship *Achille Lauro* in the Mediterranean Sea and attacks on airports led to the passage of international conventions prohibiting attacks on civilian passenger ships, civilian airports, and related targets. During the 1990s, international conventions focused on prohibiting the funding of terrorist organizations and mass-casualty bombings.

Conventions and treaties represent signatories' agreement to extradite terrorist perpetrators or to prosecute them if they are taken into custody within a signatory's borders. With few exceptions, actual definitions of terrorism are absent from conventions and treaties. The reason is that their focus is on criminal behavior rather than the motivations of perpetrators.

International conventions and treaties do not have the force of law in the United States without the passage of enabling legislation by Congress. Procedurally, the United States can become a signatory to an international convention, and then Congress must enact the requisite enabling legislation to grant it legal status within the United States. Thus, legislation such as

international conventions:
Multinational agreements between partner countries that state that specified protocols and procedures will be respected among signatories. The underlying purposes of international conventions enacted during specific historical periods reflect the quality of terrorist threats existing at the time.

Table 3.1 International Law and Cooperation: Conventions on Terrorism

International terrorist incidents have historically resulted in international cooperative treaties and conventions. The following table summarizes selected international antiterrorist conventions.

Perceived Threat	Response and Purpose	
	International Legal Response	**Purpose**
Airline hijacking	Tokyo Convention of 1963 Hague Convention of 1970 Montreal Convention of 1971	Mitigate property and human losses Extradition of hijackers Severe penalties for attacks on airports and aircraft
Attacks on embassies and diplomats	Convention to Prevent Terrorism Against Targets of International Significance Prevent & Punish Crimes Against Internationally Protected Persons (1973)	Recognize inviolability of diplomatic missions International framework for suppressing terrorist attacks
Transnational terrorist activity	Extradition treaties	Binding over terrorist suspects
Weak international legal institutions	International tribunals	Bring perpetrators of crimes against humanity to justice

the Anti-Hijacking Act of 1974 was passed by Congress to prohibit the hijacking of aircraft anywhere within the jurisdiction of the United States; it also claims jurisdiction of hijackers found in the United States regardless of where the initial hijacking occurred.

Nations enter into treaties to create predictability and consistency in international relations. When threats to international order arise, such as hijackings, kidnappings, and havens for wanted extremists, the international community often enters into multilateral agreements to manage the threat.

Table 3.1 summarizes international conventions and their purpose, and the following examples illustrate the nature of multilateral counterterrorist agreements.

Protecting Diplomats

In reply to repeated attacks on embassies and assaults on diplomats in the late 1960s and early 1970s, several international treaties were enacted to promote cooperation in combating international terrorism against diplomatic missions. These treaties included the following:

- **Convention to Prevent and Punish Acts of Terrorism Taking the Form of Crimes Against Persons and Related Extortion That Are of International Significance**. This was a treaty among members of the Organization of American States that "sought to define attacks against internationally protected persons as common crimes, regardless of motives."[4] The purpose of the agreement was to establish common ground for recognizing the inviolability of diplomatic missions.

Convention to Prevent and Punish Acts of Terrorism Taking the Form of Crimes Against Persons and Related Extortion That Are of International Significance: A treaty among members of the Organization of American States that "sought to define attacks against internationally protected persons as common crimes, regardless of motives."[4]

- **Prevention and Punishment of Crimes Against Internationally Protected Persons, Including Diplomatic Agents.** Adopted by the United Nations in 1973, this was a multilateral treaty seeking to establish an internationally recognized common framework for suppressing extremist attacks against persons protected by internationally recognized status.

Attacks against foreign diplomatic posts can have significant political ramifications domestically. An instructive case in point is the September 2012 assault on U.S. diplomatic installations in Benghazi, Libya. Chapter Perspective 3.2 discusses this incident and its political effect.

CHAPTER PERSPECTIVE 3.2

The Benghazi Attack

Islamist insurgents attacked two U.S. diplomatic facilities in Benghazi, Libya, on the night of September 11, 2012. U.S. ambassador John Christopher Stevens was killed during the attack, as was a foreign service officer (FSO) and two Central Intelligence Agency (CIA) security contractors. The insurgents first attacked and burned the U.S. diplomatic mission where the ambassador and FSO were killed and later, in the early morning, attacked the annex, where the CIA security contractors died. More than 30 American personnel were rescued from the diplomatic mission.

An international U.S. emergency response effort was activated. When the attack began, a surveillance drone was dispatched to the site. Secretary of Defense Leon Panetta and Chairman of the Joint Chiefs of Staff General Martin Dempsey met with President Barack Obama at the White House to discuss response options. Secretary Panetta ordered Marine antiterrorist teams in Spain to prepare to deploy to Benghazi and Tripoli. Special operations teams were also ordered to prepare for deployment from Croatia, the United States, and a staging area in Italy. A small security team was also deployed from the U.S. Embassy in Tripoli and arrived in Benghazi.

Initial reports were that the attack was the work of a mob angered by the promulgation of an American-made film mocking the Prophet Muhammed and Islam. It was later determined that the attackers were armed Islamists.

The political backlash about the initial assessment was intense, with calls for the resignation of the U.S. ambassador to the United Nations. Congressional hearings were held by the Senate Select Committee on Intelligence, the House Foreign Affairs Committee, the House Oversight Committee, and the Senate Foreign Relations Committee. Officials called to testify included Secretary of State Hillary Clinton. A House of Representatives Select Committee on Benghazi was established, which quickly became deeply divided along partisan political lines.

The attack and its aftermath created a partisan political crisis in the United States on the questions of what the facts of the crisis response actually were, whether information was intentionally obfuscated, and whether the congressional investigations were motivated more by partisan politics than objective fact-finding.

Discussion Questions

1. Was the response during the attack adequately managed?

2. What, in your opinion, were the causes of the subsequent political crisis?

3. Are protocols for the protection of U.S. diplomatic missions adequately adaptive to fluid security environments?

International Conventions on Hijacking Offenses

In response to the new phenomenon of airline hijackings in the late 1960s and early 1970s, the world community enacted a number of international treaties to promote cooperation in combating international terrorism directed against international travel services. These treaties included the following:

- **Tokyo Convention on Offences and Certain Other Acts Committed on Board Aircraft**. Enacted in 1963 as the first international airline crimes treaty, the Tokyo Convention required all signatories to "make every effort to restore control of the aircraft to its lawful commander and to ensure the prompt onward passage or return of the hijacked aircraft together with its passengers, crew, and cargo."[5] Partner countries therefore agreed to mitigate property and human losses, and to restore all parties to their original status prior to criminal events occurring on board an aircraft.

- **Hague Convention of 1970**. This treaty required signatories to extradite "hijackers to their country of origin or to prosecute them under the judicial code of the recipient state."[6] In this way, hijackers would be brought before legal authorities to be tried for their crimes.

- **Montreal Convention of 1971**. This treaty extended international law to cover "sabotage and attacks on airports and grounded aircraft, and laid down the principle that all such offenses must be subject to severe penalties."[7] Like hijackers brought to justice under the Hague Convention, perpetrators of crimes against airports and grounded aircraft would be prosecuted to the full extent of the law.

Hijacking incidents can have significant international political and diplomatic ramifications. An instructive case study is the September 1970 Black September hijackings and subsequent diplomatic crisis. Chapter Perspective 3.3 discusses this case.

Tokyo Convention on Offences and Certain Other Acts Committed on Board Aircraft: Enacted in 1963 as the first international airline crimes treaty, the Tokyo Convention required all signatories to "make every effort to restore control of the aircraft to its lawful commander and to ensure the prompt onward passage or return of the hijacked aircraft together with its passengers, crew, and cargo."

Hague Convention of 1970: This treaty required signatories to extradite "hijackers to their country of origin or to prosecute them under the judicial code of the recipient state."

Montreal Convention of 1971: This treaty extended international law to cover "sabotage and attacks on airports and grounded aircraft, and laid down the principle that all such offenses must be subject to severe penalties."

CHAPTER PERSPECTIVE 3.3

The Black September Crisis: Case Study and Historical Context

Palestinian hijackers attempted to hijack five airliners on September 6 and 9, 1970. Their plan was to fly all of the planes to an abandoned British Royal Air Force (RAF) airfield in Jordan, hold hostages, broker the release of Palestinian prisoners, release the hostages, blow up the planes, and thereby force the world to focus on the plight of the Palestinian people. On September 12, 255 hostages were released from the three planes that landed at Dawson's Field (the RAF base), and 56 were

kept to bargain for the release of seven Palestinian prisoners. The group then blew up the airliners.

Unfortunately for the hijackers, their actions greatly alarmed King Hussein of Jordan. Martial law was declared on September 16, and the incident led to civil war between Palestinian forces and the Jordanian army. Although the Jordanians' operation was precipitated by the destruction of the airliners on Jordanian soil, tensions had been building between the army and

(Continued)

Palestinian forces for some time. King Hussein and the Jordanian leadership interpreted this operation as confirmation that radical Palestinian groups had become too powerful and were a threat to Jordanian sovereignty.

On September 19, Hussein asked for diplomatic intervention from Great Britain and the United States when a Syrian column entered Jordan in support of the Palestinians. On September 27, a truce ended the fighting. The outcome of the fighting was a relocation of much of the Palestinian leadership and fighters to its Lebanese bases. The entire incident became known among Palestinians as Black September and was not forgotten by radicals in the Palestinian nationalist movement. One of the most notorious terrorist groups at the time took the name Black September, and the name was also used by the group led by terrorist Abu Nidal.

Discussion Questions

1. What role do you think this incident had in precipitating the PLO's cycles of violence?

2. What, in your opinion, would have been the outcome if the Jordanian government had not responded militarily to the Palestinian presence in Jordan?

3. How should the world community have responded to Black September?

Extradition Treaties

Extradition treaties require parties to bind over terrorist suspects at the request of fellow signatories. Strong extradition treaties and other criminal cooperation agreements are powerful tools in the war against terrorism.

When properly implemented, extradition agreements can be quite effective. However, these treaties are collaborative and are not easily enforceable when one party declines to bind over a suspect or is otherwise uncooperative. When this happens, there is little recourse other than to try to convince the offending party to comply with the terms of the treaty.

International Courts and Tribunals

International Court of Justice: The principal judicial arm of the United Nations. The court hears disputes between nations and gives advisory opinions to recognized international organizations.

International Criminal Court: A court established to prosecute crimes against humanity, such as genocide. Its motivating principle is to promote human rights and justice.

International Criminal Tribunal for the Former Yugoslavia (ICTY): The ICTY has investigated allegations of war crimes and genocide arising out of the wars that broke out after the fragmentation of Yugoslavia during the 1990s.

The United Nations has established several institutions to address the problems of terrorism, genocide, torture, and international crime. The purpose of these institutions is to bring the perpetrators of crimes against humanity to justice. They are international courts, and their impact can be significant when nations agree to recognize their authority. The following institutions were established to prosecute offenses in international forums:

- **International Court of Justice.** The International Court of Justice is the principal judicial arm of the United Nations. Its 15 judges are elected from among member states and sit for nine-year terms. The court gives advisory opinions to recognized international organizations and hears disputes between nations.

- **International Criminal Court (ICC).** The ICC was established to prosecute crimes against humanity, such as genocide. The promotion of human rights and justice is its motivating principle. In practice, this has meant that the ICC has issued arrest warrants for the prosecution of war criminals.

- **International Criminal Tribunal for the Former Yugoslavia (ICTY).** The fragmentation of Yugoslavia during the 1990s led to allegations of war crimes and genocide arising out of the wars that broke out afterward. These allegations were investigated by the ICTY. The fighting among Croats, Muslims, and Serbs was exceptionally brutal and occasionally genocidal. Several alleged war criminals, including former Yugoslavian president Slobodan Milosevic, have been brought before the court. Others remain at large but under indictment.

- **International Criminal Tribunal for Rwanda (ICTR).** The breakdown of order in Rwanda during the 1990s led to allegations of war crimes and genocide that resulted from internal fighting. These allegations were investigated by the ICTR. Hundreds of thousands died during the campaign of terror waged by mobs and paramilitaries. The indictments against suspected war criminals detail what can only be described as genocide on a massive scale.

International Criminal Tribunal for Rwanda (ICTR): The ICTR has investigated allegations of war crimes and genocide that resulted from the breakdown of order in Rwanda during the 1990s.

HOMELAND SECURITY: DOMESTIC STATUTORY AUTHORITY IN THE UNITED STATES

In 1996 and 2001, the U.S. Congress deliberated about and enacted counterterrorist legislation as adaptations to the newly emerging terrorist environment. The Anti-Terrorism and Effective Penalty Act of 1996 and the USA PATRIOT Act of 2001 were written into law as seminal statutes in the United States' adaptation to the modern terrorist environment. These statutes represent broad legalistic approaches to controlling emerging terrorist environments. Within the context of domestic security considerations, norms of criminal justice and legal procedures are incorporated into such legislation to investigate and punish those who commit acts of political violence. In the modern era, legislation, criminal prosecutions, and incarceration have become typical policy measures to strengthen domestic security.

Table 3.2 summarizes domestic antiterrorist laws in the United States.

PEKKA SAKKI/AFP/Getty Images

▶ **Photo 3.3** Judges for the International Criminal Tribunal for the Former Yugoslavia. Such tribunals are designed to bring terrorists and violators of human rights to justice.

Table 3.2   Domestic Laws on Terrorism

Domestic terrorist incidents often result in legislative responses and the enactment of new laws. The following table summarizes selected domestic antiterrorist laws.

Perceived Threat	Response and Purpose	
	Domestic Legal Response	Purpose
Domestic terrorism	Anti-Terrorism and Effective Death Penalty Act of 1996	First omnibus antiterrorist legislation
The new terrorism	USA PATRIOT Act of 2001 Department of Homeland Security Act of 2002 USA PATRIOT Improvement and Reauthorization Act of 2005	Comprehensive omnibus antiterrorist legislation Remedy political and operational disarray for domestic security Renewal and amendment of 2001 USA PATRIOT Act

The Anti-Terrorism and Effective Death Penalty Act of 1996

Anti-Terrorism and Effective Death Penalty Act of 1996: The United States passed its first comprehensive counterterrorism legislation, titled the Anti-Terrorism and Effective Death Penalty Act, in 1996. The purpose of the Anti-Terrorism Act was to regulate activity that could be used to mount a terrorist attack, provide resources for counterterrorist programs, and punish terrorism.

In 1996, during the administration of President Bill Clinton, the United States passed its first comprehensive counterterrorism legislation, titled the **Anti-Terrorism and Effective Death Penalty Act**. The purpose of the Anti-Terrorism Act was to provide resources for counterterrorist programs, punish terrorism, and regulate activity that could be used to mount a terrorist attack. It was an omnibus bill in the sense that it contained a multiplicity of provisions to accomplish its underlying purpose. These provisions included the following:

- A federal death penalty for deaths resulting from acts of terrorism

- The inclusion of so-called taggant agents in plastic explosives, which mark the time and place of their manufacture

- The ability to prosecute crimes against on-duty federal employees as federal (rather than state) offenses

- Funding for terrorism prevention, counterterrorism, and counterintelligence

- Stronger procedural controls on asylum, deportation, and entry into the country

- A prohibition on government and private business financial transactions with terrorist states

- Assignation of authority to the secretary of state for designating private groups as terrorist organizations and forbidding them to raise funds in the United States

The Anti-Terrorism and Effective Death Penalty Act was passed after the terrorist attack at Centennial Park during the Atlanta Olympics and the explosion of TWA Flight 800 near Long Island, New York. Although the Flight 800 disaster was later declared not to be an act of terrorism, officials considered the Anti-Terrorism Act to be a milestone in responding to domestic terrorism.

▶ **Photo 3.4** President George W. Bush delivers an address before signing the 2001 USA PATRIOT Act into law. The law represents the legal foundation for homeland security policy and procedures in the United States.

The USA PATRIOT Act of 2001

In the aftermath of the September 11, 2001, homeland attacks, the U.S. Congress quickly passed legislation with the intent to address the new security threat. On October 26, 2001, President George W. Bush signed this legislation into law. It was labeled the "Uniting and Strengthening America by Providing Appropriate Tools Required to Intercept and Obstruct Terrorism Act of 2001," commonly known as the USA PATRIOT Act. It was also an omnibus law, similar to the Anti-Terrorism and Effective Death Penalty Act, but much more comprehensive. Provisions of the USA PATRIOT Act include the following:

- Revision of the standards for government surveillance, including those pertaining to federal law enforcement access to private records

- Enhancement of electronic surveillance authority, such as the authority to tap into e-mail, electronic address books, and computers

- The use of *roving wiretaps* by investigators, which permit surveillance of any individual's telephone conversations on any phone anywhere in the country

TIM SLOAN/AFP/Getty Images

- Requiring banks to identify sources of money deposited in some private accounts and requiring foreign banks to report on suspicious transactions

- The use of nationwide search warrants

- Deportation of immigrants who raise money for terrorist organizations

- The detention of immigrants without charge for up to one week on suspicion of supporting terrorism

Debate about these and other provisions came from across the ideological spectrum. Civil liberties watchdog organizations questioned whether these provisions would erode constitutional protections. At the same time, conservatives questioned the possibility of government intrusion into individuals' privacy. To address some of these concerns, lawmakers included a *sunset provision* mandating that the USA PATRIOT Act's major provisions automatically expire unless periodically extended. Lawmakers also required the Department of Justice to submit reports on the impact of the act on civil liberties. For example, in early 2005, the House of Representatives and the U.S. Department of Justice advocated restriction of the act's ability to authorize access to certain personal records without a warrant. This resulted in the passage of the USA PATRIOT Improvement and Reauthorization Act of 2005, which is discussed later in this section.

The Department of Homeland Security Act of 2002

In the immediate aftermath of the September 11 terrorist attacks, early efforts to establish a homeland security enterprise quickly revealed a relatively fractured and turf-oriented federal bureaucracy. To remedy what quickly became highly publicized reporting of political and operational disarray within the domestic security effort, President George W. Bush, in June 2002, completely reorganized the American homeland security community by initiating a process that led to the enactment of the **Department of Homeland Security Act of 2002**. The act was signed into law by President Bush on November 25, 2002.

A large, cabinet-level Department of Homeland Security was created by the new law. Because of the apparent operational fragmentation of homeland intelligence and security— and the important fact that the original Office of Homeland Security had no administrative authority over other federal agencies—the new department absorbed the functions of several large federal agencies. The result of this massive reorganization was the creation of the third-largest federal agency, behind only the Department of Veterans Affairs and the Department of Defense in size.

Further discussion of this reorganization is provided in Chapter 5. Instructive representations of the alignment of federal agencies before and after the reorganization are also found in tables in Chapter 5.

The USA PATRIOT Improvement and Reauthorization Act of 2005

Sunset provisions were created as an integral component of congressional oversight and proactive management of renewal processes for the USA PATRIOT Act. The purpose of the sunset provisions was to counterbalance enhanced authority granted to the executive branch under the USA PATRIOT Act. Congress first renewed the USA PATRIOT Act in March 2006 after passage of the **USA PATRIOT Improvement and Reauthorization Act of 2005**. After intensive political debate, it incorporated compromise provisions that included the following:

Department of Homeland Security Act of 2002: President George W. Bush, in June 2002, initiated a process that completely reorganized the American homeland security community. This process led to the enactment of the Department of Homeland Security Act of 2002, which was signed into law on November 25, 2002. The new law created a large, cabinet-level Department of Homeland Security.

USA PATRIOT Improvement and Reauthorization Act of 2005: Congressional oversight and proactive management of renewal processes for the USA PATRIOT Act were enacted in sunset provisions to counterbalance enhanced authority granted to the executive branch under the USA PATRIOT Act. Thus, the USA PATRIOT Act was first renewed by Congress in March 2006 after passage of the USA PATRIOT Improvement and Reauthorization Act of 2005.

- Restrictions on federal agents' access to library records

- Enhanced penalties for financial support of terrorism

- Improved organizational coordination of criminal prosecutions against accused terrorists with the creation of a new position of assistant attorney general for national security within the U.S. Department of Justice

- Enhanced standards for the protection of mass transportation

- Improved information flow between law enforcement and intelligence officers

Subsequent to the USA PATRIOT Improvement and Reauthorization Act of 2005, reauthorization legislation was regularly considered and passed by the U.S. Congress and signed into law by sitting presidents. For example, reauthorization proposals were considered in 2006, 2010, and 2011.

The USA FREEDOM Act of 2015

The USA PATRIOT Act was superseded, with modifications, in June 2015 with the passage of the USA FREEDOM Act. Key provisions of the USA PATRIOT Act were restored, and other provisions were either modified or eliminated. USA FREEDOM is a colloquial acronym for "Uniting and Strengthening America by Fulfilling Rights and Ending Eavesdropping, Dragnet-collection and Online Monitoring."

Initial introduction of the USA FREEDOM Act in October 2013 was a reaction to the publication of classified National Security Agency (NSA) memoranda by defector Edward Snowden earlier in the year. The Snowden document leak revealed that the NSA engaged in bulk data collection of telecommunications records, including telephone records and Internet metadata. The NSA's program was perceived by many in Congress and elsewhere to be an example of unacceptable surveillance by intelligence agencies on the private communications of everyday Americans. The USA FREEDOM Act imposed strict limits on bulk data collection of telephone records and Internet metadata by intelligence agencies. It also limited government collection of data from specific geographic locations and specific telecommunications service providers. The act continued the practice of authorizing roving wiretaps and efforts to track possible lone-wolf terrorists.

IN PERSPECTIVE: HOMELAND SECURITY, COUNTERTERRORISM, AND THE LAW

Although the USA FREEDOM Act curtailed the surveillance authority of intelligence agencies, previous legislation (i.e., the Anti-Terrorism and Effective Death Penalty Act, USA PATRIOT Act, and Department of Homeland Security Act) effectively expanded overall executive power in the national effort to counter the threat of terrorism. Unlike previous legislation, the sweeping scope of these laws conferred enhanced powers to regulatory, security, and law enforcement agencies and sought to coordinate this authority within newly established and integrated administrative umbrellas. For example, executive agencies such as the Federal Bureau of Investigation and local law enforcement officers were granted enhanced surveillance and detention authority. Chapter Perspective 3.4 delineates examples of governmental authority that influence the policy direction of the homeland security enterprise.

CHAPTER PERSPECTIVE 3.4

Administrative Authority and the Homeland Security Enterprise

The homeland security enterprise is subject to the authority of executive, legislative, and judicial processes. These include the following examples:

Executive Orders

Declarations issued by presidents or governors that have legal authority over homeland security processes and policies. When based on prior statutory actions by the legislative branch, no further legislative consideration will occur.

Homeland Security Presidential Directives

Presidential policy directives to members of the homeland security enterprise. These are executive directives that are implemented by relevant authorities.

Judicial Decisions

Judicial decisions are rendered by federal and state courts. Such decisions have the authority of law within the jurisdiction of the rendering court. Appellate decisions likewise have constitutional authority within the scope of the rendering court's jurisdiction.

Presidential Directives

An overarching term referring to executive orders in general. Differing nomenclature may be used by different presidential administrations, but the administrative effect is generally the same. In effect, all presidential directives are orders establishing policies to be carried out by relevant executive authorities.

Statutory Authority

Legislatively driven authority granted to government agencies. Statutory authority establishes the parameters and limitations for agencies when implementing policies. These are promulgated as laws relevant to the homeland security enterprise.

Discussion Questions

1. What do you think about the scope of presidential and gubernatorial authority in ordering potentially wide-ranging administrative decisions?

2. What, in your opinion, should be the role of legislative bodies in formulating policies affecting the homeland security enterprise?

3. How should the three branches of government collaborate in assuring an efficient and effective homeland security enterprise?

A watchdog role was also conferred to the legislative branch of government over provisions of the USA PATRIOT Act through the incorporation of sunset provisions and reauthorization procedures. Congressional oversight counterbalances the enhanced executive authority contained in recent counterterrorist legislation. This process was deemed necessary because of concerns that the executive branch would be conferred unchecked authority absent periodic legislative review. Several judicial decisions have also been rendered that have checked executive authority, as indicated in the following holdings:

- *2004.* U.S. federal courts have authority to determine whether foreign terrorism suspects detained in Guantánamo, Cuba, are wrongfully imprisoned.

- *2006.* U.S. citizens arrested in the United States must be tried in the criminal court system. Also, a military tribunal system created in Guantánamo, Cuba, was declared unconstitutional because it was established without approval by Congress.

Mark Wilson/Getty Images

▶ **Photo 3.5** President George W. Bush and former Pennsylvania governor Tom Ridge. Ridge became the first secretary of the Department of Homeland Security.

- *2008*. Foreign terrorism suspects have a constitutional right to challenge their detention in Guantánamo, Cuba, in U.S. courts.

- *2013*. Customs officials must establish *reasonable suspicion* before conducting forensic examinations of laptops, cell phones, cameras, and other devices owned by U.S. citizens.

Enhanced authority is deemed necessary by supporters of homeland security legislation and, at the same time, criticized as too far-reaching by critics. Increased government authority is often viewed with skepticism and concern, usually within a political and social context where such authority is seen as being used to curtail civil liberties. However, the underlying policy rationale is that demonstrable threats posed by terrorists require coordinated action from national security agencies and criminal justice institutions. During the administrative crisis following the September 11 terrorist attacks, comprehensive legislation such as the Homeland Security Act was passed to strengthen the nation's capacity to prepare for, respond to, and recover from terrorist incidents. In the post–9/11 domestic security environment, statutory initiatives moved toward policy and administrative consolidation out of perceived necessity.

Thus, the modern era of homeland security was inaugurated by and initially defined by statutory responses to domestic security crises. Nevertheless, privacy and civil liberties considerations underlie many of the debates on and policy analyses of homeland security legislation. Further analysis of these considerations is provided in Chapter 4.

CHAPTER SUMMARY

This chapter introduced the legal foundations of homeland security. The period immediately preceding the current terrorist environment significantly influenced contemporary legal responses to terrorist threats. International terrorism and state sponsorship of terrorism during the 1970s motivated the international community to adopt cooperative conventions and treaties specifically written to address the international threat and state complicity. In the United States, enabling legislation was enacted to formally engage the international effort.

Domestically, comprehensive counterterrorist legislation was passed in response to terrorist threats in the late 1990s and early 2000s. The new laws created an extensive homeland security bureaucracy and redefined the government role in coordinating national domestic security initiatives. The Anti-Terrorism and Effective Death Penalty Act, USA PATRIOT Act, and Department of Homeland Security Act together created a new foundation for the new era of homeland security.

Thus, a statutory framework established a new homeland security environment in which significant authority was conferred to executive branch agencies. The overarching rationale for passing comprehensive homeland security legislation was to promote the creation of a coordinated administrative network that would efficiently prepare for terrorist incidents, respond to terrorist events, and restore stability afterward. This enhanced executive authority was subject to periodic legislative oversight and also became the subject of judicial decisions.

DISCUSSION BOX

The Utility of Homeland Security Legislation

This chapter's Discussion Box is intended to stimulate critical debate about the purpose of homeland security legislation.

Homeland security legislation has been integrated into the responsive process to terrorist threats. Such statutory initiatives are fundamental components for framing homeland security environments in the modern era. Some critics argue that these measures are too broad, whereas supporters insist that they are necessary for creating a legal framework to respond to verifiable terrorist threats.

Discussion Questions

1. How necessary are comprehensive legislative initiatives? Why?

2. How easily defensible are comprehensive statutes? State your position by defending such legislation.

3. How viable are criticisms of sweeping legislation? State your position by criticizing such statutes.

4. To what extent do new terrorist threats or incidents drive new legislative initiatives?

5. Should new terrorist threats or incidents drive new legislative initiatives?

KEY TERMS AND CONCEPTS

The following topics were discussed in this chapter and can be found in the glossary:

Anti-Terrorism and Effective Death Penalty Act of 1996   58

Convention to Prevent and Punish Acts of Terrorism Taking the Form of Crimes Against Persons and Related Extortion That Are of International Significance   53

Country Reports on Terrorism   49

Department of Homeland Security Act of 2002   59

foreign terrorist organizations (FTOs)   49

Hague Convention of 1970   55

international conventions   52

International Court of Justice   56

International Criminal Court   56

International Criminal Tribunal for Rwanda (ICTR)   57

International Criminal Tribunal for the Former Yugoslavia (ICTY)   56

Montreal Convention of 1971   55

Prevention and Punishment of Crimes Against Internationally Protected Persons, Including Diplomatic Agents   54

state sponsors of terrorism   49

Tokyo Convention on Offences and Certain Other Acts Committed on Board Aircraft   55

USA PATRIOT Improvement and Reauthorization Act of 2005   59

ON YOUR OWN

Get the tools you need to sharpen your study skills. SAGE edge offers a robust online environment featuring an impressive array of free tools and resources.

Access practice quizzes, eFlashcards, video, and multimedia at **edge.sagepub.com/martinhs3e**

RECOMMENDED WEBSITES

The following websites provide information about the legal foundations of homeland security as well as critical analyses of homeland security laws:

DHS Office of General Counsel: www.dhs.gov/office-general-counsel

FBI Law Enforcement Bulletin: www.fbi.gov/stats-services/publications/law-enforcement-bulletin/november-2012/supreme-court-cases-2011-2012-term

National Center for State Courts: www.ncsc.org

National Consortium for the Study of Terrorism and Response to Terrorism: www.start.umd.edu

National Counterterrorism Center: www.nctc.gov

U.S. Department of Justice: www.justice.gov/archive/ll/highlights.htm

USA PATRIOT Act: www.congress.gov/bill/107th-congress/house-bill/3162

WEB EXERCISE

Using this chapter's recommended websites, conduct online research on counterterrorist laws, including international conventions and domestic legislation.

1. What are the primary agendas of the international conventions?

2. How would you describe the differences between conventions and treaties passed before and after the September 11, 2001, terrorist attacks?

3. In your opinion, are any of these measures overly comprehensive? Not comprehensive enough?

To conduct an online search of laws and treaties, activate the search engine on your Web browser and enter the following keywords:

"Counterterrorism legislation"

"Counterterrorist treaties"

RECOMMENDED READINGS

The following publications provide discussion on the legal foundations of homeland security:

Beckman, James. 2007. *Comparative Legal Approaches to Homeland Security and Anti-Terrorism*. Aldershot, UK: Ashgate.

Shah, Niaz A. 2008. *Self-Defense in Islamic and International Law*. New York: Palgrave Macmillan.

Smith, Cary Stacy. 2010. *The Patriot Act: Issues and Controversies*. Springfield, IL: Charles C Thomas.

Walker, Clive. 2011. *The Law and Terrorism*. Oxford, UK: Oxford University Press.

CHAPTER FOUR

CIVIL LIBERTIES AND SECURING THE HOMELAND

Opening Viewpoint: Freedom of Reporting and Security Priorities

The term *homeland security* is common to the modern political lexicon and security environment. Although the term is new and originated within the context of the September 11, 2001, attacks and American policy adaptations, the underlying concept has been periodically applied during historical periods of national security and political crises. Controversial administrative measures were often implemented during these periods and were deemed necessary at the time. Restrictions on the reporting of information by the media is an example of such controversial measures.

The United States has periodically restricted media access to information about matters that affect security policy. This has occurred during times of crisis, and the logic is quite understandable: A policy decision was adopted that concluded that the war effort (or counterterrorism policy) requires limitations to be imposed to prevent information from helping the enemy and to prevent the enemy from spreading its propaganda.

The challenge for democracies is to strike a balance between governmental control over information—for the sake of national security—and unbridled propaganda. The following examples illustrate how the United States managed the flow of information during international crises:

- During the *Vietnam War*, journalists had a great deal of latitude to visit troops in the field and observe operations. Vietnam was the first "television war," so violent and disturbing images were broadcast into American homes on a daily basis. These reports were one reason why American public opinion turned against the war effort.

- During the 1991 *Persian Gulf War*, news was highly controlled. Unlike during the Vietnam War, the media received their

Chapter Learning Objectives

This chapter will enable readers to do the following:

1. Apply historical perspectives on civil liberties and homeland security

2. Describe how domestic security policies are designed to address threat environments existing at the time

3. Evaluate the importance of balancing civil liberties protections and domestic security necessities

4. Apply arguments from each side of the debate on how to balance civil liberty and homeland security

5. Analyze controversial options for promoting domestic security

The preamble to the U.S. Constitution states that its central purpose is to establish a government system providing security, peace, and welfare for the population and to secure liberty for the people: "We the People of the United States, in Order to form a more perfect Union, establish Justice, insure domestic Tranquility, provide for the common defence, promote the general Welfare, and secure the Blessings of Liberty to ourselves and our Posterity, do ordain and establish this Constitution for the United States of America."

The authors of the Constitution took great care to balance liberty and security and drafted the first ten amendments to the Constitution as a new Bill of Rights. In effect, the concepts of civil liberty, a just government, and general security are founding principles of the Republic.

There is a natural tension between the desire to safeguard human rights and the necessity of securing the homeland. This tension is reflected in spirited philosophical and political debates about how to accomplish both goals. It is also reflected in the fact that during historical periods when threats to national security existed, sweeping security measures were undertaken as a matter of perceived necessity. The implementation of these policies was often politically popular at the time—primarily because of the immediacy of the perceived threat—but questioned during objective reflection in later years.

The modern homeland security environment exists because of the attacks on September 11, 2001, and has resulted in the creation of extensive bureaucracies, the passage of new security-related laws, and the implementation of controversial counterterrorist measures. The new security environment and policies were adopted because of the immediacy of the existent threat to the homeland, yet questions nevertheless arose about the efficacy and ethics of some measures. For this reason, oversight on the potential civil liberties implications of security-related policies is embedded in the homeland security enterprise. For example, the Office for Civil Rights and Civil Liberties is an integral administrative component in the U.S. Department of Homeland Security.

The discussion in this chapter addresses the difficult balance between securing the homeland and preserving constitutionally protected civil liberties by examining the following perspectives:

- Security and liberty: The historical context

- Achieving security

- Balancing civil liberties and homeland security

SECURITY AND LIBERTY: THE HISTORICAL CONTEXT

To understand modern concerns about the correlation between achieving strong homeland security and preserving constitutionally protected civil liberties, it is necessary to evaluate this question within the context of several historical periods. Several historical eras integrated security-related policies into the fabric of domestic politics and society. The rationale was to protect the nation from perceived (and often verifiable) threats.

The United States has experienced several episodes of crisis where the American public and political leaders perceived a need to enact legally based policies to manage the crisis. Laws were passed because of fear and uncertainty precipitated by domestic or foreign threats. At the time, contemporary measures were deemed necessary and were, therefore, often quite popular. This was because domestic security measures were presented as practical necessities. However, although such policies enjoyed significant support, their implementation just as often stirred strong criticism and opposition. The constitutionality and ethics of these laws were called into question during reflection in post-crisis years.

Table 4.1 summarizes these historical periods, the perceived threat, selected counter-measures, and outcomes.

Table 4.1   Security and Civil Liberty: Historical Perspectives

Balancing the desire for domestic security while concurrently protecting civil liberties can be a complex process. The United States has experienced several historical periods in which countermeasures were implemented to thwart perceived threats to domestic security. The following table summarizes these historical periods, the perceived threat, selected countermeasures, and outcomes.

Historical Period	Perceived Threat and Responses		
	Perceived Threat	Selected Countermeasure	Outcome
Early Republic **(1797–1809)**	Enemy immigrants and critics of government	Passage of the Alien and Sedition Acts	Strong criticism and repeal of the laws
Civil War **(1860–1865)**	Confederate sympathizers in Union States	Suspension of the writ of habeas corpus	Successful suspension of the writ; 38,000 civilians detained
First Red Scare **(1919)**	Anarchist and communist terrorism and subversion	Palmer Raids, arrests, and deportations	Successful disruption of anarchist and communist organizations
Second Red Scare **(1930s to 1940s)**	Communist subversion	HUAC investigations and federal legislation	High-profile investigations and prosecutions

(Continued)

Table 4.1 (Continued)			
	Perceived Threat and Responses		
Historical Period	**Perceived Threat**	**Selected Countermeasure**	**Outcome**
Third Red Scare **(1950s)**	Communist subversion	McCarthy Senate hearings and investigations	Widespread denunciation of "McCarthyism"
Second World War **(1941–1945)**	Sympathy for Japan by Japanese Americans	Internment camps	Relocation of thousands of Japanese Americans; eventual reparations

The Early Republic and Civil War

Since its inception, the United States has periodically responded to perceived threats by empowering the federal government to restrict the liberties of legally specified groups. During the early republic, Congress passed a series of laws commonly referred to as the Alien and Sedition Acts. During the Civil War, the writ of habeas corpus was suspended under declared emergency authority assumed by the executive branch of government.

The Early Republic and the Alien and Sedition Acts

During the presidency of John Adams (1797–1801), the United States responded to growing concern over the possibility of engaging in war with France. France, which had militarily supported the United States during the American Revolution, was waging war in Europe during the early career of Napoleon Bonaparte. At this time, European monarchies were attempting to crush the French Revolution, but they fared badly when Bonaparte scored repeated victories. Some American leaders argued that the United States should assist French revolutionaries in their time of need; others opposed intervention in the European conflict.

John Adams's Federalist Party controlled Congress and supported Great Britain in its opposition to Bonaparte's France. The Democratic Republican Party supported France and enjoyed political favor from new immigrants, many of whom were French. Four laws were passed by the Federalist-controlled Congress granting authority to President Adams to suppress activism by immigrants. These statutes were known as the **Alien and Sedition Acts**, and the new laws placed significant restrictions on the liberty of political critics and immigrants.

Alien and Sedition Acts: Four laws passed during the administration of President John Adams granting Adams authority to suppress activism by immigrants. Known as the Alien and Sedition Acts, these statutes placed significant restrictions on the liberty of immigrants and political critics.

- *Alien Enemies Act.* The president was granted authority to deport or imprison citizens of "enemy" countries.

- *Alien Friends Act.* The president was granted authority to deport or imprison citizens of "friendly" countries if they were deemed to be dangerous.

- *Naturalization Act.* This act required immigrants to live in the United States for 14 years before being permitted to receive citizenship.

- *Sedition Act.* This act permitted the imprisonment of individuals for criticizing the government.

The Alien and Sedition Acts were strongly opposed as violating the First Amendment to the U.S. Constitution. During the presidential administration of Thomas Jefferson (1801–1809), who was a Democratic Republican, the four laws were amended, repealed, or allowed to expire. The Alien and Sedition Acts represent the first (but not the last) time in U.S. history when specified groups experienced the legalized abrogation of constitutional liberties. In later years, similar abrogations would be directed against specified ideological, political, and ethnic groups.

The Civil War and Suspension of Habeas Corpus

A writ of habeas corpus (Latin for "that you have the body") is an order from a judge demanding that an imprisoning authority deliver its prisoner to court for a determination on the constitutionality of the imprisonment. Should the judge determine that detention is unconstitutional, the prisoner must be released. Writs of habeas corpus represent a powerful counterweight to the authority of the state to detain individuals without bringing charges in a timely manner. Article III, Section 9 of the U.S. Constitution states, "The privilege of the writ of habeas corpus shall not be suspended, unless when in a case of rebellion or invasion the public safety may require it." Thus, any suspension of habeas corpus requires clear confirmation that rebellion or invasion is a threat to public safety, and absent such proof, the suspension is invalid.

During the Civil War, President Abraham Lincoln suspended the right to habeas corpus and ordered the military to detain an Ohio congressman, 31 legislators from Maryland, and hundreds of alleged Confederate sympathizers. President Lincoln made all detainees subject to the jurisdiction of military courts-martial rather than the civilian judiciary. The U.S. Supreme Court ruled that these measures were unconstitutional, but Lincoln ignored the Court's ruling, and during the Civil War, approximately 38,000 civilians were detained by the military. In this way, the executive branch of the government circumvented the judiciary by invoking its own interpretation of Article III, Section 9 as a justifiable suspension of habeas corpus.

The Civil War–era suspension of habeas corpus and the use of military courts-martial was the first precedent for applying such measures to alleged enemies of the state—by a similar rationale as that recently used to justify controversial detentions of modern enemy combatants in facilities such as Guantánamo Bay.

Communism and the Red Scares

"**Red Scares**" refer to several periods in U.S. history when perceived threats from anarchist, communist, and other leftist subversion created a generalized climate of political anxiety. In fact, these perceived threats were sometimes complemented by genuine security-related incidents and conspiracies by radical leftists. During these periods, sweeping security procedures were implemented that were later criticized as being overly broad or unconstitutional. Domestic terrorism in the modern era is further discussed in Chapter 8.

Anticommunist Red Scares occurred several times in the United States when national leaders sought to suppress the perceived threat of communist subversion. During the Red Scares, government officials reacted to communist activism by adopting authoritarian measures to preempt incipient sedition. The Red Scares took place during three periods in American history: first in the aftermath of the 1917 Bolshevik Revolution in Russia, again at the height of the Great Depression in the 1930s, and finally during the East–West tensions of the Cold War in the 1950s.

Red Scares: Periods in U.S. history when a generalized climate of political anxiety occurred in response to perceived threats from communist, anarchist, and other leftist subversion.

The First Red Scare

The First Red Scare began following the founding of the Communist Party USA (CPUSA) in 1919. The CPUSA was established during a global effort by communists to create an international movement to end capitalism.

In 1919, a series of letter bombs were intercepted after they were mailed to prominent officials and industrial executives. Additional bombs were detonated in several cities by violent extremists, including one directed against U.S. attorney general A. Mitchell Palmer. It was widely held at the time that leftist revolutionaries were responsible for the bombing campaign, particularly communists and anarchists. In response to these incidents, Palmer received authorization from President Woodrow Wilson to conduct a series of raids—the so-called **Palmer Raids**—against labor activists, including leftist labor groups, communists, anarchists, socialists, and mainstream American labor unions. Offices of many organizations were searched without warrants and shut down; thousands were arrested. Leftist leaders were arrested and put on trial, and hundreds of people were placed aboard ships and deported from the United States.

The legal foundations for the law enforcement crackdown against leftists were the Espionage Act of 1917 and the Sedition Act of 1918. An interesting postscript is that A. Mitchell Palmer was eventually prosecuted and convicted for misappropriation of government funds.

Palmer Raids: During a domestic bombing campaign allegedly conducted by communists and anarchists, President Woodrow Wilson authorized U.S. attorney general A. Mitchell Palmer to conduct a series of raids—the so-called Palmer Raids—against labor activists, including American labor unions, socialists, communists, anarchists, and leftist labor groups.

The Second Red Scare

The Second Red Scare began during the 1930s, at the height of the Great Depression. Communists and socialists enjoyed a measure of popularity during this period because of mass unemployment and the apparent crisis of capitalism. Fears grew that the uncertainty of the Great Depression would lead to widespread unrest encouraged by communist and socialist agitation.

Congress acted to halt the dissemination of leftist sentiment by establishing the House Un-American Activities Committee (HUAC) in 1938, which investigated alleged sedition, disloyalty, and other subversive activities by private individuals. Congress also passed the Smith Act of 1940, which made advocacy of the violent overthrow of the government a federal crime. High-profile investigations were conducted during this period, and through the late 1940s, a number of alleged communists were prosecuted. High-profile prosecutions occurred, including the investigation of Alger Hiss, a State Department official accused of being a communist spy.

McCarthyism: During the 1950s, Republican senator Joseph McCarthy of Wisconsin sought to expose communist infiltration and conspiracies in government, private industry, and the entertainment industry. The manner in which McCarthy promoted his cause was to publicly interrogate people from these sectors in a way that had never been done before: on television. Hundreds of careers were ruined, and many people were blacklisted—that is, nationally barred from employment. McCarthy was later criticized for overstepping the bounds of propriety, and the pejorative term *McCarthyism* has come to mean a political and ideological witch hunt.

The Third Red Scare

The Third Red Scare occurred during the 1950s, when Republican senator Joseph McCarthy of Wisconsin held a series of hearings to counter fears of spying by communist regimes, primarily China and the Soviet Union. The hearings reflected and encouraged a general fear that communists were poised to overthrow the government and otherwise subvert the "American way of life."

McCarthy sought to expose conspiracies and communist infiltration in the entertainment industry, government, and private industry. Hundreds of careers were ruined during the hearings, and many people were blacklisted—that is, nationally barred from employment. McCarthy was adept at promoting his cause. The manner in which he did so was to publicly interrogate people from these sectors by utilizing a mode that had never been extensively used before: on television. Critics later argued that McCarthy overstepped the bounds of propriety, and the pejorative term **McCarthyism** has come to mean an ideologically and politically motivated witch hunt.

Wartime Internment Camps

The attack on Pearl Harbor on December 7, 1941, by the Empire of Japan created a climate of fear against ethnic Japanese in the United States. Conspiracy scenarios held that domestic sympathizers would begin a campaign of sabotage and subversion on behalf of Japan. This would, in theory, be done in preparation for a Japanese invasion of the West Coast. Unfortunately, a prewar backdrop of racial prejudice against Asians in general became a focused animosity toward Asians of Japanese heritage. This combination of declared war, fear of subversion, and prejudice culminated in the relocation of ethnic Japanese from their homes to internment facilities.

Getty Images/Hulton Archive

▶ **Photo 4.1** Senator Joseph McCarthy giving testimony during the Red Scare of the 1950s. His accusations and methods were usually sensational and were widely denounced in later years.

The administration of President Franklin Delano Roosevelt established a War Relocation Authority, and the U.S. Army was tasked with moving ethnic Japanese to internment facilities on the West Coast and elsewhere. More than 100,000 ethnic Japanese, approximately two-thirds of whom were American citizens, were forced to relocate to the internment facilities. Internment facilities were operational until 1945, and most internees lost their property and businesses during the relocations. Reparation payments of $20,000 were authorized during the 1980s to be disbursed to surviving internees. In 1988, the U.S. government formally apologized for the internments by passing the **Civil Liberties Act** and declared the internment program unjust.

Civil Liberties Act: During the 1980s, reparation payments of $20,000 were authorized to be disbursed to survivors of the internment of Japanese Americans during the Second World War, and in 1988, the U.S. government passed the Civil Liberties Act, which formally apologized for the internments and declared the internment program unjust.

ACHIEVING SECURITY

Civil liberties advocates contend that a careful balance must be struck between achieving security and protecting civil liberties. To achieve security, government responses must be proportional to the perceived threat and measured in how they are implemented. At the same time, some civil liberties advocates argue that because government responses are usually reactive after a threat arises, a more permanent solution may be found by countering extremism through reform—in effect, the creation of an environment that counters conditions conducive to encouraging radical sentiment. The following discussion considers this argument within the context of balancing homeland security and civil liberty perspectives.

Practical Considerations: Civil Liberty and Government Responses

Homeland security experts must pragmatically concentrate on achieving several fundamental counterterrorist objectives. These objectives can realistically only minimize, rather than eliminate, terrorist threats, but nevertheless, they must be actively pursued. In a practical sense, counterterrorist objectives include the following:

- Disrupt and prevent domestic terrorist conspirators from operationalizing their plans.

- Deter would-be activists from crossing the line between extremist activism and political violence.

- Implement laws and task forces to create a cooperative counterterrorist environment.

- Minimize physical destruction and human casualties.

Balancing Theory and Practicality

It is clear that no single model or method for achieving security will apply across every scenario or terrorist environment. Because of this reality, the process for presenting counterterrorist models must include frameworks based on both theory and practical necessity. *Theoretical* models must reflect respect for human rights protections and balance this against options that may, out of necessity, include the use of force and law enforcement options. The *practicality* of these models requires them to be continually updated and adapted to emerging terrorist threats. With these adaptations, perhaps terrorism can be controlled to some degree by keeping extremists and violent dissidents off balance, thereby preventing them from having an unobstructed hand in planning and carrying out attacks or other types of political violence.

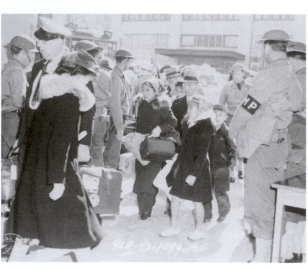

▶ **Photo 4.2** A Japanese American family in San Francisco being evacuated to an internment camp during the Second World War. The relocation program targeted Americans of Japanese ancestry in the aftermath of the attack on Pearl Harbor by the Empire of Japan.

As a matter of practical necessity, counterterrorist policy adaptations may conceivably require controlling the dissemination of information from the media or engaging in surveillance of private communications. Both options potentially challenge fundamental notions of civil liberty.

Regulating the Media

Freedom of the press is an ideal standard—and arguably an ideology—in the United States. The phrase embodies a conceptual construct that suggests that the press should enjoy the liberty to independently report information to the public, even when this information might involve national security or be politically sensitive. News editors and journalists, when criticized for their reports, frequently cite "the people's right to know" as a justification for publishing controversial information. The question is whether the right to know extends without restraint to information that may affect national security.

The counterpoint to absolute freedom of the press is regulation of the press. This issue arises when the media publish information about subjects that the public or the government would rather not consider. Regulation is theoretically a genuine option when matters of national security are at stake. When these and other concerns arise, policymakers and editors are challenged to address the following policy questions:

- Should the media be officially regulated?

- If regulation is desirable, how much regulation is acceptable?

In the United States, broad regulation of the media is not considered to be a politically viable option, and there is rarely an effort by government entities to officially suppress or otherwise regulate media content when media outlets decide to publish information. Rather, the first option by government agencies is to decline to release information by citing national security confidentiality. However, when government officials conclude that pending publication of information in the media may pose a threat to national security, the most common recourse is the U.S. judicial system, and thus, executive fiat is rarely attempted as an option.

Because of these limitations on the ability of officials to regulate the publication of information, the American media culture tends to rely on standards of journalistic self-regulation and media gatekeeping. Within the context of this system, the media will self-regulate the reporting and portrayal of potentially unsavory, controversial, or sensitive news. For example, the American media have rarely published images of casualties from terrorist incidents or the international war on terrorism. This system is sometimes not ideal, and Chapter Perspective 4.1 discusses the debate concerning national security, the reporting of terrorism, and regulation of the media.

Electronic Surveillance and Civil Liberty

Electronic surveillance has become a controversial practice in the United States and elsewhere. The fear is that civil liberties can be jeopardized by unregulated interception of telephone conversations, e-mail, telefacsimile, and other transmissions. Detractors argue that government use of these technologies can conceivably move well beyond legitimate

CHAPTER PERSPECTIVE 4.1

National Security, Reporting Terrorism, and Regulating the Media

In the United States, consensus exists that ethical standards should be observed when reporting terrorist incidents. These include the following:

> [Do not] serve as a spokesman/accomplice of the terrorists. . . .
>
> [Do not] portray terror as attractive, romantic, or heroic; honest portrayal of motives of terrorists. . . .
>
> Hold back news where there is clear and immediate danger to life and limb.
>
> Avoid . . . unchallenged terrorist propaganda. . . .
>
> Never try to solve a situation.[a]

In order to comply with these standards, one professional model is that of journalistic self-regulation. Journalistic self-regulation is sometimes referred to as media gatekeeping. If conducted under established standards of professional conduct, self-regulation obviates the need for official regulation and censorship. In theory, moral arguments brought to bear on the press from political leaders and the public will pressure them to adhere to model standards of fairness, accuracy, and objectivity.

This is, of course, an *ideal* free press environment; in reality, critics argue that journalistic self-regulation is a fluid and inconsistent process. The media report terrorist incidents using certain labels and often create a mood by spinning their reports. Some media—acting in the tabloid tradition—sensationalize acts of political violence so that very little self-regulation occurs. Critics argue that unregulated sensationalized reporting can harm national security, and therefore, outside regulation is necessary.

Internationally, many democracies occasionally regulate or otherwise influence their press community while, at the same time, advocating freedom of reporting. Some democracies selectively release information or release no information at all during terrorist incidents. The rationale is that the investigation of these incidents requires limitations to be placed on which information is made available to the public. This occurs as a matter of routine during wartime or other national crises.

Discussion Questions

1. Should the public's "right to know" take precedence over the regulation of potentially sensitive national security information?

2. Should the government rely exclusively on journalistic self-regulation to control the reporting of sensitive national security information?

3. Are adequate protections in place to control unjustifiable suppression of information by government authorities?

Note

a. See David L. Paletz and Laura L. Tawney, "Broadcasting Organizations' Perspectives," in *Terrorism and the Media*, ed. David L. Paletz and Alex P. Schmid (Newbury Park, CA: Sage, 1992), 129.

application against threats from crime, espionage, and terrorism. Absent strict protocols to rein in these technologies, a worst-case scenario envisions state intrusions into the everyday activities of innocent civilians. Should this happen, critics foresee a time when privacy, liberty, and personal security become values of the past.

Chapter Perspective 4.2 discusses an instructive case study involving the deployment of **Carnivore**, an early software surveillance tool created to monitor e-mail, and its successor, **DCS-1000**.

Carnivore: An early software surveillance tool created to monitor e-mail.

DCS-1000: The upgraded and renamed version of the Carnivore software surveillance tool.

Civil Liberty and Countering Extremism Through Reform

Ethnocentrism, nationalism, ideological intolerance, racism, and religious fanaticism are core motivations for terrorism. History has shown that coercive measures used to counter these tendencies are often only marginally successful. The reason is uncomplicated: A great deal of extremist behavior is rooted in passionate ideas, recent historical memories of conflict, and cultural tensions. Very importantly, injustice does occur, and ideologies or other expressions of identity are used to rouse opposition to injustice. Although coercion can eliminate cadres and destroy extremist organizations, sheer repression is a risky long-term solution. In fact, outright repression and the suppression of civil liberties can create a backlash in which members of the suppressed group feel justified in their violent resistance.

CHAPTER PERSPECTIVE 4.2

Carnivore: The Dawn of Internet Surveillance

In July 2000, it was widely reported that the FBI possessed a surveillance system that could monitor Internet communications. Called Carnivore, the system was said to be able to read Internet traffic moving through cooperating Internet service providers. All that was required was for Carnivore to be installed on an Internet provider's network at their facilities. Under law, the FBI could not use Carnivore without a specific court order under specific guidelines, much like other criminal surveillance orders.

The FBI received a great deal of negative publicity, especially after it was reported that the agency had evaded demands for documents under a Freedom of Information Act (FOIA) request filed by the Electronic Privacy Information Center (EPIC), a privacy rights group. Concern was also raised by critics when it was reported in November 2000 that Carnivore had been very successfully tested and that it had exceeded expectations. This report was not entirely accurate. In fact, Carnivore did not operate properly when it was used in March 2000 to monitor a criminal suspect's e-mail; it had inadvertently intercepted the e-mail of innocent Internet users. This glitch embarrassed the Department of Justice (DOJ) and angered the DOJ's Office of Intelligence Policy and Review.

By early 2001, the FBI gave Carnivore a less ominous sounding new name, redesignating the system DCS (Digital Collection System)-1000. Despite the political row, which continued well into 2002 (in part because of the continued FOIA litigation), Carnivore was cited as a potentially powerful tool in the new war on terrorism. The use of DCS-1000 after 2003 was apparently reduced markedly, allegedly because Internet surveillance was outsourced to private companies' tools. The program reportedly ended in 2005 because of the prevalence of significantly improved surveillance software.

Discussion Questions

1. Should the public be concerned that federal agencies are using surveillance technologies domestically?

2. How should ethical considerations be balanced against national security considerations in the domestic use of surveillance technologies?

3. At what point does the domestic use of surveillance technologies infringe on privacy?

Because extremism has historically originated primarily from domestic conflict (sometimes from national traumas such as invasions), efforts to counter domestic extremism must incorporate societal and cultural responses. A central consideration is that new societal and cultural norms often reflect demographic changes and political shifts. Dissent can certainly be repressed, but it is rarely a long-term solution absent preventive measures such as social reform, political inclusion, and protection of constitutional rights.

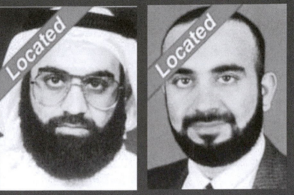

FBI/Getty Images

▶ **Photo 4.3** Al-Qaeda operative Khalid Sheikh Mohammed, who was interrogated using the waterboarding procedure during captivity in secret locations.

Case in Point: Data Mining by the National Security Agency

In June and July 2013, the British newspaper *The Guardian* published a series of articles reporting covert surveillance operations coordinated by the U.S. National Security Agency (NSA). These operations involved the acquisition of European and U.S. telephone metadata and Internet surveillance. First reports indicated that the operations were code-named **Tempora** (apparently a British operation cooperating with the NSA) and **PRISM**.[1] **Edward Snowden**, a former NSA contractor and Central Intelligence Agency (CIA) employee, became an international fugitive after leaking details of these operations to the media. The information was apparently delivered to a documentary filmmaker, the *Washington Post*, and *The Guardian*. Subsequent articles in *The Guardian* detailed another NSA operation, code-named **XKeyscore**, which apparently deployed a much more robust ability to collect

Tempora: A British covert surveillance operation conducted in cooperation with the U.S. National Security Agency.

PRISM: A covert surveillance operation coordinated by the U.S. National Security Agency.

Snowden, Edward: A former Central Intelligence Agency employee and National Security Agency contractor who leaked details of covert surveillance operations to the media prior to becoming an international fugitive. The information was apparently delivered to *The Guardian*, the *Washington Post*, and a documentary filmmaker.

XKeyscore: A covert surveillance operation, coordinated by the U.S. National Security Agency, capable of collecting real-time data on chat rooms, browsing history, social networking media, and e-mail.

Figure 4.1   XKeyscore and NSA Data Mining

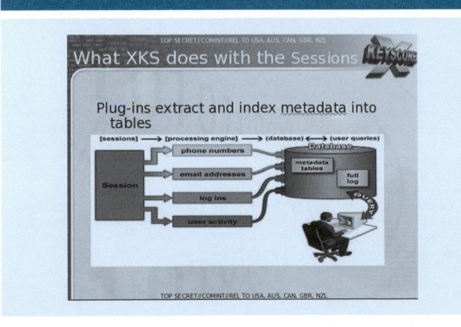

Source: Greenwald, Glenn. "XKeyscore: NSA Tool Collects 'Nearly Everything a User Does on the Internet,'" *The Guardian*, July 31, 2013.

online data.[2] According to the reports, XKeyscore was capable of collecting real-time data on social-networking media, chat rooms, e-mail, and browsing history.

These revelations began a vigorous debate in the United States and Europe about privacy, espionage, and whether the programs were justifiable. Civil libertarians questioned the legality of the extensive data-mining operations. In defense of the surveillance, intelligence officials commented that the NSA's program had thwarted approximately 50 terrorist plots in 20 countries, including at least 10 plots directed against the United States.[3]

Some independent activist organizations and outlets are keen to publish classified information without authorization. Chapter Perspective 4.3 discusses the case of the WikiLeaks organization and the unauthorized reporting of private and classified information.

Case in Point: Cultural Shifts, Inclusion, and Civil Liberty in the United States

The United States is a good subject for evaluating cultural shifts. In the aftermath of the political turmoil of the 1960s and 1970s, the United States underwent a slow cultural and ideological shift that began to promote concepts such as multiculturalism and diversity.

CHAPTER PERSPECTIVE 4.3

WikiLeaks

WikiLeaks is an independent organization founded by Australian Julian Assange and others in 2007. It maintains an international online presence through its domain wikileaks.org, operating in accordance with self-described central principles that include

the defence of freedom of speech and media publishing, the improvement of our common historical record and the support of the rights of all people to create new history. We derive these principles from the Universal Declaration of Human Rights. In particular, Article 19 inspires the work of our journalists and other volunteers. It states that everyone has the right to freedom of opinion and expression; this right includes freedom to hold opinions without interference and to seek, receive and impart information and ideas through any media and regardless of frontiers. We agree, and we seek to uphold this and the other Articles of the Declaration.[a]

WikiLeaks publishes documents and other information obtained from anonymous volunteers, including classified information not intended for release. In particular,

WikiLeaks has obtained information from volunteers who may be subject to criminal prosecution or other sanctions if their anonymity is compromised. This occurred when U.S. Army Private First Class Bradley Manning was arrested in May 2010 for allegedly delivering hundreds of thousands of classified documents to WikiLeaks.

The documents and information released by WikiLeaks have on occasion incited strong international debate about wartime policies. For example, in 2010, WikiLeaks released U.S. military reports under the titles "Afghan War Diary" and "Iraq War Logs." These releases together accounted for one of the most voluminous releases of classified information in modern U.S. history.

Other controversial releases of information included the 2012 release of U.S. Department of Defense documents concerning Guantánamo Bay detainees and several million e-mails from the private intelligence company Stratfor. In June 2015, WikiLeaks released classified reports and documents from the NSA, which WikiLeaks titled "Espionnage Élysée." It reported NSA surveillance of French government officials.

Note

a. WikiLeaks website, http://www.wikileaks.org/About.html. Accessed January 23, 2017.

These concepts have been adaptations to the United States' gradual transformation into a country in which no single demographic group will constitute a majority of the population, probably by the mid-twenty-first century. This reflects a significant cultural and societal shift from the *melting-pot ideology* of previous generations, when new immigrants, racial minorities, and religious minorities were expected to accept the cultural values of the American mainstream.

In the United States and elsewhere, grassroots efforts to promote inclusion became common features of the social and political environment (although not without political opposition). For example, private watchdog organizations monitor extremist tendencies, such as right-wing and neofascist movements. Some of these organizations, such as the Southern Poverty Law Center and the Anti-Defamation League in the United States, have implemented programs to promote community inclusiveness. In the public sector, government agencies have long been required to monitor and promote inclusion of demographic minorities and women in government-funded programs. Also in the public sector, the trend among local police forces has shifted toward practicing variants of community-oriented policing, which, in practice, means that the police operationally embed themselves as much as possible within local communities.

The theoretical outcome of these cultural tendencies would be an erosion of the root causes for extremist sentiment. In essence, the protection of civil liberties co-opts the extremists' position that the government or dominant group is uninterested in the rights of the championed group.

BALANCING CIVIL LIBERTIES AND HOMELAND SECURITY

Counterterrorist options occasionally involve controversial practices and procedures that recurrently inflame misgivings held by civil liberties advocates. This tension between homeland security necessities and civil liberty ideals is starkly highlighted when one considers several politically sensitive counterterrorist options. The underlying question is whether effective homeland security is a justifiable end for implementing controversial options. Table 4.2 summarizes selected countermeasures and their implications for civil liberty protections.

Terrorist Profiling

The American approach to domestic counterterrorism prior to the September 11, 2001, attacks was a law enforcement approach. After the attacks, the new homeland security environment called for a more security-focused approach. The FBI and other agencies created a **terrorist profile** that was similar to standard **criminal profiles** used in law enforcement investigations. Criminal profiles are descriptive composites that include the personal characteristics of suspects, such as their clothing, eye color, gender, hair color, weight, race, and height. Suspects can be administratively detained for questioning if they match these criminal profiles.

The composite of the new terrorist profile included the following characteristics: male gender, Middle Eastern heritage, young-adult age, Muslim faith, and temporary visa status. Based on these criteria—and during the serious post–9/11 security crisis—the FBI and Immigration and Naturalization Service administratively detained hundreds of men fitting this description. Material witness warrants were used from the outset to detain many of these men for questioning. This is recognized as a legally and procedurally acceptable practice as long as suspects are not differentially treated when criminal profiles are designed.

terrorist profile: A descriptive composite similar to standard criminal profiles. The modern terrorist profile includes the following characteristics: Middle Eastern heritage, temporary visa status, Muslim faith, male gender, and young-adult age.

criminal profiles: Descriptive composites that include the personal characteristics of suspects, such as their height, weight, race, gender, hair color, eye color, and clothing. Suspects who match these criminal profiles can be administratively detained for questioning.

Table 4.2 Implementing Countermeasures: Challenges to Civil Liberty

The selection of countermeasures to combat terrorism often poses challenges to the protection of civil liberties. Nevertheless, some countermeasures are deemed necessary, and the imperative for ensuring domestic security must be balanced against the need to protect civil liberties.

Selected Countermeasure	Implementation and Civil Liberty Challenges		
	Justification for Countermeasure	Mode of Implementation	Civil Liberty Consideration
Terrorist profiling	Identifying potential terrorists	Questioning, surveillance, and detention of persons fitting the profile	High potential for racial profiling
Labeling or defining the enemy	Clarifying the status of detainees	Indefinite detentions without legal recourse	Legal protections should be enforced for suspects
Extraordinary renditions	Taking suspected terrorists into custody	Covert global capturing of suspects	Illicit kidnappings and detentions violate local and international laws
Enhanced interrogations	Eliciting useful information from detainees	Redefining torture and use of coercive methods during interrogations	Torture, however defined, is fundamentally immoral

As the September 11 investigations continued and in the wake of several warnings about additional terrorist threats, the DOJ expanded the FBI's surveillance authority. New guidelines were promulgated in May 2002 that permitted field offices to conduct surveillance of religious institutions, websites, libraries, and organizations without an *a priori* (before-the-fact) finding of criminal suspicion. A broad investigative net was cast, using the rationale that verifiable threats to homeland security must be detected and preempted.

These detentions and guidelines were widely criticized by civil liberties advocates. Critics argued that the detentions were improper because the vast majority of the detainees had not been charged with violating the law and no criminal suspicion had been articulated. Critics of the surveillance guidelines contended that they gave too much power to the state to investigate innocent civilians. Many also maintained that there was a danger that these investigations could cross a conceptual threshold and become discriminatory **racial profiling,** involving the unconstitutional detention of people because of their racial heritage or ethnonational identity. Nevertheless, the new security policies continued to use administrative detentions and enhanced surveillance as counterterrorist methods.

racial profiling: The unconstitutional detention of people because of their ethnonational or racial heritage.

The Problem of Labeling the Enemy

When formulating counterterrorist policies, American homeland security experts are challenged by two problems: first, the problem of defining terrorism, and second, the problem of labeling individual suspects. The latter problem poses challenges to protecting fundamental constitutional rights.

Although defining terrorism can be an exercise in semantics and is often shaped by subjective political or cultural biases, certain fundamental elements are objectively accepted. Common features of most formal definitions include the use of illegal force, subnational actors, unconventional methods, political motives, attacks against passive civilian and military targets, and acts aimed at affecting an audience.

In comparison, official designations (labels) used to confer special status on captured insurgent and terrorist suspects have become controversial. After September 11, 2001, it became clear that official designations and labels of individual suspected terrorists are a critical legal, political, and security issue. The question of a suspect's official status when he or she is taken prisoner is central. It determines whether certain recognized legal or political protections are or are not observed.

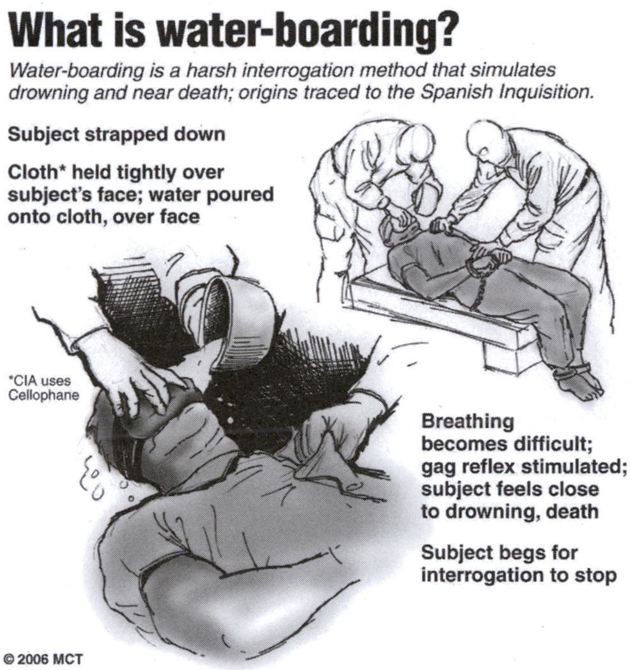

Civil Liberties and Detainees

When enemy soldiers are taken prisoner, they are traditionally afforded legal protections as *prisoners of war*. This is well recognized under international law. During the war on terrorism, many suspected terrorists were designated by the United States as *enemy combatants* and were not afforded the same legal status as prisoners of war, and were therefore technically ineligible to receive the right of legal protection. Such practices have been hotly debated among proponents and opponents.

According to the protocols of the third Geneva Convention, prisoners who are designated as prisoners of war and are brought to trial must be afforded the same legal rights in the same courts as soldiers from the country holding them prisoner. Thus, prisoners of war held by the United States would be brought to trial in standard military courts under the Uniform Code of Military Justice and would have the same rights and protections (such as the right to appeal) as all soldiers.

Suspected terrorists have not been designated as prisoners of war. Official and unofficial designations, such as *enemy combatants*, *unlawful combatants*, and *battlefield detainees*, have been used by American authorities to differentiate them from prisoners of war. The rationale is that suspected terrorists are not soldiers fighting for a sovereign nation and are, therefore, not eligible for prisoner-of-war status. When hundreds of prisoners were detained by the United States at facilities such as the American base in Guantánamo Bay, Cuba, the United States argued that persons designated as enemy combatants were not subject to the Geneva Conventions. Thus, such individuals could be held indefinitely, detained in secret, transferred at will, and sent to allied countries for more coercive interrogations. Under enemy combatant status, conditions of confinement in Guantánamo Bay included open-air cells with wooden roofs and chain-link walls. In theory, each case was to be reviewed by special military tribunals, and innocent prisoners would be reclassified as non–enemy combatants and released.

Civil liberties and human rights groups disagreed with the special status conferred on prisoners by the labeling system. They argued that basic legal and humanitarian protections should be granted to prisoners regardless of their designation.

What is water-boarding?

Water-boarding is a harsh interrogation method that simulates drowning and near death; origins traced to the Spanish Inquisition.

Subject strapped down

Cloth* held tightly over subject's face; water poured onto cloth, over face

*CIA uses Cellophane

Breathing becomes difficult; gag reflex stimulated; subject feels close to drowning, death

Subject begs for interrogation to stop

© 2006 MCT

▶ **Photo 4.4** A depiction of the waterboarding procedure. So-designated "enhanced interrogation" methods have been both condemned as unethical torture and supported as a hard necessity in the war on terror.

The Ker-Frisbie Rule and Extraordinary Renditions

In many ways, the war on terrorism is a "shadow war" that is fought covertly and beyond the attention of the public. It is also a war that employs unconventional tactics and uses resources that were hitherto either uncommon or unacceptable. One unconventional tactic adopted by the United States is known as **extraordinary rendition**, a method of covertly abducting and detaining suspected terrorists or affiliated operatives.

Extraordinary renditions were initially sanctioned during the Reagan administration in about 1987 as a means of capturing drug traffickers, terrorists, and other wanted persons. They involve an uncomplicated procedure: Find suspects anywhere in the world, seize them, transport them to the United States, and force their appearance before a state or federal court. Such compulsory appearances before U.S. courts (after forcible abductions) have long been accepted as procedurally valid and as not violating one's constitutional rights. The doctrine that permits these abductions and appearances is an old one, and it has come to be known as the **Ker-Frisbie Rule.**[4]

This practice was significantly expanded after the September 11 terrorist attacks. It became highly controversial because, unlike previous renditions in which suspects were seized and brought into the U.S. legal system, most antiterrorist abductions placed suspects in covert detention. Many abductions have been carried out by CIA operatives, who transported a number of abductees to allied countries for interrogation. The CIA also established secret detention facilities and maintained custody of suspects for extended periods of time.[5] Allegations have arisen and findings have been made that these suspects were tortured.

Western governments such as Italy, Sweden, and Germany launched investigations into alleged CIA-coordinated extraordinary renditions from their countries. In June 2005, Italy went so far as to order the arrests of 13 alleged CIA operatives for kidnapping an Egyptian cleric from the streets of Milan.[6]

Case in Point: The Torture Debate

Few counterterrorist methods garner such passionate debate as the infliction of physical and psychological pressure on terrorist suspects. The United States joined the debate in the aftermath of the invasion of Iraq. From October through December 2003, Iraqi detainees held at the U.S.-controlled Abu Ghraib prison near Baghdad were abused by American guards. The abuse included sexual degradation, intimidation with dogs, stripping prisoners naked, forcing them into human pyramids, and making them stand in extended poses in so-called stress positions. The U.S. Congress and global community became aware of these practices in April 2004 when graphic photographs were published in the media, posted on the Internet, and eventually shown to Congress. Criminal courts-martial were convened, and several guards were convicted and sentenced to prison.

Unfortunately for the United States, not only was its image tarnished, but further revelations about additional incidents raised serious questions about these and other practices. For example, in March 2005, U.S. Army and Navy investigators reported that 26 prisoners in American custody had possibly been the victims of homicide. Furthermore, detainees incarcerated in detention centers at the Guantánamo Bay Naval Base were reported to have been regularly subjected to intensive physical and psychological interrogation techniques.

A debate about the definition and propriety of torture ensued. The debate included questions of how to draw definitional lines between so-called "**enhanced interrogation**" methods and torture.

Torture is a practice that is officially eschewed by the United States, both morally and as a legitimate interrogation technique. Morally, such practices are officially held to

extraordinary renditions: A method of covertly abducting and detaining suspected terrorists or affiliated operatives.

Ker-Frisbie Rule: A doctrine that permits coercive abductions and appearances before U.S. judicial authorities, named after two cases permitting such abductions and appearances.

enhanced interrogation: Physical and psychological stress methods used during the questioning of suspects.

be inhumane and unacceptable. As an interrogation method, American officials have long argued that torture produces bad intelligence because victims are likely to admit whatever the interrogator wishes to hear. However, during the war on terrorism, a fresh debate began about how to define torture and whether physical and psychological stress methods that fall outside of this definition are acceptable.

Assuming that the application of coercion is justifiable to some degree to break the resistance of a suspect, the question becomes whether physical and extreme psychological coercion is also justifiable. For instance, do the following techniques constitute torture?

- Waterboarding, in which prisoners believe that they will drown

- Sexual degradation, whereby prisoners are humiliated by being stripped or being forced to perform sex acts

- Stress positions, whereby prisoners are forced to pose in painful positions for extended periods

- Creating a chronic state of fear

- Environmental stress, accomplished by adjusting a detention cell's temperature

- Sleep deprivation

- Inducing disorientation about one's whereabouts or the time of day

- Sensory deprivation, such as depriving suspects of sound or light

When images such as those from Abu Ghraib became public, the political consequences were serious. Nevertheless, policymakers continued their debate on which practices constitute torture and whether some circumstances warrant the imposition of as much stress as possible on suspects—up to the brink of torture. In May 2008, the DOJ's inspector general released an extensive report that revealed that FBI agents had complained repeatedly since 2002 about harsh interrogations conducted by military and CIA interrogators.

Case in Point: The Militarization of the Police

Political and social analysts of the homeland security era have referenced a policy trend toward the *militarization* of state and local police agencies. The primary reference point is an increased acquisition of surplus military hardware such as armored vehicles equipped with tear gas launchers and sniper positions. Additional reference points are the acquisition of assault rifles and submachine guns, the use of nonlethal grenades ("flash-bangs"), and an increased tendency toward deploying heavily equipped SWAT units to control crowds. Some commentators criticize this trend as overly aggressive and unnecessarily confrontational.

Although such policies frequently create political controversy in the present environment, they are not new policies. For example, SWAT units have been integral units within law enforcement agencies since the 1970s. The use of armored vehicles by some agencies likewise originated in response to urban discord during the 1960s and 1970s. The acquisition of armored vehicles and weapons with increased firepower occurred in the aftermath of incidents in which criminals wielded high-caliber and high-capacity weapons against police responders.

The militarization of the police debate intensified during the aftermath of the August 2014 police shooting of 18-year-old Michael Brown in Ferguson, Missouri. Brown was unarmed, and there was significant factual disagreement concerning police and witness accounts on the circumstances of the incident. Peaceful protests and civil disorders in Ferguson occurred for days following the fatal shooting. In November 2014, peaceful protests and civil disorders were reinflamed in Ferguson after information was leaked about grand jury testimony. Nationwide protests occurred in approximately 170 cities in the United States. Video images of the police responses in Ferguson and elsewhere gave rise to a vigorous political and policy debate about the use of armored vehicles, sniper positions, robust street clearance, and other aggressive tactics used by law enforcement agencies during the peaceful protests and civil disorders.

This chapter's Global Perspective discusses prosecutions and police investigations in the United Kingdom during an Irish Republican Army (IRA) bombing campaign. Two examples of miscarriages of criminal justice are presented.

GLOBAL PERSPECTIVE

WRONGFUL PROSECUTION IN THE UNITED KINGDOM

In the United Kingdom, where factions of the IRA were highly active in London and other cities, the British police were considered to be the front line against IRA terrorism. They usually displayed a high degree of professionalism without resorting to repressive tactics and consequently enjoyed widespread popular support. For example, London's Metropolitan Police (also known as "The Met") became experts in counterterrorist operations when the IRA waged a terrorist campaign during the 1970s.

The Met's criminal investigations bureau generally used high-quality detective work, rather than authoritarian techniques, to investigate terrorist incidents. The British criminal justice system also generally protected the rights of the accused during trials of IRA suspects. However, in the rush to stop the IRA's terrorist campaign (especially during the 1970s), miscarriages of criminal justice did occur. Examples of these miscarriages include the following examples:

- **Guildford Four.** Four people were wrongfully convicted of an October 1974 bombing in Guildford, England. Two of them were also wrongfully sentenced for a bombing attack in Woolwich. The Guildford Four served 15 years in prison before being released in 1989, when their convictions were overturned on appeal. The group received a formal apology from Prime Minister Tony Blair in June 2000 and received monetary awards as compensation. The case was made famous by the American film *In the Name of the Father*.

- **Birmingham Six.** Six men were convicted for the November 1974 bombings of two pubs in Birmingham, England, that killed 21 people and injured 168. On appeal, the court ordered the release of the Birmingham Six after it ruled that the police had used fabricated evidence. The men were released in 1991 after serving 16 years in prison.

CHAPTER SUMMARY

This chapter introduced some of the challenges inherent in the balancing of civil liberties considerations with domestic security necessities. Historical precedents indicate that the debate on this issue is as old as the nation. Several historical eras were confronted with the question of how far the nation should go to ensure domestic security.

In this regard, questions arise about whether policies deemed acceptable and appropriate at a particular time were actually, in retrospect, violations of principles of civil liberty. Modern counterterrorist practices such as extraordinary renditions and enhanced interrogation epitomize this tension.

There is often a natural tension between preserving human rights and securing the homeland. This tension is reflected in political and philosophical debates about how to accomplish both goals. Nevertheless, during historical periods when threats to national security existed, sweeping measures were undertaken as a matter of perceived necessity. The implementation of these measures was often politically popular at the time but questioned in later years. The modern homeland security environment exists because of the attacks of September 11, 2001, which resulted in the creation of extensive bureaucracies, the passage of new security-related laws, and the implementation of controversial counterterrorist measures.

DISCUSSION BOX

Civil Liberty Protections and the "Ticking Bomb" Scenario

This chapter's Discussion Box is intended to stimulate critical debate about how to balance civil liberty protections against the need to respond to an immediate homeland security threat.

It can be argued that civil liberty considerations are largely a moral debate about norms of justice in civil society—in other words, whether a just society should suspend its norms of justice when challenged by real and imminent danger from violent extremists.

Consider the following scenario: A terrorist cell has been active in several cities in the upper Midwest region of the United States. They have carried out numerous bombings and acts of sabotage, including a hostage crisis at a school that left dozens of dead and injured. There is every expectation that the cell will continue its campaign. Law enforcement officials have captured a member of the cell in an apartment apparently used as a "bomb factory," and it is evident that the suspect is a master bomb maker. The suspect refuses to reveal any information about his comrades, where completed bombs may have been sent, or where the next intended targets will be attacked. Lives are clearly at stake, and time is of the essence.

Discussion Questions

1. Which interrogation techniques are acceptable in this situation from a homeland security perspective?

2. Which interrogation techniques are *not* acceptable in this situation from a civil liberties perspective?

3. Considering the ruthlessness of the terrorists, should martial law be declared in the upper Midwest? What are the civil liberties consequences of doing so? What are the homeland security consequences of *not* doing so?

4. Should the captured suspect be afforded due process protections under the law?

5. Who should have primary custody and jurisdiction over the suspect?

KEY TERMS AND CONCEPTS

The following topics were discussed in this chapter and can be found in the glossary:

Alien and Sedition Acts   68
Carnivore   74
Civil Liberties Act   71
criminal profiles   77
DCS-1000   74
enhanced interrogation   80

extraordinary renditions   80
Ker-Frisbie Rule   80
McCarthyism   70
Palmer Raids   70
PRISM   75
racial profiling   78

Red Scares   69
Snowden, Edward   75
Tempora   75
terrorist profile   77
XKeyscore   75

ON YOUR OWN

Get the tools you need to sharpen your study skills. SAGE edge offers a robust online environment featuring an impressive array of free tools and resources.

Access practice quizzes, eFlashcards, video, and multimedia at **edge.sagepub.com/martinhs3e**

RECOMMENDED WEBSITES

The following websites provide perspectives on civil liberties and human rights issues and interventions:

- American Civil Liberties Union: www.aclu.org/national-security

- Amnesty International: www.amnesty.org

- DHS Office of Civil Rights and Civil Liberties: www.dhs.gov/office-civil-rights-and-civil-liberties

- Doctors of the World: www.doctorsoftheworld.org

- Electronic Privacy Information Center: epic.org

- Gendercide Watch: www.gendercide.org

- Human Rights Watch: www.hrw.org

- Médecins Sans Frontières: www.doctorswithoutborders.org

WEB EXERCISE

Using this chapter's recommended websites, conduct an online investigation of the civil liberty and homeland security debate. Compare and contrast each side of the debate.

1. What are the primary issues at the center of the debate?

2. How would you describe the differences between those who question homeland security policies and those who believe they are absolutely necessary?

3. In your opinion, are any of these positions more persuasive than others? Less persuasive?

To conduct an online search on research and monitoring organizations, activate the search engine on your Web browser and enter the following keywords:

"Homeland security and civil liberty"

"Homeland security and human rights"

RECOMMENDED READINGS

The following publications provide discussion on homeland security and civil liberties issues:

Begg, Moazzam and Victoria Brittain. 2006. *Enemy Combatant: My Imprisonment at Guantánamo, Bagram, and Kandahar.* New York: New Press.

Crank, John P. and Patricia E. Gregor. 2005. *Counter-Terrorism After 9/11: Justice, Security and Ethics Reconsidered.* New York: Anderson.

Daalder, Ivo H., ed. 2007. *Beyond Preemption: Force and Legitimacy in a Changing World.* Washington, DC: Brookings Institution Press.

Ginbar, V. 2008. *Why Not Torture Terrorists?* Oxford, UK: Oxford University Press.

Grey, Stephen. 2006. *Ghost Plane: The Inside Story of the CIA's Secret Rendition Programme.* New York: St. Martin's.

Heymann, Phillip B. and Juliette N. Kayyem. 2005. *Protecting Liberty in an Age of Terror.* Cambridge, MA: MIT Press.

Schawb, Stephen Irving Max. 2009. *Guantánamo, USA: The Untold History of America's Cuban Outpost.* Lawrence: University Press of Kansas.

Tsang, Steve. 2007. *Intelligence and Human Rights in the Era of Global Terrorism.* Westport, CT: Praeger Security International.

Wilson, Richard Ashby, ed. 2005. *Human Rights in the War on Terror.* New York: Cambridge University Press.

Worthington, Andy. 2007. *The Guantánamo Files.* Ann Arbor, MI: Pluto Press.

PART TWO

HOMELAND SECURITY AGENCIES AND MISSIONS

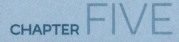
AGENCIES AND MISSIONS
Homeland Security at the Federal Level

Chapter Learning Objectives

This chapter will enable readers to do the following:

1. Compare and contrast the difference between federal bureaucracy prior to and after the September 11, 2001, terrorist attacks

2. Evaluate the roles of law enforcement and service agencies and underlying interagency challenges

3. Describe the post–9/11 bureaucratic transformation and the dawn of modern homeland security

4. Evaluate the role of the Department of Homeland Security

5. Analyze the homeland security roles of other federal agencies

6. Apply the homeland security nexus with the military

Opening Viewpoint: Homeland Security Policing in a Federal System

The question of how to create an effective law enforcement mechanism within the homeland security enterprise can be a complicated process in the United States. Federal agencies possess law enforcement authority, but they do not enforce state and local laws. This is because there is no national police force in the United States similar to the French *gendarmerie* or Mexico's *Policia Federal*. The reason for this is the constitutionally established system of federalism and reserved powers conferred to the states under the Tenth Amendment to the U.S. Constitution.

Authority to enforce state laws is constitutionally reserved for the states, and therefore, local policing authority is devolved to state agencies. However, many federal agencies are authorized to carry out law enforcement duties to support their specified missions. Federal law enforcement agencies often take the lead in investigating incidents of domestic terrorism, with other agencies performing a supportive role and assisting in resolving cases.

The federal homeland security bureaucracy is a network of specialized agencies that contribute to the overall mission of securing the United States from terrorist threats. Many of these agencies are subsumed under the direction of the secretary of homeland security, while others are directed by cabinet-level or independent officials. The U.S. Department of Homeland Security and several sector-specific agencies carry out homeland security–related bureaucratic duties assigned to them.

The role of the military is fundamentally one of defending the nation from external threats. However, in the age of the *New Terrorism*, adversaries are frequently intent on carrying out attacks domestically in the United States, and the military's homeland defense mission must be adapted to this reality in support of the homeland security enterprise.

The discussion in this chapter will review the following:

- The scope of the homeland security bureaucracy

- The Department of Homeland Security

- The homeland security missions of other federal agencies

- The role of the military

THE SCOPE OF THE HOMELAND SECURITY BUREAUCRACY

Strong proposals were made to revamp the domestic homeland security community within nine months of the September 11, 2001, attacks. This occurred because of the apparent failure of the pre–September 11 domestic security community to adapt to the new terrorist environment as well as because of highly publicized operational problems.

Conceptual Background: The Bureaucratic Context

Ideally, governments act rationally and efficiently to resolve problems. To do so, government functions are organized in operational arrangements known as *bureaucracy*. Max Weber invented the term to describe and explain rationality and efficiency in managing governments—a field of public administration known as *organizational theory*. It should be obvious that many functions of government require professional bureaucracies and trained managers to ensure social stability and the delivery of critical services. For example, efficiency in regulating interstate commerce permits the delivery of essential commodities throughout the nation. Many of these bureaucratic functions are literally life-and-death missions, such as emergency preparedness and disaster response. In terrorist environments, the consequences can be quite dire if homeland security bureaucracies are not flexible, efficient, and collaborative. Operationalizing the bureaucratic ideal in practice is often a complex and difficult task.

U.S. Federal Bureau of Investigation

▶ **Photo 5.1** Member of an FBI Special Weapons and Tactics unit. The FBI maintains ready-action units for fast deployment domestically.

The Federal Bureaucracy and the Dawn of the Homeland Security Enterprise

Prior to the domestic attacks of September 11, 2001, the United States relied on administratively separated federal law enforcement and services agencies to provide what is now referred to as *homeland security*. These agencies performed mission-specific duties that were frequently unsynchronized in the carrying out of domestic security operations.

Federal Bureau of Investigation (FBI): An investigative bureau within the U.S. Department of Justice. It is the largest federal law enforcement agency, and among its duties are domestic counterterrorism and intelligence collection.

National Strategy for Homeland Security: Published in 2002, this was the first clarification of the newly emerging homeland security culture. The document explicated the concept of homeland security, identified essential homeland security missions, and established priorities for coordinating the protection of critical domestic infrastructures.

Federal Law Enforcement Agencies

Federal law enforcement agencies are bureaus within the large cabinet agencies that are charged with enforcing federal criminal laws. The **Federal Bureau of Investigation (FBI)**; Drug Enforcement Administration; and Bureau of Alcohol, Tobacco, Firearms, and Explosives are examples of federal law enforcement agencies. Prior to the September 11 attacks, these agencies investigated security threats in the same manner that they investigated crimes: by working cases and making arrests.

Federal Services Agencies

Services agencies regulate and manage services for the general population. Federal services agencies include large cabinet agencies, regulatory agencies, and independent agencies. The U.S. Departments of Health and Human Services, Energy, and Defense are examples of services agencies. Prior to the September 11 attacks, these agencies had a variety of missions, including regulating immigration, inspecting nuclear facilities, and responding to emergencies. Table 5.1 summarizes the pre–September 11 security duties of several U.S. federal agencies.

A New Culture: The Dawn of Homeland Security

In July 2002, the first clarification of the newly emerging homeland security culture was promulgated by the Office of Homeland Security with the publication of the *National Strategy for Homeland Security*.[1] The *National Strategy for Homeland Security* identified essential homeland security missions, established priorities for coordinating the protection of critical domestic infrastructure, and explicated the concept of homeland security.

The Office of Homeland Security, placed under the administration of the White House, was dissolved after the passage of the Department of Homeland Security Act of 2002. The **Department of Homeland Security (DHS)** was created on November 25, 2002, with the enactment of the Homeland Security Act. The new department was tasked with five main areas of responsibility. These areas of responsibility reflected the underlying missions of the former security and emergency response agencies that were subsumed under the authority of the new DHS. The five areas of responsibility were as follows:

▸ **Photo 5.2** Official seal of the U.S. National Counterterrorism Center.

- Guarding against terrorism
- Securing our borders
- Enforcing our immigration laws
- Improving our readiness for, response to, and recovery from disasters
- Maturing and unifying the department[2]

Thus, the concept of homeland security incorporated counterterrorism and other domestic security and emergency response functions.

Clarifying the new homeland security enterprise is an ongoing process. In years subsequent to the 2002 publication of the *National Strategy for Homeland Security*, additional

Prior to the September 11, 2001, organizational crisis, homeland security was the responsibility of a number of federal agencies. These agencies were not centrally coordinated, and they answered to different centers of authority. Cooperation was theoretically ensured by liaison protocols, special task forces, and oversight. In reality, there was a great deal of functional overlap and bureaucratic "turf" issues.

The following table summarizes the activity profiles of several bureaus prior to the post–September 11 organizational crisis.

Agency	Activity Profile		
	Parent Organization	**Mission**	**Enforcement Authority**
Central Intelligence Agency	Independent agency	Collection and analysis of foreign intelligence	No domestic authority
Coast Guard	Department of Transportation	Protection of U.S. waterways	Domestic law enforcement authority
Customs	Department of the Treasury	Examination of people and goods entering the United States	Domestic inspection, entry, and law enforcement authority
Federal Bureau of Investigation	Department of Justice	Investigating and monitoring criminal and national security threats	Domestic law enforcement authority
Federal Emergency Management Agency	Independent agency	Responding to natural and human disasters	Coordination of domestic emergency responses
Immigration and Naturalization Service	Department of Justice	Managing the entry and naturalization of foreign nationals	Domestic inspection, monitoring, and law enforcement authority
Secret Service	Department of the Treasury	Establishing security protocols for president, vice president, and special events	Domestic protection of president and vice president and special law enforcement authority (including counterfeiting)

Department of Homeland Security (DHS): The Homeland Security Act was enacted on November 25, 2002, and created the Department of Homeland Security. The new department was tasked with five main areas of responsibility, reflecting the underlying missions of the former emergency response and security agencies subsumed under the authority of DHS.

National Security Strategy: The process of clarifying the new homeland security enterprise is ongoing, and publications regularly clarify the homeland security enterprise. One of those publications is the 2010 *National Security Strategy*.

National Strategy for Counterterrorism: The process of clarifying the new homeland security enterprise is ongoing, and publications regularly clarify the homeland security enterprise. One of those publications is the 2011 *National Strategy for Counterterrorism*.

publications regularly clarified the new homeland security enterprise. Examples of clarifications include the 2010 *National Security Strategy*,[3] the 2011 *National Strategy for Counterterrorism*,[4] the 2015 *National Preparedness Goal, Second Edition*,[5] and the 2018 *National Strategy for Counterterrorism*.[6] Table 5.2 summarizes the security duties of several U.S. federal agencies immediately after the creation of the new department.

Table 5.2 Federal Agencies and Homeland Security: After the September 11, 2001, Organizational Crisis

In the wake of the post–September 11 organizational crisis, the Bush administration subsumed the homeland security duties of several federal agencies under the jurisdiction of a new Department of Homeland Security. The goal was to coordinate operations and to end overlapping duties.

The following table is a good snapshot of a nation's reorganization of national security in response to a significant shift in a terrorist environment. It is also an example of how two security agencies that arguably precipitated the organizational crisis—the FBI and CIA—were able to maintain their independence.

Agency	Activity Profile		
	New Parent Organization	New Directorate	New Directorate's Duties
Central Intelligence Agency	No change; independent agency	No change	No change; collection and analysis of foreign intelligence
Coast Guard	Department of Homeland Security	Border and Transportation Security	Coordination of all national entry points
Customs	Department of Homeland Security	Border and Transportation Security	Coordination of all national entry points
Federal Bureau of Investigation	No change; Department of Justice	No change	No change; investigating and monitoring criminal and national security threats
Federal Emergency Management Agency	Department of Homeland Security	Emergency Preparedness and Response	Coordination of national responses to terrorist incidents
Immigration and Naturalization Service (some functions)	Department of Homeland Security	Border and Transportation Security	Coordination of all national entry points
Secret Service	Department of Homeland Security	Secret Service	Establishing security protocols for president and special events

THE DEPARTMENT OF HOMELAND SECURITY

The DHS is an extensive department in the federal government whose secretary holds cabinet-level authority. The major components of the department are a result of the consolidation of agencies with critical domestic missions in the aftermath of the September 11, 2001, attacks. Homeland security is a new concept and a new mission of the federal government. DHS is by far the largest and most mission-diverse department in the homeland security

Figure 5.1   Organization Chart of the U.S. Department of Homeland Security

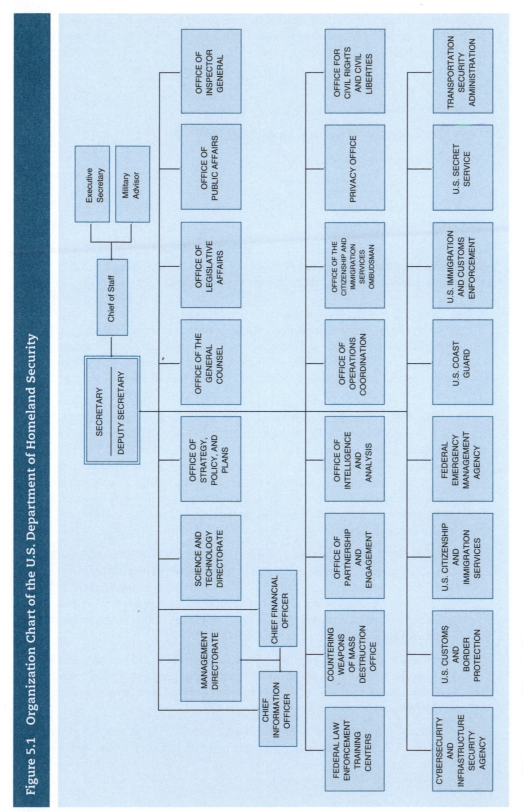

Source: U.S. Department of Homeland Security.

bureaucracy. Broadly defined, its mission is "to ensure a homeland that is safe, secure, and resilient against terrorism and other hazards."[7] More specifically, its five core missions (which have been updated) are as follows:[8]

- Prevent terrorism and enhance security

- Secure and manage our borders

- Enforce and administer our immigration laws

- Safeguard and secure cyberspace

- Ensure resilience to disasters

The Office of the Secretary

The secretary of homeland security is a cabinet-level official, the first having been former Pennsylvania governor Tom Ridge. The central administrative office of DHS is the Office of the Secretary, which operates within the following guidelines:

- The Office of the Secretary oversees Department of Homeland Security (DHS) efforts to counter terrorism and enhance security, secure and manage our borders while facilitating trade and travel, enforce and administer our immigration laws, safeguard and secure cyberspace, build resilience to disasters, and provide essential support for national and economic security—in coordination with Federal, state, local, international and private sector partners.[9]

- The Office of the Secretary carries out its administrative duties through a variety of mission-focused offices that essentially coordinate the entire DHS bureaucracy. These offices are organized as follows: *Privacy Office*. The Privacy Office addresses legitimate concerns about protecting the privacy of individuals during the course of accomplishing DHS's mission. The Privacy Office "works to preserve and enhance privacy protections for all individuals, to promote transparency of Department of Homeland Security operations, and to serve as a leader in the privacy community."[10]

- *Office for Civil Rights and Civil Liberties (CRCL)*. CRCL was created to address public concerns about protecting civil rights and to prevent infringement of civil liberties during the implementation of homeland security initiatives. CRCL "provides legal and policy advice to Department leadership on civil rights and civil liberties issues, investigates and resolves complaints, and provides leadership to Equal Employment Opportunity Programs."[11]

- *Office of Strategy, Policy, and Plans*. The Office "serves as a central resource to the Secretary and other Department leaders for strategic planning and analysis, and facilitation of decision-making on the full breadth of issues that may arise across the dynamic homeland security enterprise."[12]

- *Citizenship and Immigration Services Ombudsman*. The Citizenship and Immigration Services Ombudsman is a problem-solving office in regard to immigration issues. The Ombudsman "is dedicated to improving the quality of citizenship and immigration services delivered to the public by providing individual case assistance, as well as making recommendations to improve the administration of immigration benefits by U.S. Citizenship and Immigration Services (USCIS)."[13]

- *Office of Legislative Affairs.* Every cabinet-level agency has created an office of legislative liaison, and the Office of Legislative Affairs provides such outreach to each branch of the federal government for DHS. It "serves as primary liaison to members of Congress and their staffs, the White House and Executive Branch, and to other federal agencies and governmental entities that have roles in assuring national security."[14]

- *Office of the General Counsel.* DHS possesses a relatively large legal office in comparison to other agencies. The Office of the General Counsel "integrates approximately 1,800 attorneys from throughout the Department into an effective, client-oriented, full-service legal team. The Office of the General Counsel comprises a headquarters office with subsidiary divisions and the legal programs for eight Department components."[15]

- *Office of Public Affairs.* Clear communication to the media and public is imperative for DHS. The Office of Public Affairs "coordinates the public affairs activities of all of the Department's components and offices, and serves as the federal government's lead public information office during a national emergency or disaster."[16]

- *Office of the Executive Secretary (ESEC).* ESEC manages administrative support for chief administrators in DHS. ESEC "provides all manner of direct support to the Secretary and Deputy Secretary, as well as related support to leadership and management across the Department."[17]

- *Office of the Military Advisor.* Because of the importance of promoting coordinated management of homeland security responses and terrorist threats, the Office of the Military Advisor "is to provide counsel and support to the Secretary and Deputy Secretary in affairs relating to policy, procedures, preparedness activities, and operations between DHS and the Department of Defense (DoD)."[18]

- *Privacy Office.* The mission of the Privacy Office "is to protect individuals by embedding and enforcing privacy protections and transparency in all DHS activities."[19]

- *Partnership and Engagement.* This office "coordinates the Department of Homeland Security's outreach efforts with key stakeholders nationwide, ensuring a unified approach to external engagement."[20]

Department of Homeland Security Administrative Centers

DHS comprises a variety of administrative centers. Directorates and offices oversee administrative duties of several mission-specific bureaus. Other administrative centers comprise formerly independent agencies. The DHS's administrative centers and their multiple missions include the following:

- *National Protection and Programs Directorate.* The National Protection and Programs Directorate is a risk reduction directorate. It works "to advance the Department's risk-reduction mission. Reducing risk requires an integrated approach that encompasses both physical and virtual threats and their associated human elements."[21]

- *Science and Technology Directorate.* Research and development are important functions of DHS. The Science and Technology Directorate is "the primary research and development arm of the Department. It provides federal, state and local officials with the technology and capabilities to protect the homeland."[22]

- *Management Directorate.* The Management Directorate is the chief administrative directorate for DHS. It is "responsible for budget, appropriations, expenditure of funds, accounting and finance; procurement; human resources and personnel; information technology systems; facilities, property, equipment, and other material resources; and identification and tracking of performance measurements relating to the responsibilities of the Department."[23]

- *Countering Weapons of Mass Destruction Office (CWMD).* The mission of the CWMD Office "is to counter attempts by terrorists or other threat actors to carry out an attack against the United States or its interests using a weapon of mass destruction."[24]

- *Office of Intelligence and Analysis.* DHS has organized its own intelligence and analysis office. The Office of Intelligence and Analysis "equips the Homeland Security Enterprise with the timely intelligence and information it needs to keep the homeland safe, secure, and resilient."[25] The work of the Office of Intelligence and Analysis is discussed further in Chapter 6.

- *Office of Operations Coordination.* The Office of Operations Coordination is DHS's central office for monitoring and coordinating homeland security activities nationwide. It "provides information daily to the Secretary of Homeland Security, senior leaders, and the homeland security enterprise to enable decision-making; oversees the National Operations Center; and leads the Department's Continuity of Operations and Government Programs to enable continuation of primary mission essential functions in the event of a degraded or crisis operating environment."[26]

- *Federal Law Enforcement Training Center (FLETC).* FLETC is the hub for an interagency network of training for law enforcement professionals. FLETC "provides career-long training to law enforcement professionals to help them fulfill their responsibilities safely and proficiently."[27]

- *Domestic Nuclear Detection Office.* The Domestic Nuclear Detection Office monitors potential nuclear hazards and threats. It "works to enhance the nuclear detection efforts of Federal, state, territorial, tribal, and local governments, and the private sector and to ensure a coordinated response to such threats."[28]

- *Transportation Security Administration (TSA).* TSA "protects the nation's transportation systems to ensure freedom of movement for people and commerce."[29]

- *United States Customs and Border Protection (CBP).* The mission of CBP is "[t]o safeguard America's borders thereby protecting the public from dangerous people and materials while enhancing the Nation's global economic competitiveness by enabling legitimate trade and travel."[30]

- *United States Citizenship and Immigration Services (USCIS).* USCIS "administers the nation's lawful immigration system, safeguarding its integrity and promise by efficiently and fairly adjudicating requests for immigration benefits while protecting Americans, securing the homeland, and honoring our values."[31]

- *United States Immigration and Customs Enforcement (ICE).* ICE "promotes homeland security and public safety through the criminal and civil enforcement of federal laws governing border control, customs, trade, and immigration."[32]

- *United States Coast Guard.* The Coast Guard "is one of the five armed forces of the United States and the only military organization within the Department of Homeland Security. The Coast Guard protects the maritime economy and the environment, defends our maritime borders, and saves those in peril."[33]

- *Federal Emergency Management Agency (FEMA).* FEMA "supports our citizens and first responders to ensure that as a nation we work together to build, sustain, and improve our capability to prepare for, protect against, respond to, recover from, and mitigate all hazards."[34]

- *United States Secret Service.* The Secret Service "safeguards the nation's financial infrastructure and payment systems to preserve the integrity of the economy, and protects national leaders, visiting heads of state and government, designated sites, and National Special Security Events."[35]

THE HOMELAND SECURITY MISSIONS OF OTHER FEDERAL AGENCIES

To ensure the implementation of protective priorities, in addition to establishing the Department of Homeland Security, other federal agencies have also been assigned sector-specific homeland security missions. These agencies are known as **sector-specific agencies**. These federal agencies have been tasked with protecting critical infrastructure in the United States from terrorist attacks. Key U.S. government responsibilities for critical infrastructure are discussed in this section.

Table 5.3 summarizes the emergency support functions of sector-specific agencies for critical infrastructure/key resources (CIKR). Further discussion of CIKR is provided in Chapter 10.

The Department of Agriculture

Agricultural and food security are critical to the nation. The primary vision of the **Department of Agriculture (USDA)** is " to provide economic opportunity through innovation, helping rural America to thrive; to promote agriculture production that better nourishes Americans while also helping feed others throughout the world; and to preserve our Nation's natural resources through conservation, restored forests, improved watersheds, and healthy private working lands."[36] USDA's critical infrastructure responsibility is securing the nation's food supply and agricultural infrastructure.

sector-specific agencies: Sector-specific homeland security missions have been identified for federal agencies in addition to establishing the Department of Homeland Security. These agencies are known as sector-specific agencies.

Department of Agriculture (USDA): The primary mission of the Department of Agriculture is to ensure a "safe, sufficient and nutritious food supply for the American people."

U.S. Coast Guard

▶ **Photo 5.3** A U.S. Coast Guard rescue swimmer in heavily turbulent water. Such operations involve elite personnel and require intensive training.

Table 5.3   Relationship of Emergency Support Functions to Critical Infrastructure/Key Resources (CIKR) Duties of Sector-Specific Agencies

This table shows how the 15 emergency support functions map to the 17 critical infrastructure/ key resources sectors.

Emergency Support Function	Related CIKR Sectors
ESF Primary Agencies Coordinate Resources Support and Program Implementation for Response, Recovery, Restoration, and Mitigation programs directly related to incident management functions.	**Sector-Specific Agencies (SSAs)** Coordinate efforts to protect the Nation's CIKR from terrorist attacks and for helping to strengthen preparedness, timely response, and rapid recovery in the event of an attack, natural disaster, or other emergency.
ESF #1 – Transportation **Primary Agency:** Department of Transportation	**Transportation Systems** **SSA:** DHS/Transportation Security Administration **Postal and Shipping** **SSA:** DHS/Transportation Security Administration **Emergency Services** **SSA:** DHS/Infrastructure Protection
ESF #2 – Communications **Primary Agencies:** DHS/ Cybersecurity and Communications/ National Communications System DHS/Federal Emergency Management Agency	**Information Technology** **SSA:** DHS/Cybersecurity and Communications **Communications** **SSA:** DHS/Cybersecurity and Communications/ National Communications System **Emergency Services** **SSA:** DHS/Infrastructure Protection
ESF #3 – Public Works and Engineering **Primary Agencies:** DHS/Federal Emergency Management Agency DOD/U.S. Army Corps of Engineers	**Drinking Water and Water Treatment Systems** **SSA:** Environmental Protection Agency **Dams** **SSA:** DHS/Infrastructure Protection **Energy** **SSA:** Department of Energy **Emergency Services** **SSA:** DHS/Infrastructure Protection **Government Facilities** **SSA:** DHS/Immigration and Customs Enforcement/ Federal Protective Service **National Monuments and Icons** **SSA:** Department of the Interior

Emergency Support Function	Related CIKR Sectors
ESF #4 – Firefighting **Primary Agency:** USDA/Forest Service	**Emergency Services** **SSA:** DHS/Infrastructure Protection **Government Facilities** **SSA:** DHS/Immigration and Customs Enforcement/Federal Protective Service
ESF #5 – Emergency Management **Primary Agency:** DHS/Federal Emergency Management Agency	**Emergency Services** **SSA:** DHS/Infrastructure Protection **Government Facilities** **SSA:** DHS/Immigration and Customs Enforcement/Federal Protective Service
ESF #6 – Mass Care, Emergency Assistance, Housing, and Human Services **Primary Agency:** DHS/Federal Emergency Management Agency	**Emergency Services** **SSA:** DHS/Infrastructure Protection **Public Health and Healthcare** **SSA:** Department of Health and Human Services
ESF #7 – Logistics Management and Resource Support **Primary Agencies:** General Services Administration DHS/Federal Emergency Management Agency	All
ESF #8 – Public Health and Medical Services **Primary Agency:** Department of Health and Human Services	**Emergency Services** **SSA:** DHS/Infrastructure Protection **Public Health and Healthcare** **SSA:** Department of Health and Human Services
ESF #9 – Search and Rescue **Primary Agencies:** DHS/Federal Emergency Management Agency DHS/U.S. Coast Guard DOI/National Park Service DOD/U.S. Air Force	**Emergency Services** **SSA:** DHS/Infrastructure Protection
ESF #10 – Oil and Hazardous Materials Response **Primary Agencies:** Environmental Protection Agency DHS/U.S. Coast Guard	**Chemical** **SSA:** DHS/Infrastructure Protection **Nuclear Reactors, Materials, and Waste** **SSA:** DHS/Infrastructure Protection **Emergency Services** **SSA:** DHS/Infrastructure Protection

(Continued)

Table 5.3 (Continued)	
Emergency Support Function	**Related CIKR Sectors**
ESF #11 – Agriculture and Natural Resources **Primary Agencies:** Department of Agriculture Department of the Interior	**Agriculture and Food** **SSA:** Department of Agriculture and Department of Health and Human Services/Food and Drug Administration **National Monuments and Icons** **SSA:** Department of the Interior
ESF #12 – Energy **Primary Agency:** Department of Energy	**Energy** **SSA:** Department of Energy **Nuclear Reactors, Materials, and Waste** **SSA:** DHS/Infrastructure Protection **Dams** **SSA:** DHS/Infrastructure Protection
ESF #13 – Public Safety and Security **Primary Agency:** Department of Justice	**Emergency Services** **SSA:** DHS/Infrastructure Protection **Postal and Shipping** **SSA:** DHS/Transportation Security Administration All others as appropriate
ESF #14 – Long-Term Community Recovery **Primary Agencies:** Department of Agriculture DHS/Federal Emergency Management Agency Department of Housing and Urban Development Small Business Administration	**Banking and Finance** **SSA:** Department of the Treasury **Commercial Facilities** **SSA:** DHS/Infrastructure Protection **Drinking Water and Water Treatment Systems** **SSA:** Environmental Protection Agency
ESF #15 – External Affairs **Primary Agency:** DHS/Federal Emergency Management Agency	All

Source: U.S. Federal Emergency Management Agency.

The Department of Defense

Department of Defense (DOD): The Department of Defense is tasked to manage the armed forces of the United States.

Defending the homeland from foreign threats is of paramount importance to the overall security of the nation. The mission of the **Department of Defense (DOD)** "provides the military forces needed to deter war and ensure" the security of the United States.[37] Its critical infrastructure responsibility is to secure DOD installations, military personnel, and defense industries. The role of the military and DOD's homeland defense mission are discussed further in this chapter.

The Department of Energy

Securing energy resources, transportation, and markets requires an overarching national agenda. The mission of the **Department of Energy (DOE)** "is to ensure America's security and prosperity by addressing its energy, environmental and nuclear challenges through transformative science and technology solutions."[38] Within this context, DOE's critical infrastructure responsibility is to secure power plants, weapons production facilities, oil and gas, and research laboratories. DOE is designated a sector-specific agency for coordinating planning for the energy sector. For example, DOE handles initiatives for collaboration between the electricity subsector and the oil and natural gas subsector "to plan for and counter cybersecurity threats to energy infrastructure operations."[39]

The Department of Health and Human Services

Protecting and monitoring the health of the nation is a fundamental mission of the **Department of Health and Human Services (HHS)**. The mission of HHS is "to enhance and protect the health and well-being of all Americans. We fulfill that mission by providing for effective health and human services and fostering advances in medicine, public health, and social services."[40] The agency's critical infrastructure responsibility is to secure the nation's health care and public health system. To accomplish this responsibility, HHS is tasked to coordinate implementation of the National Health Security Strategy and Implementation Plan (NHSS). The purpose of the NHSS is "to strengthen and sustain communities' abilities to prevent, protect against, mitigate the effects of, respond to, and recover from disasters and emergencies."[41]

The Department of the Interior

The primary mission of the **Department of the Interior (DOI)** "conserves and manages the Nation's natural resources and cultural heritage for the benefit and enjoyment of the American people, provides scientific and other information about natural resources and natural hazards to address societal challenges and create opportunities for the American people, and honors the Nation's trust responsibilities or special commitments to American Indians, Alaska Natives, and affiliated island communities to help them prosper."[42] The department's critical infrastructure responsibility is to protect national monuments and lands under its jurisdiction. To accomplish this responsibility, DOI collaborates with agencies subsumed under the U.S. Department of Homeland Security. For example, in September 2010, DOI entered into an Inter-Agency Agreement with U.S. Customs and Border Protection to "fund environmental mitigation projects that will benefit several species of fish and wildlife affected by border security projects in the Southwest."[43]

The Department of the Treasury

The nation's wealth and treasure are hallmarks of the United States. The overarching mission of the **Department of the Treasury** "is to maintain a strong economy and create economic and job opportunities by promoting the conditions that enable economic growth and stability at home and abroad, strengthen national security by combating threats and protecting the integrity of the financial system, and manage the U.S. Government's finances and resources effectively."[44] Its critical infrastructure mission is to secure the U.S. financial and banking

Department of Energy (DOE): The Department of Energy is tasked to manage the nation's energy security and promote scientific research to support this task. DOE is also tasked with environmental cleaning of the nuclear weapons complex.

Department of Health and Human Services (HHS): The Department of Health and Human Services is "the United States government's principal agency for protecting the health of all Americans and providing essential human services."

Department of the Interior (DOI): The primary mission of the Department of the Interior "is to protect and provide access to our Nation's natural and cultural heritage and honor our trust responsibilities to Indian Tribes and our commitments to island communities."

Department of the Treasury: The overarching mission of the Department of the Treasury is to "maintain a strong economy and create economic and job opportunities by promoting the conditions that enable economic growth and stability at home and abroad, strengthen national security by combating threats and protecting the integrity of the financial system, and manage the U.S. Government's finances and resources effectively."

system. To accomplish this responsibility, the department was tasked to oversee ongoing security-related programs. For example, the department "performs a critical and far-reaching role in enhancing national security by implementing economic sanctions against foreign threats to the U.S., identifying and targeting the financial support networks of national security threats, and improving the safeguards of our financial systems."[45]

The Environmental Protection Agency

Environmental Protection Agency (EPA): The overarching mission of the EPA is to lead "the nation's environmental science, research, education and assessment efforts ... [and] protect human health and the environment."

Securing and preserving the nation's environment and resources is a critical component of homeland security. The overarching mission of the **Environmental Protection Agency (EPA)** is "to protect human health and the environment."[46] EPA's critical infrastructure responsibility is to secure the nation's drinking water and water treatment infrastructure. For example, the EPA assists local utilities in their efforts to comply with the Bioterrorism Act of 2002 that requires local utilities to conduct vulnerability assessments and develop emergency response plans.

THE ROLE OF THE MILITARY

Regular units of the United States armed forces are designed to defend the nation against foreign threats to national security. At the same time, states field National Guard units that serve a dual purpose: to respond when called to federal deployment and to respond when states require emergency services during domestic emergencies. Thus, the National Guard has been deployed to maintain security after natural disasters such as the Tuscaloosa–Birmingham tornadoes in 2011, Hurricane Katrina in 2005, and the Oklahoma City bombing in 1995. Federalized National Guard units have also been deployed internationally, serving under the command of the U.S. Department of Defense. The following discussion emphasizes the role of the regular armed forces and National Guard units assigned to regular deployment.

The Homeland Defense Mission of the Department of Defense

In the modern era of the New Terrorism, domestic security considerations often require the policy and operational melding of homeland security and national defense initiatives. In fact, the genesis of modern homeland security was predicated on national defense imperatives, so that homeland security, counterterrorism, and national defense institutions necessarily require clear long-term collaboration and cooperation. Further discussion of the New Terrorism is provided in Chapter 8.

Defining Homeland Defense

Although the primary mission of the U.S. Department of Defense is to prepare for global expeditionary deployment, DOD is also an integral partner in the homeland security enterprise. In this regard, DOD differentiates between its homeland defense mission and homeland security. DOD defines *homeland security* as "a concerted national effort to prevent terrorist attacks within the United States, reduce the vulnerability of the United States to terrorism, and minimize the damage and assist in the recovery from terrorist attacks."[47] *Homeland defense* is defined as "the military protection of United States territory, domestic population, and critical defense infrastructure against external threats and aggression. It also includes routine, steady state activities designed to deter aggressors and to prepare U.S. military forces for action if deterrence fails."[48]

In order to promote the homeland defense mission, DOD established an Office of Homeland Defense and Global Policy operating under the direction of an assistant secretary. DOD also created a combatant command known as Northern Command, or NORTHCOM, with an authorized mission to defend the air, land, and sea approaches to the United States. NORTHCOM's geographic responsibility includes the continental United States, Alaska, Canada, Mexico, Puerto Rico, and the U.S. Virgin Islands.

Case in Point: The Posse Comitatus Act

Debate on whether the armed forces of the United States can be deployed domestically and the conditions for such deployment is a long-standing political and legal issue. This debate centers on the concept of posse comitatus.

Posse comitatus is a very old concept in the Anglo-American legal tradition. It originally permitted a sheriff to summon the population of a county (the posse comitatus) to assist in enforcement of the law. From this medieval beginning, the concept grew to generally refer to a government's mobilization of the population or militia to enforce the law. Posse comitatus was famously applied in the United States in the Old West and elsewhere by local law enforcement officials.

The Posse Comitatus Act was enacted in 1878 during the administration of President Rutherford B. Hayes. The purpose of the law is to limit the federal government's authority to use the military to enforce domestic policies—in essence, forbidding its deployment as a posse comitatus. Relevant provisions of the U.S. Code for this limitation state the following:

> 18 U.S.C. § 1385. Use of Army and Air Force as Posse Comitatus
>
> Whoever, except in cases and under circumstances expressly authorized by the Constitution or Act of Congress, willfully uses any part of the Army or the Air Force as a posse comitatus or otherwise to execute the laws shall be fined under this title or imprisoned not more than two years, or both.

And,

> 10 U.S.C. § 375. Restriction on Direct Participation by Military Personnel
>
> The Secretary of Defense shall prescribe such regulations as may be necessary to ensure that any activity (including the provision of any equipment or facility or the assignment or detail of any personnel) under this chapter does not include or permit direct participation by a member of the Army, Navy, Air Force, or Marine Corps in a search, seizure, arrest, or other similar activity unless participation in such activity by such member is otherwise authorized by law.

The act initially only referred to the domestic deployment of the U.S. Army as a posse comitatus, but in 1956, it was amended to include the U.S. Air Force. The U.S. Navy and Marine Corps were not mentioned in the act, but those services are governed by regulations that impose the same constraints as the Posse Comitatus Act. The Posse Comitatus Act does not apply to the U.S. Coast Guard or state National Guard units, both of which may be deployed domestically to enforce the law or restore order.

Waging War in the Era of the New Terrorism

When the war on terrorism was declared in the aftermath of the September 11 attacks on the United States, it became readily apparent that this was a new kind of conflict against a new form of enemy. From the outset, policymakers understood that this war would be fought in an unconventional manner, primarily against shadowy terrorist cells and elusive leaders. It was not a war against a nation, but rather against ideas and behavior. It would be a war fought domestically and internationally. Homeland security and national defense institutions would become the front line in the new war. The mobilization of resources necessary to defend the homeland and deploy forces across the globe would require the coordination of emergency management, law enforcement, military, and intelligence assets. The mission and work of the Intelligence Community is discussed in Chapter 6.

War in the Shadows

The new security environment required the adoption of innovative counterterrorist tactics. Domestically, suspected terrorist cells were identified and dismantled by law enforcement agencies, often as a result of undercover operations by law enforcement officers. Internationally, covert operations by special military and intelligence units became the norm rather than the exception as covert operatives worked secretly around the globe. New protocols were implemented for processing suspects captured abroad, and many were detained at the U.S. naval base in Guantánamo Bay, Cuba, and other secret detention facilities. However, this war has not been fought solely in the shadows and has often involved large deployments of conventional military assets.

Overt Conflict: The Deployment of Military Assets

In contrast to the deployment of small law enforcement and covert military or intelligence assets, the U.S.-led invasions of Afghanistan and Iraq involved the commitment of large conventional military forces. In Afghanistan, reasons given for the invasion included the need to eliminate state-sponsored safe havens for al-Qaeda and other international mujahideen (holy warriors). In Iraq, reasons given for the invasion included the need to eliminate alleged stockpiles of weapons of mass destruction and alleged links between the regime of Saddam Hussein and terrorist networks. The U.S.-led operation in Iraq was symbolically named **Operation Iraqi Freedom**.

Operation Iraqi Freedom: The U.S.-led invasion of Iraq involving the commitment of large conventional military forces.

One significant challenge for waging war against extremist behavior—in this case, against terrorism—is that victory is not an easily definable condition. For example, on May 1, 2003, President George W. Bush landed on the aircraft carrier *Abraham Lincoln* to deliver a speech in which he officially declared that the military phase of the Iraq invasion had ended and that the overthrow of the Hussein government was "one victory in a war on terror that began on September 11, 2001, and still goes on."[49] Unfortunately, President Bush's declaration was premature. A widespread insurgency took root in Iraq, with the resistance employing both classic hit-and-run guerrilla tactics and terrorism. Common cause was found between remnants of the Hussein regime and non-Iraqi Islamist fighters. Thousands of Iraqis and occupation troops became casualties during the insurgency. In particular, the insurgents targeted foreign soldiers, government institutions, and Iraqi "collaborators," such as soldiers, police officers, election workers, and interpreters. Sectarian violence also spread, with Sunni and Shi'a religious extremists killing many civilians.

Is the war on terrorism being won? How can victory reasonably be measured? Assuming that the New Terrorism will continue for a time, perhaps the best measure of progress in the war is to assess the degree to which terrorist behavior is being successfully *managed*—in much the same manner that progress against crime is assessed. As the global community continues to be challenged by violent extremists during the new era of terrorism, the definition of victory is likely to continue to be refined and redefined by nations and leaders.

CHAPTER SUMMARY

Because of revelations about bureaucratic inefficiency in the aftermath of the September 11 attacks, the United States implemented a restructuring of its homeland security community. When examining homeland security agencies and missions, it is important to consider that they operate within the context of counterterrorist and antiterrorist options. Many federal agencies are participants in the overall homeland security enterprise, the Department of Homeland Security being the largest and most mission-diverse federal organization. The military performs a critical international role in securing the domestic homeland security environment. The intersection of military missions with those of domestic agencies creates a large and intricate establishment for combating terrorism domestically and internationally.

An underlying theme throughout this discussion has been that homeland security is an evolving concept. Organizational cooperation and coordination are certainly desirable, but it must be remembered that these can occur only if political and policy responses are able to adapt to changes in the terrorist environment. Homeland security in the post–September 11 era has adapted to new and emerging threats. These threats reflect the creativity and determination of those who wage terrorist campaigns against the United States and its allies. Disruption of terrorist operations requires broad cooperation and commitment to protecting the homeland from these adversaries.

DISCUSSION BOX

This chapter's Discussion Box is intended to stimulate critical thinking about the perception of homeland security training exercises by distrustful residents.

Military Training Exercises: The Politics of Jade Helm 15

Jade Helm 15 was a multistate military training exercise that began on July 15, 2015, and ended on September 15, 2015. The exercise was a training operation in unconventional warfare for members of the U.S. Army's Special Operations Command and the Joint Special Operations Command. It took place across several states, including Arizona, Florida, Louisiana, Mississippi, New Mexico, Texas, and Utah.

As planning for Jade Helm 15 progressed and the operation began, strong public opposition to the exercise arose from some quarters, including citizen groups, tabloid media, and politicians. Interestingly, much of the opposition was predicated on the notion that the exercise was not simply a training operation but was, in fact, a prelude to a federal crackdown in support of an international takeover. Elaborate conspiracy rumors included allegations of planned federal seizures of firearms, declarations of martial law, the stockpiling of

weapons for Chinese soldiers in closed Walmart stores, closed Walmart stores being prepared for FEMA detention centers, the rounding up of political opponents, and preparations made for an invasion of Texas. Other similar conspiracy rumors were also spread. The governor of Texas ordered the Texas State Guard to monitor the exercise to ensure that there were no violations of constitutional rights and liberties of residents.

In the end, Jade Helm 15 took place within its declared schedule, and none of the feared events occurred.

Discussion Questions

1. How should institutions such as the Department of Defense prepare the public for training exercises in their communities?

2. How should similar responses by local communities be addressed by homeland security officials?

3. What measures should be taken to ensure collaboration and coordination between national and local homeland security institutions?

4. Was the Texas governor's deployment of the Texas State Guard appropriate?

KEY TERMS AND CONCEPTS

The following topics were discussed in this chapter and can be found in the glossary:

Department of Agriculture (USDA) 97

Department of Defense (DOD) 100
Department of Energy (DOE) 101

Department of Health and Human Services (HHS) 101

ON YOUR OWN

Get the tools you need to sharpen your study skills. SAGE edge offers a robust online environment featuring an impressive array of free tools and resources.

Access practice quizzes, eFlashcards, video, and multimedia at **edge.sagepub.com/martinhs3e**

RECOMMENDED WEBSITES

The following websites provide information about federal homeland security agencies:

Department of Agriculture: www.usda.gov

Department of Defense: www.defense.gov

Department of Energy: www.energy.gov

Department of Health and Human Services: www.hhs.gov

Department of Homeland Security: www.dhs.gov

Department of the Interior: www.doi.gov

Department of the Treasury: www.ustreas.gov

Drug Enforcement Administration (DEA): www.usdoj.gov/dea

Environmental Protection Agency: www.epa.gov

WEB EXERCISE

Using this chapter's recommended websites, conduct an online investigation of the role of federal agencies in designing a nationwide homeland security enterprise.

1. What are the primary documents explaining the underlying purpose and missions of homeland security?

2. How would you describe the differences between the Department of Homeland Security and services agencies?

3. In your opinion, what practical options exist for coordinating national defense institutions with homeland security agencies?

To conduct an online search on research and monitoring organizations, activate the search engine on your Web browser and enter the following keywords:

"Homeland security agencies"

"Homeland security and the war on terrorism"

RECOMMENDED READINGS

The following publications provide discussions of federal agencies and their missions:

Coulson, Danny O. and Elaine Shannon. 1999. *No Heroes: Inside the FBI's Secret Counter-Terror Force.* New York: Pocket.

Dolnik, Adam and Keith M. Fitzgerald. 2007. *Negotiating Hostage Crises With the New Terrorists.* Westport, CT: Praeger.

Graff, Garrett M. 2011. *The Threat Matrix: The FBI at War in the Age of Global Terror.* New York: Little, Brown.

Pedahzur, Ami. 2010. *The Israeli Secret Services and the Struggle Against Terrorism.* New York: Columbia University Press.

Pushies, Fred J. 2009. *Deadly Blue: Battle Stories of the U.S. Air Force Special Operations Command.* New York: American Management Association.

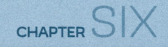
PREDICTION AND PREVENTION
The Role of Intelligence

Opening Viewpoint: Prevention: A Case of Successful International Intelligence Cooperation

An example of successful international intelligence cooperation occurred in May 2002 between American and Moroccan intelligence agencies. In February 2002, Moroccan intelligence officers interrogated Moroccan al-Qaeda prisoners held by the Americans at their naval base in Guantánamo Bay, Cuba. They received information from one of the prisoners about an al-Qaeda operative in Morocco and also received information about the operative's relatives. Moroccan officials obtained a sketched description of the man from the relatives and showed the sketch to the Guantánamo prisoner, who confirmed his likeness. The Moroccans located the suspect (a Saudi), followed him for a month, and eventually arrested him and two Saudi accomplices. The suspects eventually told the Moroccans that they were al-Qaeda operatives trained in Afghanistan and that they had escaped during the anti-Taliban campaign after receiving orders to engage in suicide attacks against maritime targets in Gibraltar. They had begun the process of inquiring about speedboats, and their ultimate targets were to be U.S. Navy ships passing through Gibraltar.

Chapter Learning Objectives

This chapter will enable readers to do the following:

1. Analyze challenges inherent in the mission of the Intelligence Community

2. Understand the organizational alignments of the Intelligence Community

3. Evaluate the types of intelligence and how intelligence is collected

4. Apply the role of intelligence collection to the context of the homeland security enterprise

5. Remember the missions of intelligence agencies

6. Understand the complexity of the intelligence craft and the roles of intelligence agencies

Intelligence refers to the collection of data. Its purpose within the context of counterterrorism is to create an informational database about terrorist movements and predict terrorist behavior. This process is not unlike that of criminal justice investigators who work to resolve criminal cases. In both contexts, the fundamental objectives of intelligence collection are prediction and prevention.

The modern **Intelligence Community** (IC) comprises mission-specific agencies representing the predictive and analytical arm of the federal government. It manages the collection and analysis of an enormous quantity of information derived from an extremely diverse array of sources. The Intelligence Community must filter this information in order to create actionable intelligence, which is critically necessary for predicting, preventing, and analyzing terrorist events. Intelligence agencies involve themselves with the collection and analysis of information. The underlying mission of intelligence agencies is to construct an accurate activity profile of terrorists. Data are collected from overt and covert sources and evaluated by expert intelligence analysts. This process—intelligence collection and analysis—is at the heart of the counterterrorist intelligence mission.

The outcome of high-quality intelligence collection and analysis can range from the construction of profiles of terrorist organizations to tracking the movements of terrorists. An optimal outcome of counterterrorist intelligence is the ability to *anticipate* the behavior of terrorists and thereby to predict terrorist incidents. However, exact prediction is relatively rare, and most intelligence on terrorist threats is generalized rather than specific. For example, intelligence agencies have had success in uncovering threats in specific cities by specific groups but less success in predicting the exact time and place of possible attacks. These considerations are summarized as elements of the overall mission of the IC:

> The Intelligence Community's mission is to collect, analyse, and deliver foreign intelligence and counterterrorist information to America's leaders so they can make sound decisions to protect our country.[1]

The discussion in this chapter addresses the role of intelligence and the mission of the Intelligence Community. Inherent in this discussion is the tension that naturally arises between the mission of the IC and the challenges of intelligence coordination, collection, and analysis. This chapter examines the following issues:

- The U.S. Intelligence Community: Mission
- The intelligence cycle
- Intelligence oversight
- Intelligence agencies
- The U.S. Intelligence Community: Challenges

THE U.S. INTELLIGENCE COMMUNITY: MISSION

Intelligence collection and analysis are important components of the homeland security enterprise. The intelligence mission is unique in the sense that it is responsible for securing the American homeland from external threats. That is, although intelligence operations have a significant effect on domestic security, their scope of operations is also outside the borders of the nation.

Background: Intelligence Collection and Jurisdiction

Federal National Security Intelligence Collection

National security intelligence collection is divided between agencies that are separately responsible for domestic and international intelligence collection. This separation is mandated by law. For example, the **Federal Bureau of Investigation (FBI)** performs domestic intelligence collection, and the **Central Intelligence Agency (CIA)** operates internationally. The FBI is a law enforcement agency that uses criminal intelligence to enforce the law and provides important assistance to state and local law enforcement agencies. However, the FBI also has primary jurisdiction over domestic counterintelligence and counterterrorist surveillance and investigations. The CIA is not a law enforcement agency and, therefore, officially performs a supportive role in domestic counterterrorist investigations.

Other federal agencies, such as the **Diplomatic Security Service**, also assist in tracking suspects wanted for acts of terrorism. The Diplomatic Security Service is a security bureau within the U.S. Department of State that, among other duties, manages an international bounty program called the **Rewards for Justice Program**. The program offers cash rewards for information leading to the arrest of wanted terrorists. The Rewards for Justice Program has successfully resulted in the capture of suspects.

State and Local Intelligence Collection

State and local intelligence collection has its origin in crime prevention and prediction. Law enforcement agencies have a long history of building criminal intelligence databases for the purpose of preventing and predicting criminal activity, and these databases are readily adaptable to providing information relevant to the national security mission of the homeland security enterprise. Modern databases are frequently linked to the FBI's criminal and forensic databases, thus creating an intertwined system of intelligence-sharing and -tracking capability. Collaborative networks and initiatives have been established to promote collaboration on intelligence sharing. Examples of these networks and initiatives include the following:

- *Homeland Security Information Network (HSIN).* "The Homeland Security Information Network (HSIN) is the trusted network for homeland security mission operations to share Sensitive But Unclassified information. . . . The Homeland Security Information Network (HSIN) provides law enforcement officials at every level of government with a means to collaborate securely with partners across geographic and jurisdictional boundaries."[2]

- *National Criminal Intelligence Sharing Plan (NCISP).* Developed in 2003, "this plan represents law enforcement's commitment to take it upon itself to ensure that the dots are connected, be it in crime or terrorism. The plan is the outcome of an unprecedented effort by law enforcement agencies, with the strong support of the Department of Justice, to strengthen the nation's security through better intelligence analysis and sharing."[3]

- *Regional Information Sharing System (RISS).* Created in 1973, RISS "offers secure information sharing and communications capabilities, critical analytical and investigative support services, and event deconfliction to enhance officer safety. RISS supports efforts against organized and violent crime, gang activity, drug activity, terrorism, human trafficking, identity theft, and other regional priorities."[4]

Evolution of the Modern Intelligence Community

The present-day IC is a successor to the missions and organizational configurations that were established during the Cold War. Rivalry between the United States and the Soviet Union, and their respective "Free World" and "Eastern Bloc" allies, necessitated the creation and funding of a global intelligence presence. At its peak during the 1980s, the IC employed

Central Intelligence Agency (CIA): The principal intelligence agency in the United States and the theoretical coordinator of American foreign intelligence collection.

Diplomatic Security Service: A security bureau within the U.S. Department of State that protects diplomats and other officials.

Rewards for Justice Program: An international bounty program managed by the U.S. Diplomatic Security Service. The program offers cash rewards for information leading to the arrest of wanted terrorists.

approximately 100,000 personnel. The IC workforce was assigned to approximately 25 agencies and elements. Each organization was tasked with performing specialized functions, often using assigned modalities of intelligence collection such as electronic surveillance or the deployment of human assets. Expansion of the Cold War–era IC necessitated a concomitant increase of fiscal resources, eventually resulting in the appropriation of approximately $30 billion for IC operations. With the end of the Cold War—dated roughly from the 1989 dismantling of the Berlin Wall—there occurred a consolidation process of hitherto discrete agency operations. Fiscal appropriations were reduced, as were the number of IC personnel, agencies, and elements.

With the post–Cold War reductions in appropriations and personnel, the IC directed much of its attention toward counterterrorist operations. This was a matter of necessity because of the following incidents:

- 1993: Vehicular bombing of the World Trade Center in New York City by Ramzi Yousef.

- 1998: Simultaneous suicide bombings of the American embassies in Nairobi, Kenya, and Dar es Salaam, Tanzania, by al-Qaeda operatives.

- 2000: Suicide attack on the destroyer USS *Cole* in Aden, Yemen.

These and other incidents indicated that determined terrorists have the ability to carry out significant attacks despite the hard work of the U.S. IC and allied intelligence agencies. The successful al-Qaeda attacks on September 11, 2001, led to the creation in November 2002 of the National Commission on Terrorist Attacks Upon the United States. Established jointly by law by Congress and President George W. Bush, it is commonly referred to as the *9/11 Commission*. The 9/11 Commission was a bipartisan panel directed to

investigate "facts and circumstances relating to the terrorist attacks of September 11, 2001," including those relating to intelligence agencies, law enforcement agencies, diplomacy, immigration issues and border control. The flow of assets to terrorist organizations, commercial aviation, the role of congressional oversight and resource allocation, and other areas determined relevant by the Commission.[5]

The final chapter of the 9/11 Commission's report is titled "How to Do It? A Different Way of Organizing the Government." In this chapter, the 9/11 Commission stressed the need for unity of effort and specifically provided detailed and pointed recommendations for restructuring the IC. It stated that "[t]he need to restructure the Intelligence Community grows out of six problems that have become apparent before and after 9/11:

- *"Structural barriers to performing joint intelligence work*. National intelligence is still organized around the collection disciplines of the home agencies, not the joint mission. The importance of integrated, all-source analysis cannot be overstated. Without it, it is not possible to 'connect the dots.' No one component holds all the relevant information.

- *"Lack of common standards and practices across the foreign-domestic divide*. The leadership of the Intelligence Community should be able to pool information gathered overseas with information gathered in the United States, holding the work—wherever it is done—to a common standard of quality in how it is collected, processed (e.g., translated), reported, shared, and analyzed. A common set of personnel standards for intelligence can create a group of professionals better able to operate in joint activities, transcending their own service-specific mind-sets.

- *"Divided management of national intelligence capabilities.* While the CIA was once "central" to our national intelligence capabilities, following the end of the Cold War it has been less able to influence the use of the nation's imagery and signals intelligence capabilities in three national agencies housed within the Department of Defense: the National Security Agency, the National Geospatial-Intelligence Agency, and the National Reconnaissance Office. One of the lessons learned from the 1991 Gulf War was the value of national intelligence systems (satellites in particular) in precision warfare. Since that war, the department has appropriately drawn these agencies into its transformation of the military. Helping to orchestrate this transformation is the Under Secretary of Defense for Intelligence, a position established by Congress after 9/11. An unintended consequence of these developments has been the far greater demand made by Defense on technical systems, leaving the Director of Central Intelligence (DCI) less able to influence how these technical resources are allocated and used.

- *"Weak capacity to set priorities and move resources.* The agencies are mainly organized around what they collect or the way they collect it. But the priorities for collection are national. As the DCI makes hard choices about moving resources, he or she must have the power to reach across agencies and reallocate effort.

- *"Too many jobs.* The DCI now has at least three jobs. He is expected to run a particular agency, the CIA. He is expected to manage the loose confederation of agencies that is the Intelligence Community. He is expected to be the analyst in chief for the government, sifting evidence and directly briefing the President as his principal intelligence adviser. No recent DCI has been able to do all three effectively. Usually what loses out is management of the Intelligence Community, a difficult task even in the best case because the DCI's current authorities are weak. With so much to do, the DCI often has not used even the authority he has.

- *"Too complex and secret.* Over the decades, the agencies and the rules surrounding the Intelligence Community have accumulated to a depth that practically defies public comprehension. There are now 15 agencies or parts of agencies in the Intelligence Community. The community and the DCI's authorities have become arcane matters, understood only by initiates after long study. Even the most basic information about how much money is actually allocated to or within the Intelligence Community and most of its key components is shrouded from public view."[6]

In recognition of the 9/11 Commission's conclusions, and to reduce the incidence of problems cited by the Commission, in December 2004, the IC was reorganized with the passage of the Intelligence Reform and Terrorism Prevention Act (IRTA). Of central importance to the IC reorganization was the creation of two new elements, the **Office of the Director of National Intelligence (ODNI)** and the **National Counterterrorism Center (NCTC)**. Members of the community were subsumed under the direction of the new ODNI. President George W. Bush appointed John Negroponte, former U.S. ambassador to Iraq, as the United States' first **Director of National Intelligence** (DNI). Officially confirmed by the Senate in April 2005, the DNI is responsible for coordinating the various components of the IC.

Office of the Director of National Intelligence (ODNI): In December 2004, the intelligence community was reorganized with the passage of the Intelligence Reform and Terrorism Prevention Act. Members of the community were subsumed under the direction of a new Office of the Director of National Intelligence, responsible for coordinating the various components of the intelligence community.

National Counterterrorism Center (NCTC): A center established to integrate the counterterrorism efforts of the intelligence community in the wake of the September 11, 2001, attacks.

Director of National Intelligence: Members of the IC are subsumed under the direction of the ODNI. President George W. Bush appointed John Negroponte, former U.S. ambassador to Iraq, as the United States' first Director of National Intelligence (DNI). The DNI is responsible for coordinating the various components of the IC.

Thus, in the post–9/11 era, the United States endeavors to advance the quality of intelligence collection and analysis by creating a coordinated and cooperative IC. This philosophy of collaboration is the primary conceptual goal of the American counterterrorist intelligence effort.

The Intelligence Community in the Post–9/11 Environment

The modern IC is comprised of agencies that function under the authority of the executive branch of government. They are administratively independent agencies that ideally cooperate in the collection and analysis of information. All agencies are tasked with providing information to the president and other relevant stakeholders on a "need to know" basis. The IC is an extensive administrative enterprise consisting of 17 elements—16 agencies and the ODNI—organized as follows:

- Office of the Director of National Intelligence
- Central Intelligence Agency
- National Security Agency
- Federal Bureau of Investigation
- Department of State (Bureau of Intelligence and Research)
- Department of Energy (Office of Intelligence and Counterintelligence)
- Drug Enforcement Administration (Office of National Security Intelligence)
- Department of Homeland Security (Office of Intelligence and Analysis)
- Department of Treasury (Office of Intelligence and Analysis)
- Defense Intelligence Agency
- Office of Naval Intelligence
- Army Intelligence and Security Command
- Marine Corps Intelligence
- Air Force Intelligence
- U.S. Coast Guard Intelligence
- National Reconnaissance Office
- National Geospatial-Intelligence Agency

Each agency must comply with mandated jurisdictional limitations on its collection of intelligence. However, because national security threats may affect multiple sectors of the homeland security enterprise, there naturally exists overlap in jurisdiction among some agencies. For example, the following problems may involve complex scenarios that activate the jurisdiction of multiple agencies:

- Threats from violent extremists
- Countering foreign intelligence operations in the United States
- Illicit weapons trafficking

- Drug trafficking

- Human trafficking

- Cyberattacks

- CBRN (chemical, biological, radiological, nuclear) threats

- Threats against infrastructure

Because of the segmentation of the IC and the complexity of its overall mission, there exists an imperative need for seamless coordination and cooperation among organizations comprising the IC. As discussed later in this chapter, the ideal of interagency collaboration is sometimes a challenging goal. Nevertheless, because of the critical need for actionable information, the IC is a central component of the homeland security enterprise.

THE INTELLIGENCE CYCLE

The Intelligence Community operates within the framework of an intelligence cycle. Ideally, the intelligence cycle represents a seamless and efficient process for providing accurate information to policymakers, who use intelligence findings to design and implement informed policies. Agencies comprising the IC are tasked to select methods for collecting desired information and to operationalize these methods. When information is successfully obtained, specialists organize, interpret, and analyze the significance of their findings. This is a dynamic process that frequently engenders new questions and new intelligence operations.

Phases of the Intelligence Cycle

The intelligence cycle involves six phases. These phases are often compartmentalized processes, consisting of the following components:

- *Planning and Direction:* "Policymakers—including the president, presidential advisors, the National Security Council, and other major departments and agencies—determine what issues need to be addressed and set intelligence priorities. The IC's issue coordinators interact with these officials to identify core concerns and information requirements."[7]

- *Collection:* "The IC uses many methods to collect information, including face-to-face meetings with human sources, technical and physical surveillance, satellite surveillance, interviews, searches, and liaison relationships. Information can be gathered through open, covert, and electronic means. All collection methods must be lawful and are subject to oversight by Congress and others. Information collected must be relevant, timely, and useful. At this state, the information is often referred to as raw intelligence, because it hasn't been thoroughly examined and evaluated yet."[8]

- *Processing:* "The collection stage of the intelligence cycle can yield large amounts of data that requires organization and refinement. Substantial resources are devoted to synthesizing this data into a form that intelligence analysts can use."[9]

- *Analysis and Production:* "Analysts examine and evaluate all the information collected, add context as needed, and integrate it into complete products. They produce *finished intelligence* that includes assessments of events and judgments about the implications of the information for the United States."[10]

- *Dissemination:* "Finished intelligence is delivered to policymakers, military leaders, and other senior government leaders who then make decisions based on the information. Finished intelligence can lead to requests for additional information, thus triggering the intelligence cycle again."[11]

- *Evaluation:* Although this is listed as a discrete step in the intelligence cycle, evaluation . . . is ongoing throughout the cycle. [The IC is] continuously evaluating . . . products for relevance, bias, accuracy, and timeliness, as well as [the] process to ensure it is efficient and thorough."[12]

Types of Intelligence Collection

The cycle of intelligence collection requires the marshaling of an integrated system of technologies, specialized agencies, professional practitioners, and collaborative government entities. This is often a complex endeavor. Nevertheless, the following six source types are routinely deployed from the IC:

SIGINT—Signal Intelligence

Intelligence collection and analysis in the modern era require the use of sophisticated technological resources. These technological resources are used primarily for the interception of electronic signals—known as **signals intelligence (SIGINT)**. SIGINT is used for a variety of purposes, such as interceptions of financial data, monitoring communications such as cell phone conversations, and reading e-mail messages. The use of satellite imagery is also commonly used by intelligence agencies, and sophisticated computers specialize in code breaking. However, the practicality of these technologies as counterterrorist options is limited in the era of the New Terrorism. Because of the cellular organizational structure of terrorist groups and their insular interactions (i.e., based on personal relationships), technology cannot be an exclusive counterterrorist resource. Human intelligence is also a critical component. Prominent SIGINT centers include the United Kingdom's Government Communications Headquarters (GCHQ) and the National Security Agency (NSA) in the United States.

HUMINT—Human Intelligence

The collection of **human intelligence**, also referred to as **HUMINT**, is often a cooperative venture with friendly intelligence agencies and law enforcement officials. This sharing of information is a critical component of counterterrorist intelligence gathering. Circumstances may also require the covert manipulation of individuals affiliated with terrorist organizations or their support groups, with the objective of convincing them to become intelligence agents. The manipulation process can include making appeals to potential spies' sense of justice or patriotism, paying them with money and other valuables, or offering them something that they would otherwise be unable to obtain (such as asylum for their family in a Western country). One significant problem with finding resources for human intelligence is that most terrorist cells are made up of individuals who know one another very well. Newcomers are not openly welcomed, and those who may be potential members are usually expected to commit an act of terrorism or other crime to prove their commitment to the cause. In other words, intelligence agencies must be willing to use terrorists to catch terrorists. This has been a very difficult task, and groups such as al-Qaeda have proven very difficult to penetrate with human assets.[13]

signals intelligence (SIGINT): Intelligence that has been collected by technological resources.

human intelligence (HUMINT): Intelligence that has been collected by human operatives rather than through technological resources.

open source intelligence (OSINT): Information collected from publicly available electronic and print outlets. It is information that is readily available to the public, but used for intelligence analysis. Examples of open sources include newspapers, the Internet, journals, radio, videos, television, and commercial outlets.

imagery intelligence (IMINT): Images are regularly collected to provide actionable intelligence. Collection technologies range from relatively routine hand-held equipment to very sophisticated means. IMINT includes intelligence information derived from the collection by visual photography, infrared sensors, lasers, electro-optics, and radar sensors.

OSINT—Open Source Intelligence

Open source intelligence (OSINT) is information collected from publicly available electronic and print outlets. It is information that is readily available to the public but used for intelligence analysis. Examples of open sources include newspapers, the Internet, journals, radio, videos, television, and commercial outlets.

IMINT—Imagery Intelligence

Images are regularly collected to provide actionable intelligence. Collection technologies range from relatively routine hand-held equipment to very sophisticated means. **Imagery intelligence (IMINT)** includes "intelligence information derived from the collection by visual photography, infrared sensors, lasers, electro-optics, and radar sensors."[14]

MASINT—Measurements and Signatures Intelligence

The use of a broad array of technical and scientific disciplines to measure the characteristics of specified subjects—for example, tracking communications signatures or measuring water and soil samples. **Measurements and signatures intelligence (MASINT)** is "intelligence information obtained by quantitative and qualitative analysis of data derived from specific technical sensors for the purpose of identifying any distinctive features associated with the source, emitter, or sender."[15]

GEOINT—Geospatial Intelligence

The collection and assessment of topography and geographical features can provide actionable intelligence regarding locations, timeframes, and other information. **Geospatial intelligence (GEOINT)** is "the all-source analysis of imagery and geospatial information to describe, assess, and visually depict physical features and geographically referenced activities on earth."[16]

The National Intelligence Priorities Framework

Intelligence policy priorities are governed by the **National Intelligence Priorities Framework** (NIPF), which "promulgates policy and establishes responsibilities for setting national intelligence priorities and translating them into action."[17]

The Director of National Intelligence is charged with overall authority to assure compliance with NIPF guidelines and is required to "approve the NIPF and the policies and processes for establishing national intelligence priorities; and adjust national intelligence priorities as necessary."[18]

This is done under consideration from, and on the recommendation of, heads of agencies that comprise the IC. It is necessary to regularly update the NIPF. Updates are intended to provide fresh direction for intelligence agencies on how best to allocate resources for intelligence collection and analysis. In theory, it is a process that promotes efficiency within the IC.

INTELLIGENCE OVERSIGHT

The Intelligence Community technically operates under the purview of the executive branch of government. However, because the work of the IC is quite often highly sensitive with potentially significant ramifications, the IC also operates under the oversight of several

measurements and signatures intelligence (MASINT): The use of a broad array of technical and scientific disciplines to measure the characteristics of specified subjects. For example, tracking communications signatures or measuring water and soil samples. MASINT is intelligence information obtained by quantitative and qualitative analysis of data derived from specific technical sensors for the purpose of identifying any distinctive features associated with the source, emitter, or sender.

geospatial intelligence (GEOINT): The collection and assessment of topography and geographical features can provide actionable intelligence regarding locations, timeframes, and other information. GEOINT is the all-source analysis of imagery and geospatial information to describe, assess, and visually depict physical features and geographically referenced activities on earth.

National Intelligence Priorities Framework: Intelligence policy priorities are governed by the National Intelligence Priorities Framework (NIPF), which promulgates policy and establishes responsibilities for setting national intelligence priorities and translating them into action.

federal policy centers. These oversight centers are drawn from the executive, legislative, and judiciary branches of government. The purpose of intelligence oversight is to confirm that the work of the IC is in compliance with relevant laws and policies. Offices possessing oversight authority within the executive branch include

- the office of the President,
- the National Security Council,
- the President's Intelligence Advisory Board,
- the Intelligence Oversight Board,
- the Office of Management and Budget, and
- the Privacy and Civil Liberties Oversight Board.

Offices possessing oversight authority within the legislative branch include

- the Senate Select Committee on Intelligence and
- the House Permanent Select Committee on Intelligence.

The Foreign Intelligence Surveillance Court is also authorized to provide oversight from within the judiciary branch. Additional oversight may originate from inspectors general operating from within each IC agency. Inspectors general conduct audits and other reviews of IC agencies.

INTELLIGENCE AGENCIES

Members of the American Intelligence Community include the following agencies and centers.

Office of the Director of National Intelligence

The ODNI was created to address concerns about the efficiency of the IC in the aftermath of the attacks on September 11, 2001. As stated on the office's website, "the mission of the ODNI is to lead and support IC integration; delivering insights, driving capabilities, and investing in the future."[19] Furthermore, "The ODNI is staffed by officers from across the IC and is organized into directorates, centers, and oversight offices that support the DNI's role as head of the IC and manager of the National Intelligence Program (NIP)."[20]

ODNI directorates "are organized around ODNI core functions to provide a more holistic view and strategic approach to intelligence integration."[21] Established director-ates include Enterprise Capacity, Mission Integration, National Security Partnerships, and Strategy and Engagement.

ODNI mission centers include the Cyber Threat Integration Center, National Counterproliferation Center, National Counterintelligence and Security Center, and the National Counterterrorism Center. The mission centers perform critical tasks for the home-land security enterprise. As explained by the ODNI,

In their roles as functional National Intelligence Managers (NIMs), the National Counterterrorism Center (NCTC), the National Counterproliferation Center (NCPC), and the National Counterintelligence and Security Center (NCSC)

also contribute to the mission of intelligence integration. For both functional and regional NIMs, the Unifying Intelligence Strategies (UIS) are critical plans for communicating priorities and achieving intelligence integration. NIMs develop UIS in line with prioritized IC requirements and are charged with leading integration across the IC by function and region.[22]

ODNI oversight offices include Civil Liberties, Privacy and Transparency; Equal Employment Opportunity and Diversity; Intelligence Community Inspector General; and Office of General Counsel. These offices function as internal controls to

ensure that the IC carries out its mission in a manner that protects privacy and civil liberties and enhances transparency; oversee equal opportunity and workforce diversity programs; conduct independent audits, investigations, inspections, and reviews; provide accurate legal guidance and counsel to ensure compliance with the Constitution, U.S. law, and corresponding regulations; and facilitate the DNI's statutory responsibility to keep the appropriate Congressional committees informed of all intelligence activities of the U.S.[23]

National Security Agency

The **National Security Agency (NSA)** is the technological arm of the U.S. Intelligence Community. Using state-of-the-art computer and satellite technologies, the NSA's primary mission is to collect communications and other signal intelligence. It also devotes a significant portion of its technological expertise to code-making and code-breaking activities. Much of this work is done covertly from secret surveillance facilities positioned around the globe.

National Security Agency (NSA): An American intelligence agency charged with signals intelligence collection, code making, and code breaking.

Central Intelligence Agency

The CIA is an independent federal agency. It is the theoretical coordinator of the Intelligence Community. The agency is charged with collecting intelligence outside of the borders of the United States, which is done covertly using human and technological assets. The CIA is legally prohibited from collecting intelligence inside the United States.

U.S. National Security Agency

▶ **Photo 6.1**

Defense Intelligence Agency

The **Defense Intelligence Agency (DIA)** is a bureau within the Department of Defense. It is the central intelligence bureau for the U.S. military. Each branch of the military coordinates its intelligence collection and analysis with the other branches through the DIA.

Defense Intelligence Agency (DIA): The central agency for military intelligence of the U.S. armed forces.

Federal Bureau of Investigation

The FBI is a bureau within the Department of Justice. It is a law enforcement agency that is charged, in part, with conducting domestic surveillance of suspected spies and terrorists. The agency also engages in domestic intelligence collection and has been deployed to American embassies around the world. Foreign counterintelligence investigations have included an FBI presence at the sites of the 1998 bombings of the U.S. embassies in Kenya and Tanzania.

▶ **Photo 6.2**

National Reconnaissance Office: Responsible for designing, building, launching, and maintaining America's intelligence satellites. NRO provides satellite reconnaissance support to the IC and Department of Defense.

DHS Office of Intelligence and Analysis: The only Intelligence Community (IC) element statutorily charged with delivering intelligence to state, local, tribal, territorial, and private-sector partners, and developing intelligence from those partners for [DHS] and the IC.

National Geospatial-Intelligence Agency

The **National Geospatial-Intelligence Agency (NGA)** is responsible for overseeing GEOINT collection and analysis. NGA "manages a global consortium of more than 400 commercial and government relationships." Furthermore, "[t]he director of NGA serves as the functional manager for GEOINT, the head of the National System for Geospatial Intelligence and the coordinator of the global Allied System for Geospatial Intelligence."[24]

National Reconnaissance Office

The **National Reconnaissance Office** (NRO) is responsible for "designing, building, launching, and maintaining America's intelligence satellites."[25] NRO provides satellite reconnaissance support to the IC and Department of Defense.

Department of Homeland Security
Office of Intelligence and Analysis

The **DHS Office of Intelligence and Analysis** (I&A) is a unique member of the IC. Unlike other agencies, "I&A is the only Intelligence Community (IC) element statutorily charged with delivering intelligence to our state, local, tribal, territorial, and private-sector partners, and developing intelligence from those partners for [DHS] and the IC."[26]

Case in Point: The International Context of Intelligence

In many democracies, intelligence collection is traditionally divided between agencies that are separately responsible for domestic and international intelligence collection. This separation is often mandated by law. For example, the following agencies roughly parallel one another's missions:

- In Great Britain, the Security Service (**MI5**) is responsible for domestic intelligence, and the Secret Intelligence Service (**MI6**) is responsible for international collection. GCHQ provides SIGINT support for both MI5 and MI6.

- In Germany, the **Bureau for the Protection of the Constitution** shares a mission similar to MI5 and the FBI, and the **Military Intelligence Service** roughly parallels MI6 and the CIA. SIGINT support is provided by several centers, including the Military Intelligence Service and the Bundeswehr's (united armed forces) Strategic Reconnaissance Command.

THE U.S. INTELLIGENCE COMMUNITY: CHALLENGES

The collection and analysis of intelligence are covert processes that do not lend themselves easily to absolute cooperation and coordination between countries or between members of domestic intelligence communities. National intelligence agencies do not readily share intelligence with allied countries; they usually do so only after careful deliberation. The same is true of intelligence communities within countries. For example, prior to the September 11, 2001, homeland attacks, dozens of federal agencies were involved in the collection of intelligence about terrorism. This led to overlapping and competing interests. A case in point is the apparent failure by the FBI and CIA to collaboratively process, share, and evaluate important intelligence between their agencies. In the case of the FBI, there was also an apparent failure of coordination between the agency's field and national offices. These problems precipitated a proposal in June 2002 by President Bush to completely reorganize the American homeland security community.

Problems of Collection and Analysis

Intelligence collection and analysis are not always exact or low-risk sciences. They can reflect only the quality and amount of data that are available. Because of the nature of counterterrorist intelligence collection and analysis, some experts in the United States have concluded that "the inherent difficulties in both collection and analysis of intelligence on terrorism mean that there will never be tactical warning of most attempted terrorist attacks, or even most major attempted attacks against U.S. targets."[27]

This observation became controversially apparent on July 7, 2004, when the U.S. Select Committee on Intelligence issued its extensive *Report on the U.S. Intelligence Community's Prewar Intelligence Assessments on Iraq.*[28] The 521-page report's findings were a scathing critique of intelligence failures regarding Iraq. For example, its first conclusion found the following:

> Most of the major key judgments in the Intelligence Community's October 2002 National Intelligence Estimate (NIE), *Iraq's Continuing Programs for Weapons of Mass Destruction*, either overstated, or were not supported by, the underlying intelligence reporting. A series of failures, particularly in analytic trade craft, led to mischaracterization of the intelligence.[29]

In another highly critical report, a presidential commission known as the Commission on the Intelligence Capabilities of the United States Regarding Weapons of Mass Destruction essentially labeled the American Intelligence Community as being dysfunctional.[30] It also said that the causes for the failure in the Iraq case continued to hinder intelligence on other potential threats, such as the nuclear programs of adversaries. The commission's 601-page report was delivered in March 2005.

Interagency Coordination and Cooperation

Among law enforcement agencies, the FBI is one of the few agencies that performs a quasi-security mission, explicitly adopting as one of its primary missions the protection of the United States from foreign intelligence and terrorist threats. The FBI does this through one of its five functional areas, the Foreign Counterintelligence functional area. The FBI also maintains missions in several U.S. embassies to coordinate its investigations of cases with international links. Among the service agencies, several bureaus perform a variety of security missions. For example, the Secret Service (part of the Department of the Treasury) protects the president, and the Federal Emergency Management Agency responds to natural and human-made disasters.

An ideal policy framework would require the FBI and CIA to coordinate and share counterterrorist intelligence in a spirit of absolute cooperation. In theory, the FBI should focus on investigating possible domestic security threats, and the CIA should pass along foreign intelligence that might affect domestic security.

Prior to the September 11, 2001, organizational crisis, homeland security was the responsibility of a number of federal agencies. These agencies were not centrally coordinated, and they answered to different centers of authority. Cooperation was theoretically ensured by liaison protocols, special task forces, and oversight. In reality, there was a great deal of functional overlap and bureaucratic "turf" issues.

One problem that became quite clear during the year following the September 11, 2001, homeland attacks was that the pre–9/11 organizational model did not adapt well to the new security crisis. This failure to adapt proved to be operationally damaging; it was politically embarrassing, and it projected an image of disarray.

Intelligence Transformation After September 11, 2001

Consolidation of the domestic security community into an efficient homeland security enterprise became a critical priority in the aftermath of the September 11 attacks. Two efforts were given particular priority: transformation of the Intelligence Community and creation of a new homeland security institutional culture.

A series of revelations and allegations called into question previous assertions by the FBI and CIA that neither agency had prior intelligence about the September 11 homeland attacks. For example, it was discovered that

- the FBI had been aware for years prior to September 2001 that foreign nationals were enrolling in flight schools, and

- the CIA had compiled intelligence data about some members of the al-Qaeda cell that carried out the attacks.

These allegations were compounded by a leak to the press of a memorandum from an FBI field agent that strongly condemned the FBI director's and headquarters' handling of field intelligence reports about Zacarias Moussaoui. Moussaoui was alleged to have been a member of the September 11, 2001, al-Qaeda cell; he had been jailed prior to the attacks. Moussaoui had tried to enroll in flying classes, in which he was apparently interested only in how to *fly* airplanes and uninterested in the *landing* portion of the classes.

Policymakers and elected leaders wanted to know why neither the FBI nor the CIA had "connected the dots" to create a single intelligence profile. Serious interagency and internal problems became publicly apparent when a cycle of recriminations, press leaks, and congressional interventions damaged the "united front" image projected by the White House. Policymakers determined that problems in the homeland security community included the following:

- Long-standing interagency rivalries

- Entrenched and cumbersome bureaucratic cultures and procedures

- No central coordination of homeland security programs

- Fragmentation of counterterrorist operations

- Poor coordination of counterterrorist intelligence collection and analysis

- Disconnect between field offices and Washington headquarters

- "Turf"-based conflict between the FBI and CIA

Subsequent commission reports led to sweeping changes in the U.S. Intelligence Community. These reports included the following:

- In July 2004, the 9/11 Commission issued its detailed report on the September 11, 2001, attacks.

- In March 2005, the Commission on the Intelligence Capabilities of the United States Regarding Weapons of Mass Destruction issued its detailed report on intelligence failures regarding the possession and proliferation of weapons of mass destruction.

The **National Counterterrorism Center (NCTC)** was established to integrate the counterterrorism efforts of the Intelligence Community. Although some jurisdictional tension existed between the NCTC and the CIA's Counterterrorism Center, the NCTC became an important component of the new homeland security culture in the United States. Clearly, the attacks of September 11, 2001, were the catalyst for a broad and long-standing reconfiguration of the American security environment.[31]

Homeland security's counterterrorist bureaucracy is conceptually an amalgamation of many functions of law enforcement and intelligence agencies as well as branches of the military. The bureaucratic ideal of rationality and efficiency requires that these sectors of the government coordinate their counterterrorist missions to promote homeland security. For example, domestic law enforcement agencies must be kept apprised of terrorist threats that may be discovered abroad by intelligence agencies or the military—the challenge is how to implement this policy in these and other scenarios.

Case in Point: Intelligence Miscalculation and the Iraq Case

One of the most disturbing scenarios involved the delivery of weapons of mass destruction (WMDs) to motivated terrorists by an aggressive authoritarian regime. This scenario was the underlying rationale given for the March 2003 invasion of Iraq by the United States and several allies.

In January 2002, U.S. president George W. Bush identified Iraq, Iran, and North Korea as the "axis of evil" and promised that the United States "will not permit the world's most dangerous regimes to threaten us with the world's most destructive weapons." In June 2002, President Bush announced during a speech at the U.S. Military Academy at West Point that the United States would engage in preemptive warfare if necessary.

Citing Iraq's known possession of weapons of mass destruction in the recent past and its alleged ties to international terrorist networks, President Bush informed the United Nations (UN) in September 2002 that the United States would unilaterally move against Iraq if the UN did not certify that Iraq no longer possessed WMDs. Congress authorized an attack on Iraq in October 2002. UN weapons inspectors returned to Iraq in November 2002. After a three-month military buildup, Iraq was attacked on March 20, 2003, and Baghdad fell to U.S. troops on April 9, 2003.

The Bush administration had repeatedly argued that Iraq still possessed a significant arsenal of WMDs at the time of the invasion, that Hussein's regime had close ties to terrorist groups, and that a preemptive war was necessary to prevent the delivery of these weapons to al-Qaeda or another network. Although many experts discounted links between Hussein's regime and religious terrorists, it was widely expected that WMDs would be found. Iraq was known to have used chemical weapons against Iranian troops during the Iran-Iraq War of 1980–1988 and against Iraqi Kurds during the Anfal Campaign of 1987.

In actuality, UN inspectors identified no WMDs prior to the 2003 invasion, nor were WMDs found by U.S. officials during the occupation of Iraq. Also, little evidence was uncovered to substantiate allegations of strong ties between Hussein's Iraq and al-Qaeda or similar networks. The search for WMDs ended in December 2004, and an inspection report submitted to Congress by U.S. weapons hunter Charles A. Duelfer essentially "contradicted nearly every prewar assertion about Iraq made by Bush administration officials."[32]

This chapter's Global Perspective discusses Israel's hunt for master bomb-maker Yehiya Ayyash, also known as "The Engineer." The manhunt is an instructive case on the response of security forces to an ongoing and imminent threat of terrorist violence.

GLOBAL PERSPECTIVE

ACTIONABLE INTELLIGENCE: ISRAEL AND THE HUNT FOR "THE ENGINEER"[a]

Yehiya Ayyash, a master bomb maker better known as "The Engineer," was a model activist within Hamas's cell-based organizational structure. Unlike PLO-style groups, Hamas required its operatives to organize themselves into small semiautonomous units. Ayyash was an al-Qassam cell (and later a "brigade") commander, but he had very few outside contacts and built his bombs in an almost solitary setting. He taught others to make bombs and how suicide bombers should position themselves for maximum effect.

The Engineer's first bomb was a Volkswagen car bomb that was used in April 1993. When Hamas began its suicide bombing campaign after the February 1994 Hebron massacre, Ayyash was the principal bomb maker. His bombs were sophisticated and custom made for each mission. They were particularly powerful compared to others previously designed by Hamas.

Ayyash was killed in January 1996. The cell phone he was using to carry on a conversation with his father had been booby-trapped by Israeli security agents and was remotely detonated. The assassination occurred as follows:

Fifty grams of RDX [plastic] explosives molded into the battery compartment of a telephone had been designed to kill only the man cradling the phone to his ear. The force of the concentrated blast caused most of the right side of Ayyash's face to implode. . . . The booby-trapped cellular phone had been . . . so target specific, that the left side of Ayyash's face had remained whole. The right hand which held the telephone was neither burnt or damaged.[b]

The Engineer had been directly and indirectly responsible for killing approximately 150 people and injuring about 500 others.

Notes

a. Primarily from Samuel M. Katz, *The Hunt for the Engineer: How Israeli Agents Tracked the Hamas Master Bomber* (New York: Fromm International, 2001).

b. Ibid., 260–61.

CHAPTER SUMMARY

The Intelligence Community occupies a central role in maintaining a viable homeland security enterprise. Intelligence agencies are charged with distinct missions within the Intelligence Community and are led by the Central Intelligence Agency, the Defense Intelligence Agency, the National Security Agency, and the Federal Bureau of Investigation. Intelligence coordination and cooperation are critically necessary to the success of homeland security, but on occasion, there have been problems and rivalries that have affected intelligence collection and analysis. Intelligence agencies perform a critical international role in securing the domestic homeland security environment. The intersection of their missions with those of domestic agencies creates a large and intricate establishment for combating terrorism domestically and internationally.

DISCUSSION BOX

This chapter's Discussion Box is intended to stimulate critical debate about the possible use, by democracies and authoritarian regimes, of antiterrorist technologies to engage in surveillance.

Toward Big Brother?

Electronic surveillance by government agencies has become a controversial practice in the United States and elsewhere. The fear is that civil liberties can be jeopardized by unregulated interception of telephone conversations, e-mail, and fax transmissions by intelligence centers. Detractors argue that government use of these technologies can conceivably move well beyond legitimate application against threats from espionage and terrorism. Absent strict protocols to rein in these technologies, a worst-case scenario envisions intelligence intrusions into the everyday activities of innocent civilians. Should this happen, critics foresee a time when privacy, liberty, and personal security become values of the past.

Discussion Questions

1. How serious is the threat from abuses in the use of information collection technologies?

2. How should information collection technologies be regulated? Can they be regulated?

3. Is it sometimes necessary to sacrifice a few freedoms to protect national security and to ensure the long-term viability of civil liberty?

4. Should the same protocols be used for domestic electronic intelligence collection and foreign collection? Why?

5. What is the likelihood that new intelligence technologies will be used as tools of repression by authoritarian regimes in the near future?

KEY TERMS AND CONCEPTS

The following topics were discussed in this chapter and can be found in the glossary:

Central Intelligence Agency (CIA)   109
Defense Intelligence Agency (DIA)   117
DHS Office of Intelligence and Analysis (I&A)   118
Diplomatic Security Service   109
Director of National Intelligence   111
geospatial intelligence (GEOINT)   115

human intelligence (HUMINT)   114
imagery intelligence (IMINT)   115
Intelligence Community (IC)   108
measurements and signatures intelligence (MASINT)   115
National Counterterrorism Center (NCTC)   111
National Intelligence Priorities Framework (NIPF)   115

National Reconnaissance Office (NRO)   118
National Security Agency (NSA)   117
Office of the Director of National Intelligence (ODNI)   111
open source intelligence (OSINT)   115
Rewards for Justice Program   109
signals intelligence (SIGINT)   114

ON YOUR OWN

Get the tools you need to sharpen your study skills. SAGE edge offers a robust online environment featuring an impressive array of free tools and resources.

Access practice quizzes, eFlashcards, video, and multimedia at **edge.sagepub.com/martinhs3e**

RECOMMENDED WEBSITES

The following websites provide information about federal homeland security agencies:

Central Intelligence Agency: www.cia.gov

Office of the Director of National Intelligence: www.dni.gov

SITE Intelligence Group (USA): www.siteintel group.org

WEB EXERCISE

Using this chapter's recommended websites, conduct an online investigation of the role of intelligence agencies.

1. What are the primary documents explaining the underlying purpose and missions of intelligence agencies?

2. How would you describe the differences between intelligence agencies?

3. In your opinion, what practical options exist for coordinating national intelligence agencies?

To conduct an online search on research and monitoring organizations, activate the search engine on your Web browser and enter the following keywords:

"Intelligence agencies"

"Intelligence and the war on terrorism"

RECOMMENDED READINGS

The following publications provide discussions of intelligence agencies and their missions:

Andrew, Christopher. 1987. *Her Majesty's Secret Service: The Making of the British Intelligence Community*. New York: Penguin.

Bamford, James. 2001. *Body of Secrets: Anatomy of the Ultra-Secret National Security Agency, From the Cold War Through the Dawn of a New Century*. New York: Doubleday.

Berentsen, Gary. 2008. *Human Intelligence, Counterterrorism, and National Leadership: A Practical Guide*. Dulles, VA: Potomac Books.

Monje, Scott C. 2008. *The Central Intelligence Agency: A Documentary History*. Westport, CT: Greenwood.

Thomas, Gordon. 2009. *Gideon's Spies: The Secret History of the Mossad*. 5th ed. New York: St. Martin's Press.

Warrick, Joby. 2011. *The Triple Agent: The al-Qaeda Mole Who Infiltrated the CIA*. New York: Doubleday.

CHAPTER SEVEN

AGENCIES AND MISSIONS
Homeland Security at the State and Local Levels

Opening Viewpoint: The Homeland Security Enterprise: State and Local Nexus

Although the federal government is positioned as the national coordinating authority for homeland security initiatives, and the Department of Homeland Security is popularly perceived to be the chief government authority for managing homeland security policies, the fact is that nearly 90,000 state and local governments are also members of the homeland security enterprise. This fact alone indicates the inherent complexity of coordinating the multivariate planning and responder policies at state and local levels of government.

As a practical matter, the federal government assists state and local homeland security initiatives in order to construct comparable policies and procedures nationally. Federal homeland security grants support state and local initiatives by reimbursing governments for related expenditures. Federal assistance to local authorities has become a multimillion-dollar feature of the homeland security enterprise.

When terrorist incidents occur, first responders to the incident site are not federal authorities but rather local agencies. Police, fire, and medical responders are all employees of state and local agencies. They take the lead in stabilizing the immediate emergency vicinity and tending to casualties in the critical minutes and hours after the incident. They are also responsible for restoring infrastructure and systems to their pre-attack status. In essence, state and local authorities bear the greatest responsibility for managing immediate homeland security events.

Chapter Learning Objectives

This chapter will enable readers to do the following:

1. Analyze state and local homeland security systems

2. Apply the implementation mission of homeland security initiatives by state executive bureaucracies

3. Evaluate local homeland security networking initiatives

4. Describe the role of law enforcement agencies in homeland security environments

5. Apply the homeland security mission of law enforcement agencies prior to and following the September 11, 2001, attacks

The domestic security environment in the United States was dramatically reordered in the aftermath of the September 11, 2001, terrorist attacks. Prior to the attacks, a great deal of responsibility for securing the homeland was decentralized, independent, and largely uncoordinated at the nonfederal government level. After the attacks, a relatively singular purpose arose in the newly reconfigured domestic security environment. Responsibility for implementing the new concept of homeland security was devolved to all levels of government and society in a new nationwide homeland security enterprise. State and local governments and agencies immediately became front-line participants in the national effort to secure the homeland and respond to verifiable terrorist threats.

The homeland security enterprise consists of collaborative initiatives at every level of society. Although the U.S. Department of Homeland Security (DHS) is the foremost coordinator of national homeland security policy, it is by no means the only partner for designing and implementing policy. DHS is the best-known homeland security agency, but homeland security operations are not limited to DHS, nor are they limited exclusively to the federal government. In addition to collaborating with the efforts of other federal agencies, DHS recognizes that state, local, and tribal governments are integrally involved in securing the implementation of homeland security efforts nationally. Every level of government is charged with homeland security responsibilities, with the consequence that thousands of jurisdictions are members of the homeland security enterprise.

Task forces and command systems have been established to coordinate multi-tiered collaboration. To develop viable collaboration, the secretary of homeland security created a National Incident Management System (NIMS). This is a foundational concept for promoting multi-tiered collaboration between federal, state, and local governments. As part of this effort, the Federal Emergency Management Agency (FEMA) and other agencies provide Incident Command System (ICS) training to state and local officials and responders; this training is instrumental for coordinating efforts to synchronize federal, state, and local response systems. Further discussion of NIMS and ICS is provided in Chapter 12.

The discussion in this chapter explains the homeland security functions of nonfederal governments and will review the following:

- State-level homeland security systems

- Local homeland security networking initiatives

- Homeland security and law enforcement agencies

STATE-LEVEL HOMELAND SECURITY SYSTEMS

The implementation of state-level homeland security systems is coordinated by each state's executive branch of government. Governors and state administrative agencies oversee the functional execution of state and local homeland security initiatives, and each state implements its own version of homeland security governance. Because of political and regional variables, there is no uniform configuration of state-level homeland security bureaucracy. However, although each state manages its requisite homeland security policies and procedures as necessitated by their unique political and administrative constraints, certain core missions and roles are similar and identifiable in virtually all state systems.

State Government and Homeland Security: Principal Roles

States manage the homeland security enterprise within the context of their preferred governmental and administrative arrangements. Bureaucratic responsibility for homeland security initiatives differs in each state, yet there are recurrent principal roles for leaders and agencies charged with implementing policy in state homeland security systems. For example, the principal roles of governors and government agencies are similar across the nation.

Emergency Leadership: The Role of the Governor

Governors serve as chief executive officers for state governments and are responsible for leading state relief efforts when terrorist attacks, natural disasters, or other emergencies occur. As chief executives, governors manage the prevention, protection, mitigation, response, and recovery systems within the jurisdiction of their states—these mission areas are presented in the 2015 *National Preparedness Goal, Second Edition,* found in Appendix B. Governors are held to be ultimately responsible for the effectiveness of state agencies and officials when they are dispatched to cope with emergencies. Successful emergency responses are deemed to reflect efficient management by governors, whereas slow or ineffective intervention can generate pointed criticism of gubernatorial administrations. These perceptions can have favorable or dire political consequences for governors, as discussed in Chapter Perspective 7.1.

CHAPTER PERSPECTIVE 7.1

Governors and the Politics of Crisis Intervention

The political fortunes of governors are frequently impacted by terrorist incidents, natural disasters, and other events that occur under the homeland security all-hazards umbrella. The severity of crises and the effectiveness of state crisis intervention efforts often serve as a political barometer for whether the public will continue to support a governor. Because governors become the "face" of state crisis intervention and recovery processes, the public's assessment of the success or failure of these processes often forms a strong influence on perceptions of the governor's competence.

Several cases in point illustrate the strength of this tendency.

- In 2005, Louisiana governor Kathleen Blanco was widely criticized for her management of the state's intervention in the aftermath of Hurricane Katrina in August and Hurricane Rita in September. Her popularity plummeted during the crises, and she declined to seek a second term in office.

- Also in 2005, Mississippi governor Haley Barbour became a local folk hero of sorts because of his perceived success in managing his state's intervention following Hurricane Katrina. Unlike Governor Blanco, his political fortunes were enhanced.

- In 2012, New Jersey governor Chris Christie's popularity among voters increased significantly during response and recovery efforts after Superstorm Sandy. When the U.S. House of Representatives adjourned prior to voting on disaster relief legislation, Governor Christie strongly criticized Congress for failing to provide needed aid for his state; he won plaudits in New Jersey for criticizing congressional members of his own political party.

- In 2013, Massachusetts governor Deval Patrick won praise for projecting determined calm following the terrorist attack on the Boston Marathon. The public widely supported his steady resolve to capture Dzhokhar Tsarnaev and cooperated with law enforcement authorities during the manhunt's door-to-door searches.

- In 2016, Florida governor Rick Scott declared the shooting of scores of patrons by Omar Mateen at the Pulse Nightclub in Orlando to be an act of terrorism. Mateen had in fact professed allegiance to ISIS, and the group later embraced him as their "soldier." In 2018, on the second anniversary of the attack, Scott designated June 12 as Pulse Remembrance Day

(Continued)

in Florida. Nevertheless, he received political criticism from some constituencies for not adequately addressing the issue of gun safety in Florida. He was also criticized politically by some members of the LGBT community for not adequately reaching out to this community.

One lesson seems clear: Events occurring under the homeland security all-hazards umbrella can become an important influence on domestic political culture.

Discussion Questions

1. To what extent should governors be held accountable for their actions during unexpected emergencies?

2. Does the possibility of political backlash have the potential to hinder gubernatorial leadership?

3. Who should ultimately receive plaudits or criticism for the success or failure of emergency response efforts?

When a homeland security–related incident occurs, governors may be required to adopt one or more critical responsibilities. For example, an official declaration of a disaster by the president is usually made after a formal request from the governor or her or his authorized surrogate. Assessments of critical needs are likewise reported either by the governor or by an authorized surrogate. In all cases, it is the expected duty of the governor to become the face of the local relief effort. Emergency orders, such as deployments or evacuations, are made either by governors or in their name.

Command Authority: Mobilizing the State National Guard

A commanding emergency power of governors under federal law is the authority to mobilize and deploy the state's National Guard units. The missions of National Guard units and reasons for deploying them range from disaster rescue relief to domestic military responses

in the event of rioting to securing critical infrastructure such as airports when terrorist attacks occur. The National Guard often serves as a "force multiplier" to state-level emergency first responders such as medical, firefighting, and law enforcement civil agencies. National Guard units may also serve as an arm of the U.S. armed forces when called into federal service by the president. Within these contexts, the National Guard may be mobilized to perform counter-drug missions, respond to domestic disasters and other emergencies as needed, enforce laws, and (when called into federal service) engage in overseas combat missions.

Thus, during emergencies, the executive authority of governors at the state level is very similar to presidential executive authority at the federal level. Chapter Perspective 7.2 summarizes historical examples of domestic deployments of the National Guard to maintain order during civil discord.

Mobilization of state National Guard units may be initiated according to various provisions of state law and the U.S. Code. For example, state codes authorize governors to mobilize the National Guard under the command and budgetary authority of state governments.

STAN HONDA/AFP/Getty Images

▶ **Photo 7.1**

Massachusetts governor Deval Patrick during the manhunt for terrorist suspects in the aftermath of the Boston Marathon bombing of April 2013.

CHAPTER PERSPECTIVE 7.2

Domestic Mobilization of the National Guard to Restore Civil Order

Many historical examples exist of domestic deployments of the National Guard to maintain order during civil discord or the threat of discord. At the time, such deployments were deemed to be of critical necessity due to the extent of unrest or the threat of further unrest. Examples in the modern era include the following incidents:

- In 1963, President John F. Kennedy federalized the Alabama National Guard to secure the racial integration of the University of Alabama after Governor George Wallace personally blocked racial integration.

- In 1965, federalized Alabama National Guard troops protected more than 3,000 civil rights marchers walking from Selma to Montgomery, the state capital.

- In 1967, thousands of National Guard and U.S. Army troops were deployed to Detroit, Michigan, during deadly rioting that left 43 people dead and damaged millions of dollars in property.

- In 1968, approximately 68,000 troops were deployed nationwide when rioting erupted in 125 cities after the assassination of the Reverend Martin Luther King Jr. by a sniper in Memphis, Tennessee.

- In 1970, President Richard Nixon ordered the National Guard to New York City to restart the delivery of mail during a nationwide strike by approximately 150,000 U.S. postal workers in more than 600 locations.

- In 1989, President George H. W. Bush deployed Army and National Guard troops to the island of St. Croix in the U.S. Virgin Islands during looting and other violence in the aftermath of Hurricane Hugo.

- In 1992, National Guard troops were deployed to Los Angeles during widespread rioting and arson after the not-guilty verdict in the trial of officers who were videotaped beating motorist Rodney King.

Discussion Questions

1. When should the National Guard be deployed during unexpected civil discord?

2. Is there a possibility of increased discord when the National Guard is deployed to domestic emergencies?

3. Who should ultimately be responsible for militarized emergency response efforts?

Federal laws likewise authorize mobilization. For example, Title 32 of the U.S. Code generally outlines the role of the National Guard and authorizes state-controlled deployment under federal budgetary authority.[1] Title 10 of the U.S. Code names the reserve components of the U.S. armed forces, including the National Guard. Under the authority of Title 10, state National Guard units may be mobilized and deployed under the command and budgetary authority of the federal government.[2]

Emergency Management: The Role of State Bureaucracies

State bureaucracies are responsible for several important assignments. Foremost among these is the implementation of state emergency plans. Depending on which contingency is activated, state bureaucracies must implement gubernatorial emergency orders by mobilizing resources and deploying state assets in a coordinated and proportional manner. The preservation of civil order during emergencies is a vital responsibility of state government, requiring the coordination of complex relief efforts and enforcement of the law.

Emergencies occur within the immediate jurisdiction of urban and local governments, and states are frequently tasked with coordinating the intervention of multiple relief agencies that report to local governments. Because efficient and collaborative responses are critical to the protection of lives and property, state agencies play an important role in assisting and intervening locally. Thus, a central function of state government is to synchronize planning and responses by local governments, especially for widespread emergencies crossing many jurisdictions.

State Government and Emergency Management: Configuration of Services

The specific configuration of homeland security and emergency management services is determined by each state. Within state governments, the bureaucratic placement of these services rarely remains static and is often subject to reorganization. Reorganization of homeland security and emergency management responsibilities usually occurs when elections are held or budget priorities change. In these cases, central oversight over homeland security and emergency services is sometimes moved to new government agencies or new command centers. One problem with the redesignation of existing reporting lines is the potential for reduced clarity about the center of authority for homeland security policies and procedures.

Because of the foregoing pattern of practice, changes in state homeland security systems are monitored by the National Emergency Management Association (NEMA, further discussed in the next section), which regularly surveys state-level administrative configurations. Survey results indicate that virtually every state establishes central oversight authority over homeland security and emergency services in specified administrative agencies. However, these centers have changed markedly over time nationally since the September 11, 2001, terrorist attacks, and reorganization of administrative placement is arguably the norm rather than an exceptional practice.

State-level administrative changes have been documented by NEMA since the inception of the modern era of homeland security. An interesting and instructive compilation is the periodic reconfiguration of state-level command responsibility for homeland security services. Table 7.1 reports NEMA data on which state-level offices were tasked with primary responsibility for providing homeland security services and how many states centered these services in the indicated offices.

National Collaboration: The National Emergency Management Association

The **National Emergency Management Association** is a professional association, essentially an umbrella organization, "*of* and *for* emergency management directors from all 50 states, eight territories, and the District of Columbia" (emphasis in original).[3] It is a nonprofit and nonpartisan organization that serves as an information and support association for emergency management professionals. NEMA resources are intended to address an all-hazards approach to homeland security and emergency management. To that end, NEMA is dedicated to "enhancing public safety by improving the nation's ability to prepare for, respond to, and recover from all emergencies, disasters, and threats to our nation's security."[4] State homeland security officials are thus offered a joint forum for receiving and sharing information on emerging issues for homeland security and emergency management.

Central oversight for homeland security and emergency services varies from state to state. The configuration of these services has changed frequently and markedly over time. These administrative changes are often the result of bureaucratic reorganizations after statewide elections.

The following table reports which government centers were responsible for primary oversight over state-level homeland security services at different points in time. It reports the number of states that centered homeland security services in the indicated administrative offices over several years following the September 11, 2001, terrorist attack.

	Governor	Adjutant General/ Military	Combined Homeland Security/ Emergency Mgt.	Public Safety	State Police	Other
2002	14	12	10	12	0	3
2008	2	8	34	7	0	0
2011	11	17	1	12	3	4
2013	9	18	8	12	3	2

NEMA's Strategic Plan outlines the association's near- and long-term visions for the provision of resources and support on a national scale for local homeland security officials and agencies. The Strategic Plan summarizes its goals as follows:

- Strengthen the nation's emergency management system.

- Provide national leadership and expertise in comprehensive, all-hazards emergency management.

- Serve as a vital emergency management information and assistance resource.

- Advance continuous improvement in emergency management through strategic partnerships, innovative programs, and collaborative policy positions. The plan is reviewed yearly by the NEMA Board of Directors and Past Presidents and is modified as needed.[5]

The National Homeland Security Consortium

NEMA established the **National Homeland Security Consortium** as a public forum for government personnel and private practitioners to "coalesce efforts and perspectives about how best to protect America in the 21st century"[6] in the aftermath of the September 11 attacks. Practitioners who founded the consortium first met in 2002 at the invitation of NEMA. The consortium now comprises 21 major national organizations representing state, local, and private professionals involved in the practice of homeland security and emergency

National Homeland Security Consortium: The National Emergency Management Association established the National Homeland Security Consortium as a public forum for government personnel and private practitioners to "coalesce efforts and perspectives about how best to protect America in the 21st century" in the aftermath of the September 11 attacks.

management. It serves as an associative assembly for presenting new and evolving policy options and concepts to a broad spectrum of state and local homeland security affiliates.

The consortium has been formally recognized by the U.S. Department of Homeland Security, and its partners represent prominent members of the homeland security enterprise. Partners include the U.S. Department of Defense, the U.S. Department of Health and Human Services, and the Centers for Disease Control and Prevention. The work of consortium partners is succinctly summarized as follows:

> The consortium provides a neutral forum for organizations to exchange ideas, have candid discussions, and galvanize input to the Federal government. The differing perspectives of member organizations do not always provide for unanimity on specifics of implementing national initiatives. Consortium members do not aim for group think but aspire to group understanding. More often than not, however, the members have found easy consensus on the consortium's major goals.[7]

Regional Collaboration: Networking and Assistance Across Jurisdictions

Councils of Government

Councils of government (COGs) are consortia of regional governing bodies that promote collaboration and cooperation among members. A large number of COGs (more than 500) have been established across the United States and are represented in every state. Among the many issues addressed by COGs, homeland security is a primary focus. Because of the large number of COGs, regional associations of councils have been formed to coordinate collaboration on regional policies and issues.

The National Association of Regional Councils

The **National Association of Regional Councils (NARC)** is an organization that promotes regionalism among governmental bodies within and across states. NARC "serves as the national voice for regionalism by advocating for regional cooperation as the most effective way to address a variety of community planning and development opportunities and issues."[8] It is a nonprofit public interest organization that promotes collaboration among regional government organizations and is an advocate at the federal level for these interests. NARC presents itself as "a recognized authority and leading advocate for regional organizations and solutions that positively impact American communities through effective interjurisdictional cooperation."[9] Thus, NARC activities provide unified advocacy on behalf of governing authorities that otherwise would be tasked with independently promoting their interests and articulating their concerns.

NARC actively promotes regional collaboration on homeland security and public safety issues. Such collaboration is intended to address an all-hazards approach to homeland security because of the need to centrally articulate the first-responder missions of state and local governing authorities. The homeland security emphasis was adopted by NARC out of concern that

> inconsistencies in preparedness efforts across state and local governments ultimately yield greater vulnerability to the terror hazard and less effective homeland security policy. For this reason, a major effort is needed to establish or

fund planning on a multi-jurisdictional basis. Experience has shown that it takes the coordinated efforts of numerous jurisdictions to successfully protect America's cities and counties, and the metropolitan and rural regions of the nation.[10]

In support of its efforts, NARC has cited the 2007 *National Homeland Security Strategy*, published by the White House, and the *National Response Framework*, published by DHS, as explanatory documents on the necessity for building coordinated regional homeland security initiatives.

The Emergency Management Assistance Compact

Congress established the **Emergency Management Assistance Compact (EMAC)** in 1996 to serve as a national disaster relief network.[11] As signatories of EMAC, members are committed to assist cosignatories with disaster relief assistance following disaster declarations, and it is greatly advantageous.

> EMAC offers assistance during governor-declared states of emergency through a responsive, straightforward system that allows states to send personnel, equipment, and commodities to help disaster relief efforts in other states. Through EMAC states can also transfer services. . . . The strength of EMAC and the quality that distinguishes it from other plans and compacts lie in its governance structure; its relationship with federal organizations, states, counties, territories, and regions; the willingness of states and response and recovery personnel to deploy; and the ability to move any resource one state wishes to utilize to assist another state.[12]

All states and U.S. territories are members of EMAC. Members agree to render assistance to each other, and to reimburse providers for their assistance. This framework is legally binding on members, thus ensuring compliance and restitution when disaster relief is provided. An important advantage of EMAC membership is that the compact is capable of leveraging and managing the disbursement of federal financial resources.

Although EMAC promotes critical networking and other collaborative assistance, it is important to understand that local agencies and other entities often rely upon mutual assistance compacts among themselves, especially between neighboring counties and municipalities. Thus, local homeland security networking initiatives often serve as a primary mechanism for preparing response and recovery when an emergency event occurs.

Emergency Management Assistance Compact (EMAC): A national disaster relief network established by Congress in 1996. Membership in EMAC commits signatories to assist cosignatories with disaster relief assistance following disaster declarations.

LOCAL HOMELAND SECURITY NETWORKING INITIATIVES

Local governments numbered more than 89,004 in 2012, according to census data released that year.[13] Many of these governments allocate substantial resources to homeland security systems within their jurisdiction. In some jurisdictions, the commitment of financial and human resources can be quite significant. Although some financial assistance is received from the federal government, federal assistance does not always fully reimburse local initiatives. Because local resources are finite and very limited, homeland security systems often draw support away from other local priorities. This is a dilemma that is widely experienced and understood, but it nevertheless represents an added challenge to the effective administration of local government.

Local planning committees and national associations representing local government entities actively advocate increased support for local homeland security initiatives. In many ways, these committees and associations are the only centralized means of communication between local governments, and they are also conduits for communication between local governments and federal authorities. At the county and municipal levels, Local Emergency Planning Committees coordinate collaborative initiatives among local officials and agencies. Other prominent local government advocacy associations include the National Association of Counties, the National League of Cities, and the U.S. Conference of Mayors. In order to promote overarching collaboration on a macro level, some local jurisdictions also engage with the International Association of Emergency Managers. All of these organizations have access to national government officials and elected representatives.

▶ **Photo 7.2** Official seal of the Cook County, Illinois, Homeland Security and Emergency Management Agency. States have established administrative mechanisms to carry out homeland security duties.

Local Emergency Planning Committees

Local Emergency Planning Committees (LEPCs) are community-based associations that allow local planners to prepare for emergency scenarios. In particular, LEPCs are tasked pursuant to the Emergency Planning and Community Right-to-Know Act of 1986 to prepare for emergencies involving hazardous materials. The LEPC missions and membership are summarized by the U.S. Environmental Protection Agency as follows:[14]

> Under the Emergency Planning and Community Right-to-Know Act (EPCRA), Local Emergency Planning Committees (LEPCs) must develop an emergency response plan, review the plan at least annually, and provide information about chemicals in the community to citizens. Plans are developed by LEPCs with stakeholder participation. There is one LEPC for each of the more than 3,000 designated local emergency planning districts. The LEPC membership must include (at a minimum):

- Elected state and local officials
- Police, fire, civil defense, and public health professionals
- Environment, transportation, and hospital officials
- Facility representatives
- Representatives from community groups and the media

Local Emergency Planning Committees (LEPCs): Community-based associations that allow local planners to prepare for emergency scenarios. LEPCs are tasked pursuant to the Emergency Planning and Community Right-to-Know Act of 1986 to prepare for emergencies involving hazardous materials.

The National Association of Counties

The **National Association of Counties (NACo)** represents county governments and counts more than 2,000 counties as members. It provides extensive services and includes policy reports, advocacy to the federal government, newsworthy publications, technical assistance, surveys, and research initiatives. NACo's activities have been recognized as an important influence for promoting county government interests.

Homeland security reporting and advocacy are included among NACo's support activities for county initiatives. NACo responded to the September 11 attacks by promptly engaging in advocacy and research on behalf of county governments. Its post–9/11 efforts are ongoing and have included the following initiatives:

National Association of Counties (NACo): An organization representing county governments. Its services include research initiatives, surveys, technical assistance, newsworthy publications, policy reports, and advocacy to the federal government.

- NACo convened a Homeland Security Task Force consisting of 43 members, which, in October 2001, proposed 20 recommendations regarding homeland security.

- In 2002, NACo published a policy paper titled *Counties and Homeland Security: Policy Agenda to Secure the People of America's Counties*[15] explaining its position that counties will be the first responders to terrorist attacks.

- In February 2004, NACo reported the results of a survey it conducted of selected *core counties* in the United States, consisting of counties from potentially threatened urban locations as identified in the DHS Urban Areas Security Initiative (UASI). It reported core county participation in the UASI and assessed how the core counties perceived the effectiveness of UASI participation.[16]

- Funding issues are a central component of NACo's homeland security advocacy, and in May 2011, the organization detailed the impact of budget cuts in a press release titled "NACo Fights Massive Cuts to Homeland Security."[17] This was followed by an article in *NACo County News: The Voice of America's Counties* titled "Homeland Security Assistance Faces Severe Reductions."[18]

The organization continues to publish periodic assessments and proposals on homeland security issues relevant to county governments. NACo publications are used by county governments as guidelines for implementing homeland security policies and procedures.

The National League of Cities

Advocacy support on behalf of over 19,000 cities, towns, and villages is provided by the **National League of Cities (NLC)**. It was founded as the first national association representing municipalities, and it continues to be the largest association. Its advocacy has been effective at the federal level. More than 2,000 municipalities are active dues-paying members of the NLC network.

Homeland security–related issues are an important component of NLC's advocacy initiatives. For example, because city personnel serve as first responders to domestic security events, NLC has lobbied extensively for enhanced federal funding to support first responders. NLC has been successful in many of these efforts; in May 2012, NLC advocacy was partly responsible for persuading congressional appropriations officials to halt efforts to consolidate 16 state and local homeland security grant programs into one state-focused grant.[19] NLC also posts resources online for members, such as links to homeland security regulatory updates and federal policy statements. These resources and documents allow members to remain current with national policy.

The U.S. Conference of Mayors

Advocacy for issues affecting major cities and other urban governments is provided by the **United States Conference of Mayors (USCM)**. Because large population concentrations characterize its membership, USCM engages in advocacy and publicizing of issues on behalf of governments representing a large proportion of the population of the United States. USCM

National League of Cities (NLC): An organization providing advocacy support on behalf of over 19,000 cities, towns, and villages.

HENNY RAY ABRAMS/AFP/Getty Images

▶ **Photo 7.3** New York City mayor Rudolph Giuliani meets with members of the New York Fire Department during the aftermath of the September 11, 2001, terrorist attacks.

United States Conference of Mayors (USCM): An organization representing urban areas on issues affecting major cities and other urban governments. USCM engages in advocacy and publicizing of issues on behalf of governments representing a large proportion of the population of the United States.

summarizes its mission as follows: "The United States Conference of Mayors (USCM) is the official non-partisan organization of cities with populations of 30,000 or more. There are 1,296 such cities in the country today. Each city is represented in the Conference by its chief elected official, the mayor."

The primary roles of the U.S. Conference of Mayors are to do the following:

- Promote the development of an effective national city and metro area focused policy;

- Strengthen federal–city relationships;

- Ensure that federal policy meets urban needs;

- Provide mayors with leadership and management tools that allow them to do their jobs better and make them more effective as leaders; and

- Create a forum in which mayors can share ideas, information, and best practices.[20]

USCM actively engages in homeland security appropriations advocacy on behalf of its membership. Because federal funding is usually disbursed to cities as reimbursement payments, USCM lobbies extensively for homeland security funds to mitigate the economic impact of expenditures by city governments.

The International Association of Emergency Managers

International Association of Emergency Managers (IAEM): "[A] non-profit educational organization dedicated to promoting the 'Principles of Emergency Management' and representing those professionals whose goals are saving lives and protecting property and the environment during emergencies and disasters."

The **International Association of Emergency Managers (IAEM)** "is a non-profit educational organization dedicated to promoting the 'Principles of Emergency Management' and representing those professionals whose goals are saving lives and protecting property and the environment during emergencies and disasters."[21] IAEM holds events and conferences that allow for sharing of emergency management policies and experiences for an international membership of professionals. In this regard, some local-level U.S. emergency management agencies participate in IAEM-sponsored events and initiatives.

Case in Point: The Challenge of Funding Initiatives

Funding homeland security initiatives can be an expensive responsibility for local authorities, and the cost to local governments is often a point of contention with the federal government. Some communities agree that they receive adequate budgetary subsidies from federal sources, but other communities argue that their budgets are strained in the effort to comply with homeland security mandates. Smaller communities advocate an even distribution of federal subsidies, but larger communities (especially urban areas) advocate for a distribution system that is proportional to the potential occurrence of an attack. The issue of oversight for proper use of subsidies can also pose a challenge to distribution programs. For example, local priorities may not reflect federal intentions without proper oversight.

HOMELAND SECURITY AND LAW ENFORCEMENT AGENCIES

Prior to the September 11 attacks, state and local law enforcement agencies in the United States operated within the parameters of traditional policing roles. These included criminal investigations, order maintenance, routine patrols, search and rescue, and crime laboratory analysis. After September 11, a new homeland security model was integrated into operational procedures at all levels of law enforcement and by all law enforcement agencies. As a result, state and local law enforcement agencies were transformed into important members of the homeland security enterprise.

The Police Mission in Homeland Security Environments

Prior to the formation of the modern homeland security environment, the overarching mission of police agencies was to serve in a *law enforcement* capacity, with relatively few resources being devoted to preparing for and responding to domestic security emergencies. After the new homeland security environment became established, the law enforcement mission of the police was augmented by an *internal security* mission. Within the context of this mission, state and local police became front-line members of a new homeland security enterprise responsible for planning and initiating civil protection operations.

The Internal Security Imperatives of Law Enforcement Agencies

When homeland security missions occur, internal security imperatives require law enforcement units to be stationed at strategic locations and to perform security-focused patrols rather than crime-focused operations. These responsibilities are sometimes threat specific and carried out in response to plausible terrorist risks. For example, because of the threat of airline attacks and hijackings, the United States commonly places law enforcement officers aboard aircraft to act as sky marshals. Other security duties include securing airports, patrolling borders, looking out for fugitives and other suspects, and conducting surveillance of groups and people who fit terrorist profiles.

When considering internal security imperatives, local police officers are arguably the domestic front line in the war on terrorism because they are among the first officials to respond to terrorist incidents. For example, 37 Port Authority of New York and New Jersey Police Department officers and 23 New York City Police Department officers died during the September 11 attacks when they were deployed to the World Trade Center site. Local police agencies are also the first to stabilize the immediate vicinity around attack sites and are responsible for maintaining long-term order in scenarios where cities suffer under terrorist campaigns. Thus, the role of law enforcement agencies varies in scale and mission depending on the characteristics of the threatening incident. Modern law enforcement internal security missions include the following:

- *Traditional police work*, in which criminal investigations are carried out by detective bureaus
- *Specialized services*, in which duties require special training (for example, defusing and removing explosive ordnance by bomb disposal units)
- *Order maintenance*, in which attack sites are secured and terrorist environments are stabilized with large and visible deployments of police personnel
- *Paramilitary deployment*, in which highly trained and specially equipped units are sent to crisis locations. Such engagement includes the deployment of hostage rescue units and Special Weapons and Tactics (SWAT) teams.

Los Angeles Police Department

▶ **Photo 7.4** A homeland security warning notice distributed by the Los Angeles Police Department. Many large urban law enforcement agencies have well-established homeland security procedures.

The Operational Impact of Terrorism: Adaptations by Law Enforcement Agencies

The operations of law enforcement agencies were immediately affected by the September 11 terrorist attacks and required extensive operational adjustments. The scope of these operational adjustments was studied in a 50-state survey of law enforcement agencies conducted in 2004 by the Council of State Governments and Eastern Kentucky University.[22] The survey reported and evaluated the impact of terrorism on law enforcement agencies in the immediate aftermath of the September 11 attacks. Key findings from the survey indicate that after September 11, a rapid and systemic sea change occurred in American law enforcement. Among its findings were the following statistics:

- Roughly 75 percent of state agencies said they either [had] a great amount of involvement with or actually serve[d] as their state's leader for terrorism-related intelligence gathering, analysis and dissemination.

- More than 70 percent of state agencies agreed or strongly agreed that their individual officers and investigators [had] significant new responsibilities in terrorism-related intelligence gathering, investigations and emergency response.

- Large local agencies (more than 300 uniformed personnel) [were] much more likely than small local agencies to report that their officers [had] significant new responsibilities in conducting terrorism-related intelligence gathering and investigations.

- In comparison with the period before Sept. 11, more than 70 percent of state agencies reported allocating more or many more resources for critical infrastructure protection; special event security; protection of dignitaries; intelligence gathering, analysis and sharing; and terrorism-related investigations.

- State police [were] more likely than local agencies to say that since Sept. 11 they [had] allocated more or many more resources for border and port security; commercial vehicle enforcement; high-tech/computer crime investigation; and preventive patrols.

- Local agencies [were] more likely to report allocating additional resources to airport security; community policing; traffic safety; drug enforcement and investigation; and traditional criminal investigations.

- More than 60 percent of state agencies reported that local agency requests for their assistance in high-tech/computer crime investigations and general training and technical assistance [had] increased or significantly increased since Sept. 11.

- More than 75 percent of state agencies reported that their assignment of personnel to Federal task forces such as the [Joint Terrorism Task Forces had] increased or significantly increased since Sept. 11.

- Only 28 percent of small local agencies [said] they [were] interacting more with the FBI, compared with 87 percent of large local agencies.[23]

Case in Point: Joint Terrorism Task Forces

Joint Terrorism Task Forces (JTTFs) have become a ubiquitous feature of law enforcement cooperation and have been established nationally in dozens of jurisdictions. JTTFs are "small cells of highly trained, locally based, passionately committed investigators, analysts, linguists, SWAT experts, and other specialists from dozens of U.S. law enforcement and intelligence

Joint Terrorism Task Forces (JTTFs): "Small cells of highly trained, locally based, passionately committed investigators, analysts, linguists, SWAT experts, and other specialists from dozens of U.S. law enforcement and intelligence agencies. [This] is a multi-agency effort led by the Justice Department and FBI designed to combine the resources of Federal, state, and local law enforcement."

agencies. [This] is a multi-agency effort led by the Justice Department and FBI designed to combine the resources of Federal, state, and local law enforcement."[24]

JTTFs existed prior to the September 11 attacks, but they were greatly expanded in the aftermath of those events. In the modern homeland security environment, there are approximately 4,400 JTTF members (quadruple the pre–9/11 number) from more than 600 state and local agencies and 50 federal agencies.[25] JTTFs are based in more than 100 cities, with at least one JTTF being housed in each FBI field office.[26]

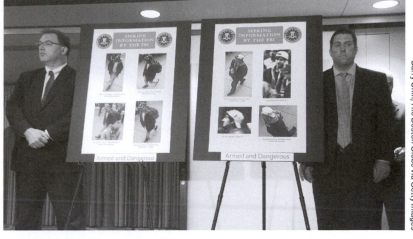

▶ **Photo 7.5** An FBI-led briefing by the Joint Terrorism Task Force seeking assistance in apprehending the Tsarnaev brothers during the aftermath of the April 2013 Boston Marathon bombing.

The purpose of JTTFs is to broadly coordinate planning and responses to terrorist incidents. When an incident occurs, members of the local JTTF immediately convene to coordinate emergency response and investigation efforts. In such cases, the collaborative effort is intended to be an efficient and purposeful adaptation to each incident. In essence, the benefits of JTTF collaboration are that "they enable a shared intelligence base across many agencies . . . [t]hey create familiarity among investigators and managers before a crisis . . . [a]nd perhaps most importantly, they pool talents, skills, and knowledge from across the law enforcement and intelligence communities into a single team that responds together."[27]

National coordination of JTTFs is accomplished through the **National Joint Terrorism Task Force**. The Federal Bureau of Investigation has been tasked with managing this interagency office from FBI headquarters. The National Joint Terrorism Task Force shares information with local JTTFs so that the national collaborative effort is conducted effectively. Table 7.2 compares and contrasts the mission and purpose of JTTFs in relation to local fusion centers.

National Joint Terrorism Task Force: The Federal Bureau of Investigation has been tasked with coordinating JTTFs nationally through the National Joint Terrorism Task Force, an interagency office operating from FBI headquarters.

Table 7.2   Operational Comparison: Joint Terrorism Task Forces and Fusion Centers

Fusion centers are owned and operated by state and local entities and are uniquely situated to empower front-line law enforcement, public safety, fire service, emergency response, public health, and private sector security personnel to lawfully gather and share threat-related information. The FBI created, coordinates, and manages JTTFs, which primarily focus on terrorism-related investigations. Both rely on expertise and information derived from all levels of government to support their efforts.

Fusion Centers	Joint Terrorism Task Forces
Focus on terrorism, criminal, and public safety matters in support of securing communities and enhancing the national threat picture	Focus primarily on terrorism and other criminal matters related to various aspects of the counterterrorism mission

(Continued)

Table 7.2 (Continued)	
Fusion Centers	**Joint Terrorism Task Forces**
Receive, analyze, gather, produce, and disseminate a broad array of threat-related information and actionable intelligence to appropriate law enforcement and homeland security agencies	Conduct counterterrorism investigations and provide information for assessments and intelligence products that are shared, when appropriate, with law enforcement and homeland security agencies
Owned and operated by state and local authorities and include federal; state, local, tribal, and territorial (SLTT); and private sector partners from multiple disciplines (including law enforcement, public safety, fire service, emergency response, public health, and critical infrastructure)	Multijurisdictional task forces managed by the FBI and include other federal and SLTT law enforcement partners that together act as an integrated force to combat terrorism on a national and international scale

Source: U.S. Department of Homeland Security.

GLOBAL PERSPECTIVE

PARAMILITARY DEPLOYMENT OF POLICE UNITS BY WESTERN DEMOCRACIES

Many nations have special units within their police forces that participate in counterterrorist operations. In several examples, these units have been used in a semimilitary role.

France. The GIGN (Groupe d'Intervention Gendarmerie Nationale) is recruited from the French *gendarmerie*, the military police. GIGN is a counterterrorist unit with international operational duties. In an operation that foiled what was arguably a precursor to the September 11 attacks, the GIGN rescued 173 hostages from Air France Flight 8969 in December 1994. Four Algerian terrorists had landed the aircraft in Marseilles, intending to fly to Paris to crash or blow up the plane over the city. GIGN assaulted the plane in a successful and classic operation.

Germany. GSG-9 (Grenzschutzgruppe 9) was organized after the disastrous failed attempt to rescue Israeli hostages taken by Black September at the 1972 Munich Olympics. It is a paramilitary unit that has been used domestically and internationally as a counterterrorist and hostage rescue unit. GSG-9

first won international attention in 1977 when the group freed hostages held by Palestinian terrorists in Mogadishu, Somalia. The Mogadishu rescue was heralded as a flawless operation.

Israel. The Police Border Guards are an elite force that is frequently deployed as a counterterrorist force. Known as the Green Police, it operates in two subgroups. YAMAS is a covert group that has been used extensively during the Palestinian *intifada*. It has been used to neutralize terrorist cells in conjunction with covert IDF operatives. YAMAM was specifically created to engage in counterterrorist and hostage rescue operations.

Spain. In 1979, the Spanish National Police organized a counterterrorist and hostage rescue force called GEO (Grupo Especial de Operaciones). Its training has allowed it to be used in both law enforcement and counterterrorist operations. Most of the latter—and a significant proportion of its operations—have been directed against the Basque Fatherland and Liberty (ETA) terrorist movement.

United States. At the national level, the United States has organized several units that have counterterrorist capabilities, all paramilitary groups that operate under the administrative supervision of federal agencies and perform traditional law enforcement work. These units are prepared to perform missions similar to Germany's GSG-9, Spain's GEO, and France's GIGN. Perhaps the best known is the FBI's Hostage Rescue Team. Not as well known but very important is the Department of Energy's Emergency Search Team. Paramilitary capabilities have also been incorporated into the Department of Justice's Bureau of Alcohol, Tobacco, Firearms and Explosives and the Department of Homeland Security's Secret Service. At the local level, American police forces also deploy units that have counterterrorist capabilities. These units are known by many names, but the most commonly known designation is SWAT.

Discussion Questions

1. To what extent should police units engage in paramilitary operations?

2. Are paramilitary operations by police units necessary? Why?

3. Who should have operational control over paramilitary police operations?

CHAPTER SUMMARY

Because state and local governments are likely to deploy first responders in the event of a terrorist attack, state and local agencies in the United States fully participate in homeland security initiatives. When examining non-federal governmental collaboration on homeland security issues, several national associations of state and local government have provided collaborative forums for establishing national response strategies implemented at the local level.

Law enforcement agencies throughout the homeland security enterprise have accepted homeland security duties as an integral component of their overall mission. The crossing of local homeland security initiatives with those of other jurisdictions has created a large and collaborative network for combating terrorist threats to the homeland. In particular, the FBI's coordination of the Joint Terrorism Task Force network has been instrumental in ensuring effective responses to terrorist threats.

DISCUSSION BOX

This chapter's Discussion Box is intended to stimulate critical thinking about interagency collaboration within the context of the Joint Terrorism Task Force model.

Operational Utility of the Joint Terrorism Task Force Model

Joint Terrorism Task Forces (JTTFs) are a nationwide network of policymakers and responders who collaborate on terrorism-related homeland security issues. They have effectively thwarted verifiable terrorist conspiracies prior to the initiation of attacks. Examples of successful investigations include the following:

- Breaking up cells like the Portland Seven, the Lackawanna Six, and the Northern Virginia jihad

- Foiled an attack on the Fort Dix army base in New Jersey

- Foiled an attack on JFK International Airport in New York City

Successes such as the foregoing are cited as examples of how interagency collaboration is an effective counterterrorist homeland security model.

Discussion Questions

1. How can the relevance of the JTTF model be kept up to date?

2. How should responsibilities be shared among state and local members of JTTFs?

3. When should federal leadership defer to local issues and considerations raised by local JTTFs?

4. What measures should be taken to assure collaboration and coordination when a terrorist incident occurs?

5. Should some agencies have priority authority in JTTF collaboration? If so, which ones? If not, why not?

Source: Federal Bureau of Investigation, "Protecting America From Terrorist Attack: Our Joint Terrorism Task Forces," http://www.fbi.gov/about-us/investigate/terrorism/terrorism_jttfs, accessed April 16, 2013.

KEY TERMS AND CONCEPTS

The following topics were discussed in this chapter and can be found in the glossary:

councils of government (COGs)   132

Emergency Management Assistance Compact (EMAC)   133

International Association of Emergency Managers (IAEM)   136

Joint Terrorism Task Forces (JTTFs)   138

Local Emergency Planning Committees (LEPCs)   134

National Association of Counties (NACo)   134

National Association of Regional Councils (NARC)   132

National Emergency Management Association (NEMA)   130

National Homeland Security Consortium   131

National Joint Terrorism Task Force   139

National League of Cities (NLC)   135

United States Conference of Mayors (USCM)   135

ON YOUR OWN

Get the tools you need to sharpen your study skills. SAGE edge offers a robust online environment featuring an impressive array of free tools and resources.

Access practice quizzes, eFlashcards, video, and multimedia at **edge.sagepub.com/martinhs3e**

RECOMMENDED WEBSITES

The following websites provide homeland security information about state and local agencies and organizations:

Department of Justice Joint Terrorism Task Force: www.justice.gov/jttf

National Association of Counties: www.naco.org

National Association of Regional Councils: narc.org/about-narc/about-the-association

National Emergency Management Association: www.nemaweb.org

National League of Cities: www.nlc.org

U.S. Census Bureau: www.census.gov/govs/cog2012

U.S. Conference of Mayors: usmayors.org/about/overview.asp

WEB EXERCISE

Using this chapter's recommended websites, conduct an online review of law enforcement agencies that engage in homeland security work on the state and local levels of government. Compare and contrast these agencies.

1. What are the primary missions of these agencies?

2. How would you describe the differences between state and local law enforcement agencies?

3. In your opinion, are any of these organizations devoting enough resources to their homeland security initiatives? Too many resources?

To conduct an online search on research and monitoring organizations, activate the search engine on your Web browser and enter the following keywords:

"Law enforcement homeland security agencies"

"State homeland security agencies"

RECOMMENDED READINGS

The following publications provide discussion on the homeland security missions of state and local agencies:

Aelehrt, Barbara. 2010. *Emergency Medical Responder: First Responder in Action*. 2nd ed. New York: McGraw-Hill.

Bergeron, David J., Gloria Bizjack, Chris LeBadour, and Keith Wesley. 2008. *First Responder*. 8th ed. New York: Prentice Hall.

International Fire Service Training Association. 2011. *Hazardous Materials for First Responders*. 4th ed. New York: Author.

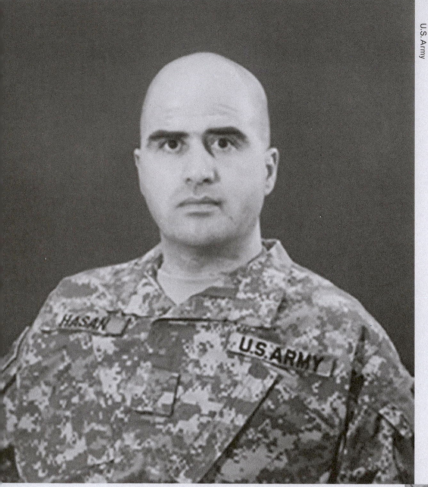

Major Nidal Malik Hasan and Anwar al-Awlaki. American-born jihadists who represent the plausible threat from homegrown sympathizers of radical international movements. Major Hasan was responsible for the mass shooting at Fort Hood, Texas, on November 5, 2009, and al-Awlaki became a central organizer and propagandist for al-Qaeda in Yemen. Intelligence agencies confirmed that Hasan corresponded with al-Awlaki via e-mail during late 2008 to mid-2009, stating in one message to al-Awlaki, "I can't wait to join you" after death. Hasan also inquired about the acceptability of killing innocents and the proper conditions for waging jihad.

PART THREE

THE TERRORIST THREAT AND HOMELAND SECURITY

CHAPTER EIGHT

SEA CHANGE

The New Terrorism and Homeland Security

Chapter Learning Objectives

This chapter will enable readers to do the following:

1. Define and discuss attributes of the New Terrorism

2. Analyze the terrorist environment prior to the September 11, 2001, terrorist attacks

3. Describe plausible scenarios stemming from the New Terrorism

4. Evaluate possible counterterrorist policy options

Opening Viewpoint: The New Terrorism and Soft Targets

The New Terrorism is characterized by the threat of weapons of mass destruction, asymmetrical tactics, indiscriminate targeting, and intentionally high casualty rates—as occurred in the attacks of September 11, 2001, in the United States; the March 11, 2004, bombings in Madrid; the July 7, 2005, attacks in London; the November 26–29, 2008, assault in Mumbai; the September 21, 2013, assault on a shopping mall in Nairobi; the April 2, 2015, assault at Garissa University College in Kenya; the November 13, 2015, assault in Paris; the July 14, 2016, vehicular attack against pedestrians in Nice; and the August 17, 2017, vehicular attack against pedestrians in Barcelona. The use of these weapons and tactics against civilians is indefensible, no matter what cause is championed by those who use them.

Terrorists select targets that have maximum effect on enemy audiences. From the terrorists' perspective, the symbolism of potential targets is as important—and perhaps more important—than the amount of actual damage an attack would have on infrastructure or the victim's economy. Thus, for homeland security planning purposes, a basic understanding of target selection requires recognition that terrorism is

- politically motivated violence;

- usually directed against civilian and administrative government targets; and

- with an intention to affect (terrorize) a target audience.

Modern terrorists select targets that are both symbolic and "soft." Soft targets include civilians and passive military targets, which are unlikely to offer resistance until after the terrorists have inflicted casualties or other destruction. Such target selection is sometimes quite effective in the short term and occasionally successful in forcing targeted interests to grant concessions. Vulnerable public venues and relatively unprotected buildings and events are naturally considered to be desirable targets.

Those who practice the New Terrorism have regularly selected soft targets that symbolize enemy interests. These targets are chosen, in part, because of their symbolic value but also because they are likely to result in significant casualties.

Internationally, suicide bombers have become particularly adept at maximizing casualties. For example, 14 people (11 of them French workers) were killed in May 2002 by a suicide bomber outside a hotel in downtown Karachi, Pakistan. In Syria and Iraq, hundreds of people have been killed by suicide bombers in dozens of sectarian attacks. Domestically, homegrown assailants have selected venues where large numbers of people congregate. The November 2009 attack by Major Nidal Hasan at Fort Hood, the May 2013 Boston Marathon bombings, and the June 2016 Pulse nightclub attack in Orlando were selected, in part, by a desire to inflict maximum casualties in an open location.

The **New Terrorism** is a principal attribute of the modern era and has become a destabilizing influence in the affairs of the global community. It signifies a marked escalation in lethality by dedicated extremists, as well as continuing progression in utilizing modern networking and technological resources by extremist movements. Terrorists in the modern era are quite willing to use modern weapons technologies in ways guaranteed to increase the likelihood of mass casualties. Acceptable weapons technologies include weapons of mass destruction and innovative stealth weapons, such as suicide bombs and improvised explosive devices. Modern terrorists are no longer "surgical" in the deployment of such weapons; entire groups of people are declared to be the enemy and, therefore, attacked as legitimate targets. Thus, while terrorists in the past would justify civilian casualties by declaring that they represented inadvertent "collateral damage" to the intended target, terrorists in the present era declare that civilian populations *are* the intended and justifiable targets.

The New Terrorism also represents an unprecedented period of globalized terrorism. "In this new era, advanced communications technologies such as the Internet and cable news outlets confer an unprecedented ability for terrorists to influence the international community quickly, cheaply, and with little risk to the extremists themselves."[1] Terrorists understand the value of disseminating their message to a global audience, and they carefully adapt their tactics to manipulate modern information technologies to their benefit. The New Terrorism signifies the increasing sophistication of modern extremists, who often demonstrate proficiency in coordinating their tactics by using globalized information technologies. Furthermore, independent stateless terrorist groups operate "within the context of modern integrated economies and regional trade areas."[2] In this milieu, the globe has become a rich battlefield for committed extremists.

The discussion in this chapter will review the following:

- Defining terrorism: An ongoing debate

- Case in point: The Political Violence Matrix

- Historical context: The "Old Terrorism" prior to September 11, 2001

New Terrorism: A typology of terrorism characterized by a loose, cell-based organizational structure; asymmetrical tactics; the threatened use of weapons of mass destruction; potentially high casualty rates; and usually a religious or mystical motivation.

- September 11, 2001, and the New Terrorism

- Plausible scenarios: The New Terrorism and new modes of warfare

- Policy options

DEFINING TERRORISM: AN ONGOING DEBATE

The effort to formally define terrorism is a critical one because government antiterrorist policy calculations must be based on criteria that determine whether a violent incident is an act of terrorism. Governments and policymakers must piece together the elements of terrorist behavior and demarcate the factors that distinguish terrorism from other forms of conflict.

There is some consensus among experts—but no unanimity—on what kind of violence constitutes an act of terrorism. Governments have developed definitions of terrorism, individual agencies within governments have adopted definitions, private agencies have designed their own definitions, and academic experts have proposed and analyzed dozens of definitional constructs. This lack of unanimity, which exists throughout the public and private sectors, is an accepted reality.

In Europe, countries that endured terrorist campaigns have written official definitions of terrorism. The British have defined terrorism as "the use or threat, for the purpose of advancing a political, religious or ideological cause, of action which involves serious violence against any person or property."[3] In Germany, terrorism has been described as an "enduringly conducted struggle for political goals, which are intended to be achieved by means of assaults on the life and property of other persons, especially by means of severe crimes."[4] And the European interior ministers note that "terrorism is . . . the use, or the threatened use, by a cohesive group of persons of violence (short of warfare) to effect political aims."[5]

The United States has not adopted a single definition of terrorism as a matter of government policy, instead relying on definitions that are developed from time to time by government agencies. These definitions reflect the United States' traditional law enforcement approach to distinguishing terrorism from more common criminal behavior. The following definitions are a sample of the official approach.

The U.S. Department of Defense defines terrorism as "the unlawful use of violence or threat of violence, often motivated by religious, political, or other ideological beliefs, to instill fear and coerce governments or societies in pursuit of goals that are usually political."[6] The U.S. Code defines terrorism as illegal violence that attempts to "intimidate or coerce a civilian population; . . . influence the policy of a government by intimidation or coercion; or . . . affect the conduct of a government by assassination or kidnapping."[7] The Federal Bureau of Investigation has defined terrorism as "the unlawful use of force or violence against persons or property to intimidate or coerce a Government, the civilian population, or any segment thereof, in furtherance of political or social objectives."[8] The State Department has defined terrorism as "premeditated, politically motivated violence perpetrated against non-combatant targets by subnational groups or clandestine agents."[9]

Using these definitions, common elements can be combined to construct a composite American definition, which is as follows:

> Terrorism is a premeditated and unlawful act in which groups or agents of some principal engage in a threatened or actual use of force or violence against human or property targets. These groups or agents engage in this behavior, intending the purposeful intimidation of governments or people to affect policy or behavior, with an underlying political objective.

CASE IN POINT: THE POLITICAL VIOLENCE MATRIX

To properly conceptualize the New Terrorism, one must understand the qualities and scales of violence that define terrorist violence. The Political Violence Matrix is a tool that aids in this conceptualization.

Experts have identified and analyzed many terrorist environments. These environments include state, dissident, religious, ideological, international, and criminal terrorism. One distinguishing feature within each model is the relationship between the *quality of force* used by the terrorists and the *characteristics of the intended target* of the attack. Figure 8.1 depicts how the relationship between quality of force and target characteristics often defines the type of conflict between terrorist and victim.

Combatants, Noncombatants, and the Use of Force

Definitional and ethical issues are not always clearly drawn when one uses terms such as *combatant target*, *noncombatant target*, *discriminate force*, and *indiscriminate force*. Nevertheless,

Figure 8.1   The Political Violence Matrix

The purpose of the Political Violence Matrix is to create a framework for classifying and conceptualizing political violence. This classification framework is predicated on two factors: **Force** and **Intended Target**.

When force (whether conventional or unconventional) is used against *combatant* targets, it occurs in a warfare environment. When force is used against *noncombatant* or *passive military* targets, it often characterizes a terrorist environment. Violent environments can be broadly summarized as follows:

- **Total War**. Force is indiscriminately applied to destroy the military targets of an enemy combatant to absolutely destroy them.
- **Total War/Unrestricted Terrorism**. Indiscriminate force is applied against noncombatant targets without restraint, either by a government or by dissidents.
- **Limited War**. Discriminating force is used against a combatant target, either to defeat the enemy or to achieve a more limited political goal.
- **State Repression/Restricted Terrorism**. Discriminating force is directed against noncombatant targets either as a matter of domestic policy or as the selective use of terrorism by dissidents.

The following figure summarizes factors to be considered when evaluating the application of different scales of force against certain types of targets.

Indiscriminate force, Combatant target	Total war (WWII Eastern Front)	Limited war (Korean War)	*Discriminate force, Combatant target*
Indiscriminate force, Noncombatant target	Total war (WWII bombing of cities) Unrestricted terrorism (Rwandan genocide)	State repression (Argentine "Dirty War") Restricted terrorism (Italian Red Brigade)	*Discriminate force, Noncombatant target*

Source: Adapted from Peter C. Sederberg, *Terrorist Myths: Illusion, Rhetoric, and Reality* (Englewood Cliffs, NJ: Prentice Hall, 1989), 34.

the association of these concepts and how they are applied to each other are instructive references for determining whether a violent incident may be defined as terrorism.

Combatant and Noncombatant Targets

The term *combatants* certainly refers to conventional or unconventional adversaries who engage in armed conflict as members of regular military or irregular guerrilla fighting units. The term *noncombatants* obviously includes civilians who have no connection to military or other security forces. There are, however, circumstances in which these definitional lines become blurred. For example, in times of social unrest, civilians can become combatants. This has occurred repeatedly in societies where communal violence (e.g., civil war) has broken out between members of ethnonational, ideological, or religious groups. Similarly, noncombatants can include off-duty members of the military in nonwarfare environments.[10] They become targets because of their symbolic status.

Indiscriminate and Discriminate Use of Force

Indiscriminate force is the application of force against a target without attempting to limit the level of force or the degree of destruction of the target. *Discriminate force* is a more surgical use of limited force. Indiscriminate force is considered to be acceptable when used against combatants in a warfare environment. However, it is regularly condemned when used in *any* nonwarfare environment, regardless of the characteristics of the victim.[11] There are, however, many circumstances when adversaries define "warfare environment" differently. When weaker adversaries resort to unconventional methods (including terrorism), they justify them by defining them as necessary during a self-defined state of war. Discriminate force is considered to be a moral use of force when it is applied against specific targets with the intention to limit so-called collateral damage, or unintended destruction and casualties.

HISTORICAL CONTEXT: THE "OLD TERRORISM" PRIOR TO SEPTEMBER 11, 2001

Terrorism is, in many ways, a reflection of global politics so that global and domestic terrorist environments are dominated at different times by different terrorist typologies. For example, during historical periods of nationalist insurrection, terrorist campaigns were conducted in solidarity with nationalist causes. During historical periods of ideological polarization and conflict, terrorist campaigns were conducted on behalf of a preferred ideological worldview. The intensity of terrorist campaigns ebbed and flowed with the degree of extremist fervor emanating from the championed group or ideology.

The Pre–September 11 Terrorist Environment

The global terrorist environment prior to the September 11, 2001, terrorist attacks was characterized by a terrorist operational profile that was relatively uncomplicated and predictable. This "**Old Terrorism**" profile typically included the following elements:

- Clearly identifiable organizations or movements
- Use of conventional weapons, usually small arms and explosives
- Explicit grievances championing specific classes or ethnonational groups

"Old Terrorism": Prior to the September 11, 2001, terrorist attacks, the global terrorist environment was characterized by a relatively uncomplicated and predictable terrorist operational profile.

- Use of ideological or nationalist justifications for revolutionary violence

- Relatively "surgical" selection of targets

Within the context of these elements, the global terrorist environment progressed through several phases in the modern era, roughly summarized as predominantly left-wing, ethnonational, Palestinian nationalist, and religious phases.[12] The last phase—religious—heralded the dawn of the New Terrorism. The genesis of these phases is found in the hostile ideological conflict of this era: the Cold War and its modus of indirect confrontation, also known as *proxy war*. During this period, there also existed some collaboration among similarly motivated revolutionaries, as well as replication of tactics between movements.

The replication, or copycatting, phenomenon continues and is ongoing in the modern era. Chapter Perspective 8.1 discusses this phenomenon, referred to as the **contagion effect**, within the contexts of the "Old Terrorism" and New Terrorism.

Proxy Warfare: Ideological Conflict and the Old Terrorism

The Cold War and the competition between East and West defined the terrorist environment in the period immediately prior to September 11, 2001. Many military and intelligence practitioners, scholars, and commentators have argued that the terrorism in this era was predictable and logical.

During the Cold War rivalry between the United States and the Soviet Union, each superpower displayed a persistent pattern of international behavior. The Soviets tended to side with nationalist insurgencies in the developing world, whereas the United States supported the embattled established governments. These insurgencies took on the characteristics of Marxist revolutions, and the embattled governments became, from the perspective of the United States, bulwarks against the spread of communism. The Soviets and, to a lesser extent, the Chinese armed and financed many of these **wars of national liberation** against U.S.-supported regimes.

Nationalism and Marxism were synthesized repeatedly by twentieth-century revolutionaries in the developing world. Three of these conflicts—led by Mao Zedong in China, Ho Chi Minh in Vietnam, and Fidel Castro in Cuba—came to symbolize the new phenomenon of leftist nationalism. Left-wing revolutionaries in the West drew on these examples and developed theories of solidarity with Marxist nationalists to justify acts of violence in Western democracies. Many New Left revolutionaries in the West were particularly receptive to the theory that their terrorist campaigns in Western democracies were linked to the nationalist armed insurgencies in the developing world. From their perspective, all of these struggles were part of a worldwide war against capitalism, imperialism, and exploitation.

Chapter Perspective 8.2 discusses contemporary Western theory about the origins of terrorism during the Cold War.

Hulton Archive/Keystone/Getty Images

▶ **Photo 8.1** Prisoner of the Red Army Faction. A photograph of West German industrialist Hanns-Martin Schleyer, taken during his captivity by the leftist Red Army Faction (also known as the Baader-Meinhof Gang). The RAF later executed Schleyer.

contagion effect: Copycat terrorism in which terrorists imitate each other's behavior and tactics. This theory is still debated.

wars of national liberation: A series of wars fought in the developing world in the postwar era. These conflicts frequently pitted indigenous guerrilla fighters against European colonial powers or governments perceived to be pro-Western. Insurgents were frequently supported by the Soviet bloc or China.

CHAPTER PERSPECTIVE 8.1

The Contagion Effect

The contagion effect refers to the theoretical influence of media exposure on the future behavior of other like-minded extremists. In theory, when terrorists successfully garner wide exposure or a measure of sympathy from the media and their audience, other terrorists may be motivated to replicate the tactics of the first successful incident. This may be especially true if concessions have been forced from the targeted interest. Assuming that contagion theory has merit (the debate on this point continues), the question becomes the extent to which the contagion effect influences behavior.

Examples of the contagion effect arguably include the following cycles:

- Diplomatic and commercial kidnappings for ransom and concessions in Latin America during the 1960s and 1970s

- Hijackings on behalf of Middle East–related causes (usually Palestinian) from the late 1960s to the 1980s

- Similarities in the tactics of left-wing Western European ideological terrorists during their heyday from the late 1960s to the 1980s

- The taking of Western hostages in Lebanon during the 1980s

- The taking of hostages and the use of broadcasted beheadings in the Middle East during the 2000s

- High-casualty urban attacks in Western democracies during the 2000s

Assessments of the contagion effect produced some consensus that the media do have an effect on terrorist cycles. For example, empirical studies have indicated a correlation between media coverage and time lags between terrorist incidents. These studies have not definitively *proven* that contagion is a behavioral fact, but they do suggest that the theory may have some validity.

The era of the New Terrorism arguably presents an unprecedented dynamic for contagion theory because transnational cell-based movements are a new model for—and may suggest new assessments of—the theory. Transnational organizations such as al-Qaeda engage in a learning process from the lessons of attacks by their operatives around the world. The advent of communications technologies such as social networking media, cellular telephones, e-mail, text messaging, and the Internet—especially in combination with focused manipulation of the media—means that the terrorists' international learning curve can be quick and efficient. Hence, in theory, the contagion effect may be enhanced within New Terrorist movements on a global scale.

Discussion Questions

1. To what extent do you think the contagion effect occurs?

2. If cycles of terrorist methods are verifiable, what kind of cycles do you think will occur in the future?

3. If the contagion effect is a fact, are the Internet and modern terrorist networking arrangements cause for concern?

The Recent Past: The Left-Wing Terrorist Environment Prior to September 11, 2001

From the 1960s through the early 1980s, left-wing terrorists figured prominently in domestic and international incidents. Latin American and Western European groups in particular waged terrorist campaigns of significant intensity.

In Latin America, armed leftist activism in the postwar era posed serious challenges to many established governments. Armed Marxist guerrillas operated in the countryside of several nations, and urban terrorists appeared repeatedly in Latin America. Although leftist nationalism did occur in the postwar era, most leftist insurgents were dedicated Marxists.

CHAPTER PERSPECTIVE 8.2

Cold War Terrorist-Networking Theory

During the Cold War rivalry between the Western allies and the Eastern bloc, many experts in the West concluded that the communist East was responsible for sponsoring an international terrorist network. The premise was that the Soviet Union and its allies were at least an indirect—and often a direct—source of most international terrorism. Under this scenario, state-sponsored terrorism was a significant threat to world security and was arguably a manifestation of an unconventional World War III. It represented a global network of state-sponsored revolutionaries whose goal was to destabilize the democratic West and its allies in the developing world (referred to at that time as the *Third World*).

The Western democracies were able to cite evidence to support their claim that global terrorism related back to a Soviet source. One source of evidence was the fact that the Soviets never denied that they supported revolutionary groups. However, they labeled them as freedom fighters waging wars of national liberation rather than as terrorists. Another source of evidence was the truism that the West was the most frequent target of international terrorism; Soviet interests were rarely attacked. Perhaps the most credible evidence was that a number of regimes were clearly implicated in supporting terrorist movements or incidents. Many of these regimes were pro-Soviet in orientation or at least recipients of Soviet military aid. Thus, when Soviet- or Chinese-manufactured weapons were found in terrorist arms caches, it was clear that these regimes were conduits for Soviet support for radical movements.

Despite these indications, one significant problem with the Soviet sponsorship scenario was that most of the evidence was circumstantial and inconclusive. For instance, many of the world's terrorist movements and extremist governments were either non-Marxist in orientation or only secondarily Marxist. They were comfortable with accepting assistance from any willing donor, regardless of the donor's ideological orientation.

Although some nationalist movements, such as the Palestine Liberation Organization, certainly had Marxist factions that received training and support from the Soviets and although some governments, such as those of Cuba and Syria, received Soviet military and economic aid, it is questionable how much actual *control* the Soviets had over their clients. And, very significantly, many dissident movements and state sponsors of terrorism, such as the Provisional Irish Republican Army and revolutionary Iran, were completely independent of Soviet operational or ideological control.

Thus, the belief that terrorism was part of a global conflict between democracy and communism (and hence an unconventional World War III) was too simplistic. It did not take into account the multiplicity of ideologies, motivations, movements, or environments that represented international terrorism. Having said this, there was, without question, a great deal of state sponsorship of terrorism that emanated at least indirectly from the communist East. Ideological indoctrination, material support, and terrorist training facilities did provide revolutionary focus for extremists from around the world. Therefore, although there was not a communist-directed terrorist network, and the Soviets were not a "puppet master" for a global terrorist conspiracy, they did actively inflame terrorist behavior.

Discussion Questions

1. Assuming Cold War networking theories were accurate, in what way was the terrorist environment more "rational" for counterterrorist operations?

2. How was state sponsorship of terrorism during the Cold War different from modern terrorist sponsorship?

3. What was the motivation behind Soviet-bloc sponsorship of terrorism?

In Europe prior to the opening of the Berlin Wall in 1989, very little terrorism of any kind occurred in the communist Eastern bloc. However, democratic Western Europe experienced a wave of leftist terrorism that began during widespread student and human rights activism in the 1960s. This wave of political violence occurred in numerous Western countries. Although some nationalist terrorism occurred—for example, in Northern Ireland,

Corsica, and Spain—most terrorist groups were ideologically motivated. Western European groups frequently attacked international symbols in solidarity with defined oppressed groups. Only a few leftist groups remain in the present terrorist environment, and they no longer pose a significant threat to Latin American, Asian, or European nations or to the United States.

The Recent Past: The Ethnonational Terrorist Environment Prior to September 11, 2001

Throughout the postwar era, ethnonational terrorism occupied an important presence in domestic and international arenas. Its incidence has ebbed and flowed in scale and frequency, but ethnonationalist violence has never completely disappeared. Nationalism is a concept that promotes the aspirations of groups of people distinguished by their cultural, religious, ethnic, or racial heritage. The guiding motivation behind these movements is national identity.

By the late 1980s, violent leftist sympathizers in the West had all but disappeared. By the late 1990s, ethnonationalist groups were operating primarily inside their home countries but were continuing to occasionally attack international symbols to bring attention to their domestic agendas.

Ongoing Communal Conflict: The Palestinian Terrorist Environment

Beginning in the late 1960s, Palestinian nationalists were arguably the leading practitioners of international terrorism. Participating in their struggle were Western European and Middle Eastern sympathizers who struck targets in solidarity with the Palestinian cause. For example, the Popular Front for the Liberation of Palestine established strong links with European terrorist groups. This loose alliance inspired other extremists to imitate their tactics on the world's stage.

By the late 1990s, with the creation of the governing authority on the West Bank and Gaza, Palestinian-initiated terrorism focused primarily on targets inside Israel and the occupied territories. Since September 28, 2000, Palestinian resistance has periodically taken on the characteristics of a broad-based uprising as well as communal terrorism. On that date, Israeli general Ariel Sharon visited the Temple Mount in Jerusalem. After Sharon's visit (which was perceived by Palestinians to be a deliberate provocation), enraged Palestinians began a second round of massive resistance—the "shaking off," or *intifada*. The new dissident environment included violent demonstrations, street fighting, and suicide bombings. The violence was regularly characterized by bombings, shootings, and other attacks against civilian targets. Thus, the Palestinian nationalist movement entered a phase distinguished by the acceptance of communal dissident terrorism as a strategy.

Military strikes have also been launched against Israel, often with significant intensity. For example, during July and August 2014, hundreds of rockets were fired from Gaza into Israel. In November 2018, another round of bombardment occurred, with hundreds of rockets being fired from Gaza—of the nearly 400 rockets launched, Israel's Iron Dome missile defense system reportedly intercepted 100. The communal aspect of Palestinian resistance also continued, with dozens of attempted assaults by individual Palestinians against Israelis during 2015. Many of these attacks were carried out by youths and others wielding nothing more than edged weapons or motorized vehicles.

The Religious Terrorist Environment and the New Terrorism

Although vestiges of nationalist and ideological terrorism continued in some regions of the world at the end of the twentieth century, they were usually limited to pockets of violent revolutionary sentiment. They did not compare in scale to the international revolutionary fervor that had existed earlier in the twentieth century, nor to the wave of urban terrorist warfare that came to the West toward the end of the century. And although a few nationalist and ideological insurgencies continued to be fought in the countryside of some countries, they likewise did not compare in scale or frequency to similar wars in the postwar era.

By the beginning of the twenty-first century, the most prominent practitioners of international terrorism were religious extremists. Although Islamist movements, such as al-Qaeda and ISIS offshoots, were the most prolific international religious terrorists by a considerable degree, extremists from other major religions also operated on the international stage.

Prior to September 11, 2001, the most serious domestic incident of religious terrorism in the United States occurred on February 26, 1993, when Ramzi Yousef detonated a vehicular bomb at the World Trade Center in New York City. Internationally, in August 1998, the U.S. embassies in Nairobi, Kenya, and Dar es Salaam, Tanzania, were bombed, with more than 200 fatalities. In October 2000, the U.S. destroyer USS *Cole* was nearly sunk by a boat bomb in Aden, Yemen, resulting in the deaths of 17 sailors and 39 wounded. These incidents heralded a gradual transition from the previous terrorist environment toward the new environment.

Bryn Colton/Getty Images

▶ **Photo 8.2** The remains of Pan Am Flight 103 after being bombed in the skies over Lockerbie, Scotland. The attack led to economic sanctions against Libya and the trial of two members of the Libyan security service.

SEPTEMBER 11, 2001, AND THE NEW TERRORISM

It is clear from human history that terrorism is deeply woven into the fabric of social and political conflict. This quality has not changed, and in the modern world, states and targeted populations are challenged by the New Terrorism, which is characterized by the following:

- Loose, cell-based networks with minimal lines of command and control
- Desired acquisition of high-intensity weapons and weapons of mass destruction
- Politically vague, religious, or mystical motivations
- Asymmetrical methods that maximize casualties
- Skillful use of the Internet and manipulation of the media

Table 8.1 compares several key characteristics of the "Old Terrorism" and New Terrorism.

New information technologies and the Internet create unprecedented opportunities for terrorist groups, and violent extremists have become adept at bringing their wars into the homes of literally hundreds of millions of people. Those who specialize in suicide bombings, car bombs, or mass-casualty attacks correctly calculate that carefully selected targets will attract the attention of a global audience. Thus, cycles of violence not only disrupt normal routines but also produce long periods of global awareness. Such cycles can be devastating.

Religion and Terrorism in the Modern Era

Religious terrorism is political violence conducted by groups of religious "true believers" who have fervent faith in the sacred righteousness of their cause. Any behavior carried out in the defense of this sacred cause is considered to be not only justifiable but blessed.

Table 8.1 Supplanting the Old With the New

The "Old Terrorism" was, in many ways, symmetrical and predictable. It was not characterized by terrorist environments exhibiting massive casualty rates or indiscriminate attacks. Its organizational profile was also characterized by traditional organizational configurations.

The following table contrasts selected attributes of the activity profiles for the "Old Terrorism" and the New Terrorism.

Terrorist Environment	Activity Profile				
	Target Selection	Casualty Rates	Organizational Profile	Tactical/Weapons Selection	Typical Motives
"Old Terrorism"	Surgical and symbolic	Low and selective	Hierarchical and identifiable	Conventional and low to medium yield	Leftist and ethnocentric
New Terrorism	Indiscriminate and symbolic	High and indiscriminate	Cellular	Unconventional and high yield	Sectarian

Most major religions—in particular, Christianity, Islam, Judaism, and Hinduism—possess extremist adherents, some of whom have engaged in terrorist violence. Smaller religions and cults have similar adherents. Among the ubiquitous principles found among religious extremists is their conviction that they are defending their faith from attack by nonbelievers or that their faith is an indisputable and universal guiding principle that must be advanced for the salvation of the faithful. These principles are manifested in various ways and to varying degrees by religious extremists, but they are usually at the core of their belief system.

The new millennium began with a resurgence of religious terrorism. Unlike previous terrorist environments, the new era of terrorism is largely shaped by the international quality of this resurgence; in essence, modern religious terrorism is a global phenomenon affecting every member of the international community. The current ideological profile of this development is one of activism and momentum among radical Islamists. Although extremist members of other faiths certainly strike periodically, the Islamist tendency continues to attract new cadres of *jihadis* who oppose secular governments and Western influence in the Middle East. Chapter Perspective 8.3 provides some clarification on the concept of jihad.

Religion is a central feature of the New Terrorism, which is characterized by asymmetrical tactics, cell-based networks, indiscriminate attacks against "soft" targets, and the threatened use of high-yield weapons technologies. Al-Qaeda and other Islamist organizations pioneered this strategy, and it serves as a model for similarly motivated individuals and groups. Religious extremists understand that by adopting these characteristics, their agendas and grievances will receive extensive attention, and their adversaries will be sorely challenged to defeat them.

U.S. Department of State

▶ **Photo 8.3** The rubble of Khobar Towers. An indication of the destruction in the aftermath of a 1996 terrorist attack against the American military facility in Dharan, Saudi Arabia.

CHAPTER PERSPECTIVE 8.3

Jihad: Struggling in the Way of God

The concept of jihad is a central tenet in Islam. Contrary to misinterpretations common in the West, the term literally means a sacred "struggle" or "effort" rather than an armed conflict or fanatical holy war.[a] Although a jihad can certainly be manifested as a holy war, it more correctly refers to the duty of Muslims to personally strive "in the way of God."[b]

This is the primary meaning of the term as used in the Qur'an, which refers to an internal effort to reform bad habits in the Islamic community or within the individual Muslim. The term is also used more specifically to denote a war waged in the service of religion.[c]

Regarding how one should wage jihad,

The **greater jihad** refers to the struggle each person has within himself or herself to do what is right. Because of human pride, selfishness, and sinfulness, people of faith must constantly wrestle with themselves and strive to do what is right and good. The **lesser jihad** involves the outward defense of Islam. Muslims should be prepared to defend Islam, including military defense, when the community of faith is under attack.[d] (boldface added)

Thus, waging an Islamic jihad is not the same as waging a Christian Crusade; it has a broader and more intricate meaning. Nevertheless, it is permissible—and even a duty—to wage war to defend the faith against aggressors. Under this type of jihad, warfare is conceptually defensive in nature. In contrast, the Christian Crusades were conceptually *offensive* in nature. Those who engage in armed jihad are known as *mujahideen,* or holy warriors. Mujahideen who receive "**martyrdom**" by being killed in the name of the faith will find that

awaiting them in paradise are rivers of milk and honey, and beautiful young women. Those entering paradise are eventually reunited with their families and as martyrs stand in front of God as innocent as a newborn baby.[e]

The precipitating causes for the modern resurgence of the armed and radical jihadi movement are twofold: the revolutionary ideals and ideology of the 1979 Iranian Revolution and the practical application of jihad against the Soviet Union's occupation of Afghanistan.

Some radical Muslim clerics and scholars have concluded that the Afghan jihad brought God's judgment against the Soviet Union, leading to the collapse of its empire. As a consequence, radical jihadis fervently believe that they are fighting in the name of an inexorable force that will end in total victory and guarantee them a place in paradise. From their perspective, their war is a just war.[f]

Discussion Questions

1. Is the concept of the lesser jihad a reasonable rationale for violent behavior?

2. Should those who practice the lesser jihad always be labeled as terrorists?

3. In what ways do you think the concept of martyrdom appeals to would-be terrorists?

Notes

a. Karen Armstrong, *Islam: A Short History* (New York: Modern Library, 2000), 201.

b. Josh Burke and James Norton, "Q&A: Islamic Fundamentalism: A World-Renowned Scholar Explains Key Points of Islam," *Christian Science Monitor,* October 4, 2001.

c. Armstrong, 201.

d. Burke and Norton.

e. Walter Laqueur, *The New Terrorism: Fanaticism and the Arms of Mass Destruction* (New York: Oxford University Press, 1999), 100.

f. See Evan R. Goldstein, "How Just Is Islam's Just-War Tradition?" *The Chronicle Review,* April 18, 2008.

The New Terrorism and Globalization

Political and economic integration has created a new field of operations for international terrorists; in effect, globalization accommodates the operational choices of committed extremists. Global trade and political integration permit extremists to provoke the attention of targeted audiences far from their home territories. In many respects, because of globalized information and integration, terrorists are able to operate on a virtual battlefield and cross virtual borders to strike their enemies. Globalized political and economic arrangements offer terrorists the capability of affecting the international community much faster and much more intensely than previous generations of terrorists could. Technologies are quite capable of broadcasting visual images and political interpretations of attacks to hundreds of millions of people instantaneously.

The globalization of political violence is a manifestation of new information technologies, mass audiences, and the features of the New Terrorism. "These potentialities, if skillfully coordinated, provide unprecedented opportunities for small groups of violent extremists to broadly influence targeted audiences."[13]

PLAUSIBLE SCENARIOS: THE NEW TERRORISM AND NEW MODES OF WARFARE

The New Terrorism is different from previous environments because it is characterized by vaguely articulated political objectives, indiscriminate attacks, attempts to achieve maximum psychological and social disruption, and the potential use of weapons of mass destruction. It is also distinguished by an emphasis on building horizontally organized, semiautonomous cell-based networks.

Asymmetrical Warfare

asymmetrical warfare: The use of unconventional, unexpected, and nearly unpredictable methods of political violence. Terrorists intentionally strike at unanticipated targets and apply unique and idiosyncratic tactics.

The concept of **asymmetrical warfare** has been adapted to the characteristics of contemporary political violence.[14] Asymmetrical warfare refers to the use of unexpected, nearly unpredictable, and unconventional methods of political violence. Terrorists apply unique and idiosyncratic tactics and intentionally strike at unanticipated targets. This way, they can seize the initiative, redefine the international security environment, and overcome the traditional protections and deterrent policies that societies and the international community use.

Although it is an old practice, asymmetrical warfare has become a core feature of the New Terrorism. In the modern era of asymmetrical warfare, terrorists can theoretically acquire and wield new high-yield arsenals, strike at unanticipated targets, cause mass casualties, and apply unique and idiosyncratic tactics. The dilemma for victims and for counterterrorism policymakers is that by using these tactics, terrorists can win the initiative and redefine the international security environment. In this way, the traditional protections and deterrent policies used by societies and the global community can be surmounted by dedicated terrorists.

Asymmetrical Warfare and the Contagion Effect: The Case of Motorized Vehicle Attacks

Motorized vehicle attacks by terrorists have occurred with some frequency in Western countries. Such attacks involve the use of unarmored nonexplosive vehicles against soft civilian targets. Tactical considerations and selection of vehicles are uncomplicated: Terrorists use readily obtainable civilian vehicles that blend in completely with urban traffic patterns and use these vehicles to run down pedestrians.

The deployment of civilian motorized vehicles by violent extremists is a classic case of asymmetrical warfare. Such attacks are unpredictable, locations for attacks cannot be easily predetermined, and law enforcement and other security agencies cannot monitor every purchased or rented vehicle. Stolen vehicles can be readily deployed against pedestrian targets in a very short period of time.

Motorized vehicle attacks also arguably illustrate the viability of the contagion effect, as indicated by the following incidents:

- October 20, 2014: St.-Jean-sur-Richelieu, Quebec, Canada, 1 killed, 1 injured

- July 14, 2016: Nice, France, 84 killed, over 200 injured

- November 28, 2016: Columbus, USA, 11 injured

- December 19, 2016: Berlin, Germany, 12 killed, 48 injured

- January 8, 2017: Jerusalem, Israel, 4 killed, 10 injured

- March 22, 2017: London, UK, 5 killed, many injured

- April 7, 2017: Stockholm, Sweden, 5 killed, 12 injured

- June 3, 2017: London, UK, 8 killed, 40 injured

- August 12, 2017: Charlottesville, USA, 1 killed, 19 injured

- August 17, 2017: Barcelona, Spain, 14 killed, over 100 injured

- October 31, 2017: New York City, USA, 8 killed, 12 injured

These attacks represent a deliberate asymmetrical tactic that is encouraged by extremist organizations. For example, in 2010, al-Qaeda published the following tactical advice in its online magazine:

> To achieve maximum carnage, you need to pick up as much speed as you can while still retaining good control of your vehicle in order to maximize your inertia and be able to strike as many people as possible in your first run.[15]

Chapter Perspective 8.4 further discusses the appeal of asymmetrical warfare.

Maximum Casualties and the New Terrorism

Terrorist violence is, at its core, symbolic in nature. With notable exceptions, methods and targets during the "Old Terrorism" era tended to be focused and relatively surgical, and they were modified to accommodate terrorists' definitions of who should be labeled as a championed group or as an enemy. However, the redefined morality of the New Terrorism opens the door for methods to include high-yield weapons and for targets to include large concentrations of victims. Symbolic targets and concentrations of enemy populations can now be hit much harder than in the past; all that is required is the will to do so.

Why would terrorists deliberately use high-yield weapons with the goal of maximizing casualties? What objectives would they seek? Depending on the group, many reasons have been suggested, including the following general objectives:

- *Attracting attention.* No one can ignore movements that carry out truly devastating attacks. This is the ultimate manifestation of armed propaganda and propaganda by the deed.

- *Pleasing God.* Divinely inspired terrorists seek to carry out what they believe to be a mandate from God. For example, Christian terrorists believing in the inevitability of the apocalypse might wish to hasten its arrival by using a weapon of mass destruction.

- *Damaging economies.* This could be accomplished by the contamination of food or other consumer products. A few poisoning events or other acts of consumer-focused sabotage could damage an economic sector.

- *Influencing enemies.* Terrorists may be moved to wield exotic weapons as a way to influence a large population. After using these weapons, their demands and grievances would receive serious scrutiny.[16]

In the era of the New Terrorism, terrorists may strike with the central objective of killing as many people as possible. They are not necessarily interested in overthrowing governments or changing policies as their primary objectives. Rather, their intent is simply to deliver a high body count and thereby terrorize and disrupt large audiences.

Netwar: A New Organizational Theory

The New Terrorism incorporates maximum flexibility into its organizational and communications design. Semiautonomous cells are either prepositioned around the globe as sleepers (such as the November 14, 2015, terrorists in France), or they travel to locations where an attack is to occur (such as the September 11, 2001, hijackers in the United States), or they are lone wolves inspired by a foreign cause (such as the November 5, 2009, mass shooter in Fort Hood, Texas). They communicate using new cyber and digital technologies. The netwar theory is an important concept in the new terrorist environment. It refers to "an emerging mode of conflict and crime . . . in which the protagonists use network forms of organization and related doctrines, strategies, and technologies attuned to the information age. These protagonists are likely to consist of dispersed small groups who communicate, coordinate, and conduct their campaigns in an internetted manner, without a precise central command."[17]

The new *internetted* movements have made a strategic decision to establish virtual linkages via the Internet and other technologies. They represent modern adaptations of the following organizational models:

- *Chain networks.* People, goods, or information move along a line of separated contacts, and end-to-end communication must travel through the intermediate nodes.

- *Star, hub, or wheel networks.* A set of actors is tied to a central node or actor and must go through that node to communicate and coordinate.

- *All-channel networks.* There is a collaborative network of small militant groups, and every group is connected to every other group.[18]

The Internet, Social Networking Media, and Other Technologies

The Internet, encryption applications, and social communication technologies are now used extensively by many terrorist groups and extremist movements. It is not uncommon for websites to be visually attractive, user friendly, and interactive. Music, photographs, videos, and written propaganda are easily posted on web pages or disseminated via social networking media. Encryption technologies allow extremists to communicate directly, including in encrypted group chats. Extremists evidently can also use online gaming technologies that allow groups to communicate while participating in an online game. Further discussion on cybersecurity is provided in Chapter 10.

Tweeting, texting, and other social networking media platforms are used to record incidents (often graphically) and tout claimed successes. The fundamental attraction of social networking media is that it affords the capability to send messages and images live as they occur. This is an advantage for media outlets or individuals who wish to be the first to break a story. Activists quickly adapted to this capability, first most notably during the Arab Spring in 2011 when protesters tweeted and texted videos and other information during antigovernment demonstrations. Extremists also took advantage of social networking media by recording and disseminating real-time images of fighting, executions, beheadings, and casualties.

netwar: An important concept in the new terrorist environment is the netwar theory, which refers to "an emerging mode of conflict and crime . . . in which the protagonists use network forms of organization and related doctrines, strategies, and technologies attuned to the information age. These protagonists are likely to consist of dispersed small groups who communicate, coordinate, and conduct their campaigns in an internetted manner, without a precise central command."

Many Internet postings portray the sense of a peaceful and rich culture of a downtrodden group. Graphic, gory, or otherwise moving images are skillfully posted, sometimes as photo essays that "loop" for continuous replay. Bloggers have posted links to hundreds of websites where viewers may obtain jihadist videos.[19] E-mail addresses, mailing addresses, membership applications, and other means to contact the movement are also common, so that a virtual world of like-minded extremists thrives on the Internet.[20] An example of the anonymity and scope of the Internet was the activities of a purported member of the Iraqi resistance who called himself Abu Maysara al-Iraqi. Al-Iraqi regularly posted alleged updates and communiqués about the Iraqi resistance on sympathetic Islamic websites. It proved to be very difficult to verify his authenticity or even whether he was (or they were) based in Iraq because he was accomplished in the skillful substitution of new online accounts.[21]

As a counterpoint to such online postings, organizations independently monitor extremist websites for their origin and content. For example, the Search for International Terrorist Entities Intelligence Group, also known as the SITE Intelligence Group, maintains a website dedicated to identifying Web postings by several extremist organizations and terrorist groups.

CASES IN POINT: THE AL-QAEDA NETWORK AND ISIS

The Al-Qaeda Network

The modern era's prototypical Islamist revolutionary organization is al-Qaeda ("The Base"), which seeks to unite Muslims throughout the world in a holy war. Founded by Saudi national Osama bin Laden, al-Qaeda is not a traditional hierarchical revolutionary organization, nor does it call for its followers to do much more than engage in terrorist violence in the name of the faith. Al-Qaeda is best described as a cell-based movement or a loose network of like-minded Sunni Islamist revolutionaries. Compared to other movements in the postwar era, it is a different kind of network because central al-Qaeda

- holds no territory,
- does not champion the aspirations of an ethnonational group,
- has no top-down organizational structure,
- has virtually nonexistent state sponsorship,
- promulgates political demands that are vague, and
- is completely religious in its worldview.

Chapter Perspective 8.5 discusses the ideology of al-Qaeda.

Al-Qaeda has inspired Sunni Islamic fundamentalist revolutionaries and terrorists in a number of countries. It became a significant source for the financing and training of thousands of *jihadis*. The network is essentially a nonstate catalyst for transnational religious radicalism and violence.

Experts do not know how many people count themselves as al-Qaeda operatives, but estimates range from 35,000 to 50,000. Of these, perhaps 5,000 received training in camps in Sudan and Afghanistan soon after the founding of the organization. Others are new recruits from around the Muslim world and Europe, and many others are veteran Afghan Arabs who fought in the jihad against the Soviets and later against the post–September 11, 2001, American-led coalition forces in Afghanistan. These numbers fluctuated markedly during

CHAPTER PERSPECTIVE 8.5

The Ideology of Al-Qaeda

Prior to his death in May 2011, Osama bin Laden designed al-Qaeda to symbolize the globalization of terrorism in the twenty-first century. The network is perceived by many to represent a quintessential model for small groups of like-minded revolutionaries who wish to wage transnational insurgencies against strong adversaries. Although al-Qaeda certainly exists as a loose network of relatively independent cells, it has also evolved into an idea—an ideology and a fighting strategy—that has been embraced by sympathetic revolutionaries throughout the world.

Al-Qaeda leaders such as the late Osama bin Laden and his successor as leader, Ayman al-Zawahiri, consistently released public pronouncements of their goals, often by delivering audio and video communiqués to international news agencies such as Al Jazeera in Qatar. Based on these communiqués, the following principles frame the ideology of al-Qaeda:[a]

- The struggle is a clash of civilizations. Holy war is a religious duty and necessary for the salvation of one's soul and the defense of the Muslim nation.

- Only two sides exist, and there is no middle ground in this apocalyptic conflict between Islam and the forces of evil. Western and Muslim nations that do not share al-Qaeda's vision of true Islam are enemies.

- Violence in a defensive war on behalf of Islam is the only course of action. There cannot be peace with the West.

- Because this is a just war, many of the theological and legal restrictions on the use of force by Muslims do not apply.

- Because U.S. and Western power is based on their economies, large-scale mass-casualty attacks that focus on economic targets are a primary goal.

- Muslim governments that cooperate with the West and do not adopt strict Islamic law are apostasies and must be violently overthrown.

- Israel is an illegitimate nation and must be destroyed.

These principles have become a rallying ideology for Islamist extremists who have few, if any, ties to al-Qaeda. Thus, the war on terrorism is not solely a conflict against an organization but is also a conflict against a belief system.

Discussion Questions

1. What is the ideology of al-Qaeda?

2. Why did a network of religious revolutionaries evolve into a potent symbol of global resistance against its enemies?

3. Which underlying commonalities appeal to motivated Islamist activists?

Note

a. Adapted from Office of the Coordinator of Counter Terrorism, *Country Reports on Terrorism 2009* (Washington, DC: U.S. Department of State, 2010).

---

the coalition campaign in Afghanistan because of the deaths or capture of many mid- and upper-level personnel, including the death of Osama bin Laden in 2011. However, with a presence in an estimated 50 to 60 countries, it is likely that new recruits will continue to join the al-Qaeda cause or other al-Qaeda-inspired causes.

Al-Qaeda's religious orientation is a reflection of Osama bin Laden's sectarian ideological point of view. Bin Laden's worldview was created by his exposure to Islam-motivated armed resistance. As a boy, he inherited between $20 million and $80 million from his father, with some estimates ranging as high as $300 million. When the Soviets invaded Afghanistan in 1979, bin Laden eventually joined with thousands of other non-Afghan Muslims who traveled to Peshawar, Pakistan, to prepare to wage jihad. However, his main contribution to the holy war was to solicit financial and matériel contributions from wealthy Arab sources. He apparently excelled at this. The final leg of his journey toward international Islamic terrorism occurred when he and thousands of other Afghan veterans—the Afghan Arabs—returned to their countries to carry on their struggle in the name of Islam. Beginning in 1986, bin Laden organized a training camp that grew in 1988 into the al-Qaeda group.

After the Gulf War, bin Laden and a reinvigorated al-Qaeda moved to its new home in Sudan for five years. It was there that the al-Qaeda network began to grow into a self-sustaining financial and training base for promulgating jihad.

When al-Qaeda moved to Afghanistan, its reputation as a financial and training center attracted many new recruits and led to the creation of a loose network of cells and "sleepers" in dozens of countries. Al-Qaeda also became an inspiration for Islamist insurgent groups such as the al-Nusra Front in Syria, the Haqqani Network in Pakistan and Afghanistan, and al-Shabaab in Somalia. It became closely allied with the Taliban, who provided sanctuary and other support for the group. Significantly, aboveground radical Islamist groups with links to al-Qaeda took root in some nations and overtly challenged authority through acts of terrorism. For example, regional insurgent groups include al-Qaeda in the Arabian Peninsula and al-Qaeda in the Islamic Maghreb (West Africa).

ISIS

In early 2014, the international community was surprised when a little-known Sunni insurgent movement overran significant swaths of territory in northern Iraq. The international community was further alerted when these insurgents seized major population centers in Iraq—particularly the cities of Mosul and Tikrit (Saddam Hussein's home city)—as the Iraqi army and other security forces were routed. The movement was the Islamic State of Iraq and the Levant, also known as the Islamic State of Iraq and al-Sham, or ISIS (in Arabic the acronym is *Daesh*).

ISIS was founded as the de facto successor to Abu Musab al-Zarqawi's Al-Qaeda Organization for Holy War in Iraq (AQI), which had waged an intensive Islamist insurgency from about 2005 to 2006 against U.S.-led occupation forces, the Iraqi army, and the Shi'a Badr Brigade. After al-Zarqawi's death in 2006, AQI renamed itself the Islamic State in Iraq. In April 2013, Islamic State in Iraq subsequently announced the formation of Islamic State of Iraq and the Levant (Islamic State of Iraq and al-Sham) during the Syrian civil war that began in the aftermath of the 2011 Arab Spring protests.

As internecine warfare escalated in Syria and Iraq, ISIS in effect declared that the new operational scope of the movement would transcend the borders of neighboring countries. Thus, the central tenets of ISIS are twofold: first, its refusal to recognize the borders of Syria, Iraq, and other nations; and second, waging war to achieve the avowed goal of establishing a renewed caliphate (Islamic state) transcending these borders and eventually encompassing the Muslim world. In June 2014, ISIS leader Abu Bakr al-Baghdadi announced the formation of an Islamic caliphate called the Islamic State, with himself at its head as the new caliph. The new caliphate was to be governed in accordance with a harsh fundamentalist interpretation of Islamic *shari'a* law. These events and declarations brought the movement into political opposition with al-Qaeda's central leadership, in particular leader Ayman al-Zawahiri, who disavowed ISIS when the group refused to limit its operations only to Syria. This disavowal did nothing to diminish ISIS's operations on the battlefield.

ISIS adopted brutal tactics from its inception, especially in the manner in which it governed captured territory and how it prosecuted its war. The movement regularly executed captured soldiers and police officers, imprisoned and tortured civilians, kidnapped and executed Western civilians, and imposed draconian *shari'a* law and order in the territory it occupied. ISIS routinely recorded and broadcast beheadings, crucifixions, prisoner burnings, and massacres via social networking media. The group engaged in extreme repression and ethnic/religious cleansing of Christians, Yazidis, and Shi'a Muslims in areas they occupied. ISIS formally instituted legalized enslavement of women and children captured in their campaigns, selling and abusing victims under self-instituted slavery laws and regulations.

As knowledge of ISIS's tactics spread, thousands of members of the Iraqi army and security forces literally shed their uniforms and abandoned weapons when relatively small numbers of ISIS fighters advanced during the initial offensive in 2014. Equipment and weapons captured by ISIS allowed the group to wage conventional warfare against Iraqi and Syrian opponents, thus consolidating territory under its control. This began to be reversed in 2016 and 2017, when counteroffensives in Iraq and Syria recaptured ISIS-occupied cities and territory during heavy combat. By 2019, territory previously occupied by ISIS had been overrun.

ISIS successfully inspired thousands of international fighters to join the movement in Syria and Iraq, many of whom were volunteers from North Africa, Chechnya, Europe, central Asia, and the United States. A stated goal by ISIS was to have these volunteers return to their home countries to wage jihad domestically. The group also embedded operatives among refugees migrating to Western countries. In conjunction with these stratagems, ISIS encouraged fellow believers in the West and elsewhere to carry out attacks in their host countries. In this regard, ISIS claimed responsibility for the following incidents:

- In January 2015, an ISIS lone-wolf sympathizer opened fire at a Jewish market in Vicennes, France, killing five people, including a policewoman.

- In November 2015, two ISIS suicide bombers killed 43 people and wounded 239 in Beirut, Lebanon.

- In November 2015, ISIS operatives attacked several sites in Paris, France, using firearms and explosives, killing approximately 130 people and wounding 350 others.

- In December 2015, two ISIS sympathizers shot and killed 14 people and wounded 22 in San Bernardino, California, in the United States.

- In March 2016, three ISIS suicide bombers killed 32 people and wounded more than 300 in Brussels, Belgium.

- In June 2016, an ISIS lone wolf shot and killed 49 people and wounded 53 in Orlando, Florida, in the United States.

- In December 2016, an ISIS operative drove a truck through a Christmas market in Berlin, Germany, killing 12 and injuring 56.

Like al-Qaeda, the ISIS "brand" inspired other regional insurgent groups to profess allegiance to the ISIS cause of creating a pan-Islamic state. Boko Haram in northeastern Nigeria and bordering regions declared allegiance in 2015. Another interesting case is that of the Abu Sayyaf Group in the Philippines. Abu Sayyaf for some time claimed affiliation with al-Qaeda and later professed allegiance to ISIS.

CASE IN POINT: ASYMMETRICAL WARFARE IN PARIS AND BRUSSELS

The Paris Attacks

In 2015, Paris, France, was the site of two significant terrorist incidents carried out by Islamist extremists. From January 7 to 9, 2015, several incidents occurred. On January 7, two brothers wielding AK-47 assault rifles opened fire inside the offices of Charlie Hebdo magazine, a popular French satirical publication. Charlie Hebdo had published satirical cartoons of Islam and other religions, including depictions of the Prophet Muhammed, and the assault was conducted in retaliation for these depictions. The gunmen killed 12 people, including eight cartoonists and journalists and two police officers. The next day, a police officer was killed outside Paris, and on January 9, the same shooter took four people hostage at a Jewish kosher market, eventually killing four hostages before being killed by French police. The kosher market shooter was an acquaintance of the two Charlie Hebdo attackers, who were killed on January 9 by French police.

▶ **Photo 8.4** Death in Kenya. A Kenyan man is lifted out of the rubble of the U.S. embassy in Nairobi, Kenya. On August 7, 1998, the U.S. embassy in Nairobi was bombed by an al-Qaeda cell, which coordinated the assault with an attack conducted by another cell against the U.S. embassy in Dar es Salaam, Tanzania.

On the night of November 13, 2015, several gunmen methodically attacked several sites in the heart of Paris. The assault began when a suicide explosion occurred near the Stade de France, where a Germany–France soccer match was being held. French President François Hollande was in attendance at the match. Elsewhere, approximately 15 people were killed at a bar and a restaurant when men opened fire from an automobile with AK-47 assault rifles. A second suicide explosion occurred outside the Stade de France, and nearly simultaneously, more people were killed in a separate attack from an automobile at another bar in Paris. A few minutes later, at least 19 people were killed at a café by gunfire from an automobile. A third suicide explosion occurred at another restaurant, and nearly simultaneously, three armed men opened fire and seized hostages during a concert at the Bataclan concert hall. The Bataclan attackers calmly and methodically executed scores of patrons with gunfire and hand grenades. As this was occurring at the concert hall, a fourth suicide bomb was detonated at the Stade de France—the third at the stadium. French police successfully stormed the Bataclan concert hall and killed all of the assailants but not before they had executed more than 100 patrons.

On November 18, 2015, French police and soldiers killed Abdelhamid Abaaoud, the alleged mastermind of the attack, during a raid conducted on a fortified dwelling in the Saint Denis neighborhood. A fifth suicide bomb was detonated during the raid.

The Brussels Attacks

On March 22, 2016, three suicide bombs were detonated in Brussels, Belgium—two in Zaventem at the Brussels Airport, and one on a train at the Maalbeek metro station near the European Commission headquarters. A fourth bomb had been placed at the airport, but failed to detonate. Thirty-two victims died, more than 300 were wounded, and three terrorists were killed. Islamic State of Iraq and the Levant declared they were responsible for the attacks. At the time, the assault was the most lethal terrorist attack in Belgian history.

Belgium's status in the international terrorist environment represents an instructive case in point, especially regarding its vulnerability prior to the attack. The Belgian Muslim community is mostly concentrated in Brussels, and represents more than 20% of the city's population. There exists some pro-jihadist sentiment among members of the community, as evidenced by several terrorist attacks and plots in the early 2000s and immediately prior to the March 2016 attack. Significantly, as a proportion of its population, more Belgian nationals traveled abroad to join violent Islamist movements such as the self-proclaimed Islamic State than volunteers from other Western countries. Within this environment, the Belgian security community was notably weak because of the existence of independent intelligence agencies with competing policy priorities. This resulted in uncoordinated contingency planning, inefficient investigative practices, and overall difficulty in countering potential security threats.

In the immediate aftermath of the November 2015 Paris attacks, investigators determined that a clear affiliation existed between the assailants in Brussels and Paris. In fact, members of the cell that carried out the attack in Paris were based in the Molenbeek suburb of Brussels. In the Schaerbeek district of Brussels, investigators later found a "bomb factory" in a residence used by the cell. The cell was apparently initially activated and tasked to conduct several attacks in Paris. However, their plans were revised because of the arrest of an accomplice on March 18, 2016, in Brussels. The arrest motivated members of the cell to carry out alternate strikes in Brussels, resulting in the bombings on March 22, 2016.

POLICY OPTIONS

Proactive policies specifically seeking to eliminate terrorist groups and environments form the definitional foundation for **counterterrorism**. The ultimate goal of counterterrorism is clear regardless of which policy is selected: to save lives by proactively preventing or decreasing the number of terrorist attacks. As a corollary, **antiterrorism** refers to enhanced security, target hardening, and other defensive measures seeking to prevent or deter terrorist attacks.

Counterterrorist experts in the modern era are required to concentrate on achieving several traditional counterterrorist objectives. These include the following:

- Disrupting and preventing terrorist conspiracies from operationalizing their plans

- Deterring would-be terrorist cadres from crossing the line between extremist activism and political violence

- Implementing formal and informal international treaties, laws, and task forces to create a cooperative counterterrorist environment

- Minimizing physical destruction and human casualties

Countering Extremism

Policymakers generally grasp the limitations of exclusive reliance on coercive counterterrorist methods to counter extremist ideologies and appreciate the necessity to develop alternative measures. Operations other than war include conciliatory options, if feasible. These options may provide long-term solutions to future extremism. Past reliance on conciliatory options, such as peace processes, negotiations, and social reforms, had some success in resolving both immediate and long-standing terrorist crises. If skillfully applied, adaptations of these options could present extremists with options other than political violence. In the past, these options were usually undertaken with the presumption that some degree of coercion would be kept available should the conciliatory options fail; this is a pragmatic consideration.

counterterrorism: Proactive policies that specifically seek to eliminate terrorist environments and groups. The ultimate goal of counterterrorism is to save lives by proactively preventing or decreasing the number of terrorist attacks.

antiterrorism: Target hardening, enhanced security, and other defensive measures seeking to deter or prevent terrorist attacks.

Countering the New Terrorism

Counterterrorism must adapt to the fact that the New Terrorism reflects the ability of extremists to operate within emerging political environments. During this adaptation process, it is important to remember that "[counterterrorist] instruments are complementary, and the value of using them should be—and generally [is]—more than just the sum of the parts. If the process is not properly managed, the value may be less than the sum of the parts, because of the possibility of different instruments working at cross purposes."[22]

Thus, as some terrorist environments continue to be characterized by the New Terrorism, models must be flexible enough to respond to new environments and must avoid stubborn reliance on methods that "fight the last war." This reality is particularly pertinent to homeland security and the war against terrorism. Unlike previous wars, the new war has been declared against *behavior* as much as against terrorist groups and revolutionary cadres. The fronts in the new war are amorphous and include homeland security measures, covert *shadow wars*, counterterrorist financial operations, global surveillance of communications technologies, and identification and disruption of transnational terrorist cells and support networks.

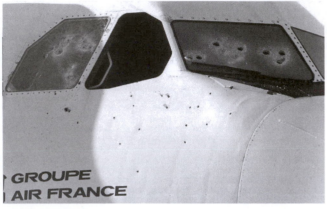

Georges GOBE/AFP/Getty Images

▶ **Photo 8.5** Air France Flight 8969 after the rescue of hostages by elite French GIGN commandos.

- *Homeland security measures.* These are required for hardening targets, deterring attacks, and thwarting conspiracies. Internal security requires the extensive use of nonmilitary security personnel, such as customs officials, law enforcement agencies, and immigration authorities.

- *Covert shadow wars.* These are fought outside of public scrutiny using unconventional methods. Shadow wars require the deployment of military, paramilitary, and coercive covert assets to far regions of the world.

- *Counterterrorist financial operations.* These operations are directed against bank accounts, private foundations, businesses, and other potential sources of revenue for terrorist networks. Intelligence agencies can certainly hack into financial databases, but a broad-based coalition of government and private financial institutions is necessary for this task.

- *Global surveillance of communications technologies.* This requires surveillance of technologies including telephones, cell phones, and e-mail. Agencies specializing in electronic surveillance, such as the U.S. National Security Agency, are the institutions most capable of carrying out this mission.

- *Identification and disruption of transnational terrorist cells and support networks.* This requires international cooperation to track extremist operatives and "connect the dots" on a global scale. Primary responsibility for this task lies with intelligence communities and law enforcement agencies.

The foregoing new fronts in the new war clearly highlight the need to continuously upgrade physical, organizational, and operational counterterrorist measures; flexibility and creativity are essential. Failure to do so is likely to hinder adaptation to the dynamic terrorist environment of the 2000s. Thus, for example, the inability to control and redress long-standing bureaucratic and international rivalries could be disastrous in the new environment.

GLOBAL PERSPECTIVE: DIGITAL, VIDEO, AND AUDIO TERRORISM

This chapter's Global Perspective explores a tactic adopted by insurgent groups and hostage takers of recording their victims and promulgating their images in the mass media, on social networking media, and on the Internet.

GLOBAL PERSPECTIVE

TACTICAL HORROR: DIGITAL, VIDEO, AND AUDIO TERRORISM

With the advent of the Internet, cable news networks, and social media, terrorists now possess unprecedented access to global audiences. Communications technologies quickly and cheaply bring symbolic images and extremist messages to the attention of policymakers and civilians around the world. Terrorists have adapted their tactics to these new technologies, and many utilize them to broadcast their messages and operations.

During the early millennium, hostage takers discovered that the plight of their victims would garner intensive global attention so long as their images were promulgated to noteworthy cable news networks.

The typical pattern was for an international figure—often a foreign worker—to be seized by extremists, followed by a communiqué claiming credit for the abduction. A video or series of videos would be delivered to a news outlet, with images of the victim pleading for his or her life while seated before a flag and surrounded by hooded and armed terrorists. The outcome was sometimes satisfactory, with the hostage being granted freedom. At other times, the video incidents ended horrifically.

The first noted incident was the kidnapping and videotaped murder of American journalist Daniel Pearl in Pakistan in January 2002. Since then, Islamist insurgents in Iraq and Syria, terrorists in Saudi Arabia and Libya, and violent jihadists elsewhere have either issued Internet, cable news, or social media communiqués; videotaped their hostages; executed them; or committed all of these

actions. After the Daniel Pearl murder, a gruesome cycle of beheadings occurred, as illustrated by the following incidents from Iraq:

- Victims represent the international community and have included citizens from Bulgaria, Pakistan, South Korea, Nepal, Norway, the United States, Great Britain, Turkey, and Iraq.

- Al-Qaeda in Iraq and other Islamist or other sectarian movements appeared to be responsible for most of the kidnappings and murders.

- A number of hostages were beheaded, sometimes on video recordings that were posted on the Internet.

In the aftermath of the initial cycle of broadcasts, subsequent cycles of media-oriented terror have included social media images of killings and other incidents, such as mass executions of prisoners by the Islamic State of Iraq and the Levant (ISIS). ISIS became very adept at posting skillfully made propaganda videos on sympathetic websites. The videos portray heroic fighters marching into battle, parading in captured territory, and proclaiming their cause in speeches to listeners. This is balanced against broadcasts of executions, including beheadings and immolations.

CHAPTER SUMMARY

This chapter discussed the threat from the New Terrorism, including its objectives, methods, and targets. In the era of the New Terrorism, objectives have become characterized by vagueness, and methods have included indiscriminate attacks, intentional mass casualties, and the use of high-yield weapons. Modern terrorist methods reflect the changing global political environment and are characterized by asymmetrical warfare and new cell-based organizational models. Sophisticated use of the Internet and social networking media serve to spread extremist messages and images.

Countering the New Terrorism requires adapting to new fronts in the war against terrorism. These new fronts require creative use of homeland security measures, intelligence, and overt and covert operations, as well as cooperation among counterterrorist agencies. Disruption of terrorist financial operations and communications are among the most important and challenging priorities in the new war.

DISCUSSION BOX

The One True Faith

This chapter's Discussion Box is intended to stimulate critical debate about faith-motivated terrorism within major religions.

Most religious traditions have produced extremist movements whose members believe that their faith and value system is superior to other beliefs. This concept of the "one true faith" has been used by many fundamentalists to justify violent intolerance on behalf of their religion. Religious terrorists are modern manifestations of historical traditions of extremism within the world's major faiths.

Modern religious extremism is arguably rooted in faith-based natural law. Natural law is a philosophical "higher law" that is theoretically discoverable through human reason and references to moral traditions and religious texts. In fact, most religious texts have passages that can be selectively interpreted to encourage extremist intolerance. To religious extremists, it is God's law that has been revealed to and properly interpreted by the extremist movement.

In the United States, government response to the problem of religious extremism is fraught with politically and socially sensitive issues. This is because the free exercise of religion is protected by the First Amendment to the U.S. Constitution. Nevertheless, local law enforcement agencies, such as the New York Police Department, and federal agencies, such as the FBI, have engaged in monitoring activities directed against some religious institutions.

Discussion Questions

1. Is faith-motivated activism a constructive force for change?

2. At what point does the character of faith-motivated activism become extremist and terrorist?

3. Does faith-based natural law justify acts of violence?

4. Why do religious traditions that supposedly promote peace, justice, and rewards for spiritual devotion have so many followers who piously engage in violence, repression, and intolerance?

5. What, if any, policies should be adopted by law enforcement agencies to monitor religious institutions?

KEY TERMS AND CONCEPTS

The following topics were discussed in this chapter and can be found in the glossary:

ON YOUR OWN

Get the tools you need to sharpen your study skills. SAGE edge offers a robust online environment featuring an impressive array of free tools and resources.

Access practice quizzes, eFlashcards, video, and multimedia at **edge.sagepub.com/martinhs3e**

RECOMMENDED WEBSITES

The following websites provide general information about extremism, historical terrorism, and the New Terrorism. Included are links to dissident revolutionary organizations and movements.

al-Fatah: www.microsofttranslator.com/bv.aspx?ref=SERP&br=ro&mkt=en-US&dl=en&lp=AR_EN&a=http%3a%2f%2fwww.fateh.ps%2f

al-Qaeda "Training Manual": www.justice.gov/ag/manualpart1_1.pdf

CIA World Factbook: www.cia.gov/library/publications/the-world-factbook/index.html

Foreign Terrorist Organizations: www.state.gov/j/ct/rls/other/des/123085.htm

Hezbollah (Islamic Resistance): www.microsofttranslator.com/bv.aspx?ref=SERP&br=ro&mkt=en-US&dl=en&lp=AR_EN&a=http%3a%2f%2fwww.moqawama.org%2f

Intifada.com: www.intifada.com

Iraq Coalition Casualty Count (iCasualties.org): icasualties.org

Irish Northern Aid (Noraid): inac.org

Islamic Propagation Organization: www.al-islam.org/short/jihad

Muslim Brotherhood Movement: www.ikhwanweb.com

Naval Postgraduate School (Terrorism): www.nps.edu/Library/Research%20Tools/Subject%20Guides%20by%20Topic/Special%20Topics/Terrorism/TerrorismSubjectGuideEssay.html

Radio Islam: www.radioislam.org

RAND Corporation: www.rand.org

Timeline of Terrorism: www.timelineofterrorism.com

WEB EXERCISE

Using this chapter's recommended websites, conduct an online investigation of the New Terrorism and terrorist objectives, methods, and targets.

1. What common patterns of behavior and methods can you identify across regions and movements?

2. Conduct a search for other websites that offer advice on organizing terrorist cells and carrying out terrorist attacks. Do you think that the online terrorist manuals and weapons advice are a danger to global society?

3. Compare the websites for the monitoring organizations. How would you describe the quality of their information? Are they providing a useful service?

To conduct an online search on the New Terrorism, activate the search engine on your Web browser and enter the following keywords:

"New Terrorism"

"Religious terrorism"

RECOMMENDED READINGS

The following publications provide discussions on the New Terrorism:

Bergen, Peter L. 2001. *Holy War, Inc.: Inside the Secret World of Osama bin Laden*. New York: Simon & Schuster.

Berko, Anat. 2009. *The Path to Paradise: The Inner World of Suicide Bombers*. Westport, CT: Praeger Security International.

Cragin, Kim, Peter Chalk, Sara A. Daly, and Brian A. Jackson. 2007. *Sharing the Dragon's Teeth: Terrorist Groups and the Exchange of New Technologies*. Santa Monica, CA: RAND.

Dolnik, Adam. 2007. *Understanding Terrorist Innovation: Technology, Tactics and Global Trends*. London: Routledge.

Frantz, Douglas and Catherine Collis. 2007. *The Nuclear Jihadist: The True Story of the Man Who Sold the World's Most Dangerous Secrets—And How We Could Have Stopped Him*. New York: Twelve Books.

Gerges, Fawaz A. 2009. *The Far Enemy: Why Jihad Went Global*. Cambridge, UK: Cambridge University Press.

Glucklich, Ariel. 2009. *Dying for Heaven: Holy Pleasure and Suicide Bombers*. New York: HarperOne.

Gunaratna, Rohan. 2002. *Inside Al Qaeda: Global Network of Terror*. New York: Columbia University Press.

Hafez, Mohammed M. 2007. *Suicide Bombers in Iraq: The Strategy and Ideology of Martyrdom*. Washington, DC: United States Institute of Peace Press.

Katz, Samuel M. 2001. *The Hunt for the Engineer: How Israeli Agents Tracked the Hamas Master Bomber*. New York: Fromm International.

Kegley, Charles W., Jr., ed. 2002. *The New Global Terrorism: Characteristics, Causes, Controls*. New York: Prentice Hall.

Levi, Michael. 2009. *On Nuclear Terrorism*. Cambridge, MA: Harvard University Press.

Lutz, James M. and Brenda J. Lutz. 2008. *Global Terrorism*. 2nd ed. New York: Routledge.

Mueller, Robert. 2009. *Atomic Obsession: Nuclear Alarmism From Hiroshima to Al Qaeda*. New York: Routledge.

Oliver, Anne Marie and Paul F. Steinberg. 2005. *The Road to Martyrs' Square: A Journey Into the World of the Suicide Bomber*. New York: Oxford University Press.

Pape, Robert Anthony. 2005. *Dying to Win: The Strategic Logic of Suicide Terrorism*. New York: Random House.

Powell, William. 1971. *The Anarchist Cookbook*. New York: Lyle Stuart.

Ranstorp, Magnus and Magnus Normark, eds. 2009. *Unconventional Weapons and International Terrorism: Challenges and a New Approach*. New York: Routledge.

Sageman, Marc. 2008. *Leaderless Jihad: Terror Networks in the Twenty-First Century*. Philadelphia: University of Pennsylvania Press.

Scheuer, Michael. 2011 *Osama bin Laden*. New York: Oxford University Press.

Siniver, Asaf, ed. 2010. *International Terrorism Post-9/11: Comparative Dynamics and Responses*. London: Routledge.

Stern, Jessica. 1999. *The Ultimate Terrorists*. Cambridge, MA: Harvard University Press.

Thornton, Rod. 2007. *Asymmetric Warfare: Threat and Response in the Twenty-First Century*. Cambridge, UK: Polity Press.

Tucker, Jonathan B., ed. 2000. *Toxic Terror: Assessing Terrorist Use of Chemical and Biological Weapons*. Cambridge, MA: MIT Press.

Zubay, Geoffrey et al., eds. 2005. *Agents of Bioterrorism: Pathogens and Their Weaponization*. New York: Columbia University Press.

CHAPTER NINE

THE THREAT AT HOME
Terrorism in the United States

Opening Viewpoint: Lynching— Vigilante Communal Terrorism in the United States

Lynchings were public communal killings. On most occasions, they were racially motivated hangings or burnings of African American males. Lynch mobs would typically abduct the victim, drag him to the place of execution, physically abuse him (often gruesomely), and then publicly kill him. Lynchings exhibited the following profile:

- White mobs
- Killings of African Americans (usually men) and others
- Physical abuse, including torture, mutilations, and the taking of "souvenirs" from the corpses (bones, toes, etc.)
- Symbolic protection of the white community
- Symbolic warnings to the African American community

Photography was commonly used to record lynchings, and it was not uncommon for members of lynch mobs to pose proudly next to the corpses. This is significant because the use of the camera to memorialize lynchings testified to their openness and to the self-righteousness that animated the participants. Not only did photographers capture the execution itself but also the carnival-like atmosphere and the expectant mood of the crowd.[a]

The term *lynching* comes from Charles Lynch, a colonial-era Virginia farmer who, during the American Revolution, acted as a judge who hanged outlaws and pro-British colonials (Tories). From 1882 to 1968, nearly 5,000 African Americans are known to have been lynched. Some had been accused of crimes, but most were simply innocent sacrificial victims.

Note

a. Leon F. Litwack, "Hellhounds," in *Without Sanctuary: Lynching Photography in America*, ed. James Allen, Hilton Als, John Lewis, and Leon Litwack (Santa Fe, NM: Twin Palms, 2000), 10–11.

Chapter Learning Objectives

This chapter will enable readers to do the following:

1. Describe extremist ideologies and behavior in the United States

2. Analyze differences between left-wing and right-wing extremist movements

3. Analyze domestic terrorism in the United States

4. Discuss international terrorism in the United States

5. Evaluate lone-wolf terrorism in the United States

U nlike many terrorist environments elsewhere in the world, where the designations of *left* and *right* are not always applicable, most political violence in the United States falls within these designations. Even nationalist and religious sources of domestic political violence have tended to reflect the attributes of U.S. leftist or rightist movements. It is only when we look at the international sources of political violence that the left and right designations begin to lose their precision in the United States. Table 9.1 shows groups responsible for terrorist incidents in the United States, from September 12, 2001, to the most recent data available from 2017.

Table 9.1	Groups Responsible for Most Terrorist Attacks in the United States, 2001–2017			
Rank	Organization	Number of Attacks	Number of Fatalities	Number of Injured
1	Unknown	145	29	209
2	Earth Liberation Front (ELF)	38	0	0
3	Antigovernment extremists	31	72	888
4	Jihadi-inspired extremists	30	106	186
5	Animal Liberation Front (ALF)	29	0	2
6	Anti-Muslim extremists	19	3	2
7	Anti-abortion extremists	17	4	9
8	White extremists	13	24	8
8	Muslim extremists	13	15	295
9	Anti-police extremists	7	12	14
10	Anti-white extremists	6	10	10

Source: National Consortium for the Study of Terrorism and Responses to Terrorism (2018).

Within American culture, mainstream values include free enterprise, freedom of speech, and limited government. Depending on one's ideological perspective, the interpretation of these mainstream values can be very different. Consider the following examples:

- Free enterprise may be viewed with suspicion by the far left but considered sacrosanct (untouchable) by the far right.

- Freedom of speech would seem to be a noncontroversial issue, but the right and left disagree about which kinds of speech should be protected and which should be regulated.

- The role of government is a debate dating back to the time of the American Revolution. The right and left disagree about the degree to which government should have a role in regulating private life.

▶ **Photo 9.1** Communal terrorism in America. The lynchings of Tommy Shipp and Abe Smith in Marion, Indiana, on August 7, 1930. The crowd is in a festive mood, including the young couple holding hands in the foreground.

Lawrence Beitler/by Hulton Archive/Getty Images

Also, mainstream American values of past generations, such as Manifest Destiny and racial segregation, have been rejected by later generations as unacceptable extremist ideologies. Thus, conceptualizing the political left, center, and right will shift during changes in political and social culture.

The discussion in this chapter will review the following:

- Extremism in America

- Left-wing terrorism in the United States

- Right-wing terrorism in the United States

- The New Terrorism in the United States

- Lone-wolf terrorism in the United States

EXTREMISM IN AMERICA

Domestic terrorism in the United States is rooted in extremist ideologies and behaviors emanating from the political left and right and from international sources. These attributes reflect political conditions unique to the United States, including American policies that are opposed from abroad by international actors. The notion of violent extremism growing from American political interaction is often misunderstood. To facilitate your appreciation of the unique qualities of the American case, it is instructive to briefly survey the American left, the American right, and international terrorism in the United States. Table 9.2 compares the championed groups, methodologies, and desired outcomes of typical political environments on the left, at the center, and on the right. Chapter Perspective 9.1 summarizes what is meant by *extremist* ideology and behavior.

Table 9.2 Typical Political Environments

Activism on the left, right, and center can be distinguished by a number of characteristics. A comparison of these attributes is instructive. The representation here compares their championed groups, methodologies, and desired outcomes.

	Left Fringe	Far Left	Liberalism	Moderate Center	Conservatism	Far Right	Fringe Right
Championed groups	Class/nationality	Class/nationality	Demographic groups	General society	General society	Race, ethnicity, nationality, and religion	Race, ethnicity, nationality, and religion
Methodology/process	Liberation movement	Political agitation	Partisan democratic processes	Consensus	Partisan democratic processes	Political agitation	"Order" movement
Desired outcome	Radical change	Radical change	Incremental reform	Status quo or slow change	Traditional values	Reactionary change	Reactionary change

CHAPTER PERSPECTIVE 9.1

Understanding Extremism

Extremism is a radical expression of one's political values. Both the *content* of one's beliefs and the *style* in which one expresses those beliefs are basic elements for defining extremism. Thus, a fundamental definitional issue for extremism is *how* one expresses an idea, in addition to the question of *which* belief one acts upon. It is characterized by what a person's beliefs are as well as how a person expresses his or her beliefs. Both elements—style and content—are important for an investigation of fringe beliefs and terrorist behavior.

Thus, no matter how offensive or reprehensible one's thoughts or words are, they are not by themselves acts of terrorism. Only those who violently act out their extremist beliefs are terrorists. Extremists who cross the line to become terrorists always develop noble arguments to rationalize and justify acts of violence directed against enemy nations, people, religions, or other interests.

In essence, extremism is a precursor to terrorism; it is an overarching belief system terrorists use to justify their violent behavior. It is characterized by intolerance toward opposing interests and divergent opinions, and it is the primary catalyst and motivation for terrorist behavior.

Left-Wing Extremism in the United States

The American left traditionally refers to political trends and movements that emphasize group rights. Several trends characterize the American left: labor activism, people's rights movements, single-issue movements, and antitraditionalist cultural experimentation. Examples include the following:

- *Labor activism.* Historically, labor activism and organizing have promoted ideals frequently found on the left. The labor movement of the late nineteenth and early twentieth centuries was highly confrontational, with violence emanating from management, the unions, and the state. Socialist labor activists such as Samuel Gompers were quite active in organizing workers. However, the mainstream American labor movement was distinctive, in comparison with European labor movements, in that the dominant labor unions generally rejected Marxist and other socialistic economic ideologies.[1]

- *People's rights*. There have been a number of people's rights movements on the American left. In the modern era, activism on the left has generally promoted the interests of groups that have historically experienced discrimination or a lack of opportunity. Examples of people's rights movements include the civil rights, women's rights, Black Power, gay rights, and New Left movements.

- *Single issue*. Single-issue movements such as the environmentalist and peace movements have also been common on the left.

- *Questioning traditions*. One facet of the left has been a tendency toward antitraditionalist cultural trends. Manifestations of this tendency have included experimentation with alternative lifestyles and the promotion of countercultural directions such as drug legalization.[2]

On the far and fringe left, one finds elements of Marxist ideology and left-wing nationalist principles. Terrorist violence from the left has usually been ideological or ethnonationalist in nature. It has typically been carried out by covert underground organizations or cells that link themselves (at least ideologically) to leftist rights movements. Although there have been human casualties as a direct result of leftist terrorism, most violence has been directed at nonhuman symbols, such as unoccupied businesses, banks, or government buildings. Law enforcement officers have also occasionally been targeted, usually by ethnonationalist terrorists. The heyday of leftist terrorism in the United States was from the late 1960s to the mid-1980s.

In sum, *left-wing extremism* is future oriented, seeking to reform or destroy an existing system prior to building a new and just society. To the extent that leftists champion a special group, it is usually one that is perceived to be unjustly oppressed by a corrupt system or government. This group is commonly a class or ethnonational category that, in the leftists' belief system, must receive the justice and equality that has been denied it. In championing their cause, leftists believe that either reform of the system or revolution is needed to build a just society. In this sense, left-wing extremism is *idealistic*.

Right-Wing Extremism in the United States

The American right traditionally encompasses political trends and movements that emphasize conventional and nostalgic principles. On the mainstream right, traditional values are emphasized. Examples include family values, educational content, and social order ("law and order") politics. It is also common on the American right (unlike in the European and Latin American rightist movements) to find an infusion of fundamentalist or evangelical religious principles.

On the far and fringe right, one finds that racial, mystical, and conspiracy theories abound; one also finds a great deal of antigovernment sentiment, with some fringe extremists opting to separate themselves from mainstream society. Terrorist violence has usually been racial, religious, or antigovernment in nature. With few exceptions, terrorism from the right has been conducted by self-isolated groups, cells, or individual lone wolves. Unlike most leftist attacks, many of the right's targets have intentionally included people and occupied symbolic buildings. Most ethnocentric **hate crimes**—regardless of whether one considers them to be acts of terrorism or aggravated crimes—come from the far and fringe right wing. This type of ethnocentric violence has a long history in the United States. "Since the middle of the nineteenth century, the United States has witnessed several episodic waves of xenophobia. At various times, Catholics, Mormons, Freemasons, Jews, blacks, and Communists have been targets of groups . . . seeking to defend 'American' ideals and values."[3] Chapter Perspective 9.2 discusses the differences between hate crimes and acts of terrorism.

CHAPTER PERSPECTIVE 9.2

Are Hate Crimes Acts of Terrorism?

Hate crimes refer to behaviors that are considered to be bias-motivated crimes but that, at times, seem to fit the definition of acts of terrorism. Hate crimes are a legalistic concept in the United States that embody (in the law) a criminological approach to a specific kind of deviant behavior. These laws focus on a specific motive for criminal behavior—crimes that are directed against protected classes of people because of their membership in these protected classes. Thus, hate crimes are officially considered to be a law enforcement issue rather than one of national security.

The separation between hate crimes and terrorism is not always clear because "hate groups at times in their life cycles might resemble gangs and at other times paramilitary organizations or terrorist groups."[a] They represent "another example of small, intense groups that sometimes resort to violence to achieve their goals by committing . . . vigilante terrorism."[b] Among experts, the debate about what is or is not *terrorism* has resulted in a large number of official and unofficial definitions. A similar debate has arisen about how to define hate crimes because "it is difficult to construct an exhaustive definition of the term. . . . Crime—hate crime included—is relative."[c] In fact, there is no agreement on what label to use for behaviors that many people commonly refer to as *hate crimes*. For example, in the United States, attacks by white neo-Nazi youths against African Americans, gays, and religious institutions have been referred to with such diverse terms as *hate crime, hate-motivated crime, bias crime, bias-motivated crime,* and *ethno-violence.*[d]

Are hate crimes acts of terrorism? The answer is that not all acts of terrorism are hate crimes, and not all hate crimes are acts of terrorism. For example, dissident terrorists frequently target a state or system with little or no animus against a particular race, religion, or other group. Likewise, state terrorism is often motivated by a perceived need to preserve or reestablish the state's defined vision of social order without targeting a race, religion, or other group. On the other hand, criminal behavior fitting federal or state definitions of hate crimes in the United States can have little or no identifiable political agenda other than hatred toward a protected class of people.

It is when *political* violence is directed against a particular group—such as a race, religion, nationality, or generalized "undesirable"— that these acts possibly fit the definitions of both hate crimes and terrorism. Terrorists often launch attacks against people who symbolize the cause that they oppose. In the United Kingdom, Germany, the United States, and elsewhere, many individuals and groups act out violently to promote an agenda that seeks to "purify" society. These crimes are committed by groups or individuals who are "dealing in the artificial currency of . . . 'imagined communities'—utopian pipe dreams and idealizations of ethnically cleansed communities."[e]

Notes

a. Steven E. Barkan and Lynne L. Snowden, *Collective Violence* (Boston: Allyn & Bacon, 2001), 105.

b. Ibid., 106.

c. Barbara Perry, *In the Name of Hate: Understanding Hate Crimes* (New York: Routledge, 2001), 8.

d. Mark S. Hamm, "Conceptualizing Hate Crime in a Global Context," in *Hate Crime: International Perspectives on Causes and Control,* ed. Mark S. Hamm (Cincinnati, OH: Anderson, 1994), 174.

e. Robert J. Kelly and Jess Maghan, *Hate Crime: The Global Politics of Polarization* (Carbondale: Southern Illinois University Press, 1998), 6.

Right-wing terrorism has occurred within different political and social contexts, from Ku Klux Klan violence during the civil rights movement of the 1950s and 1960s to neo-Nazi violence in the 1980s to antigovernment and **single-issue terrorism** in the 1990s.

In sum, *right-wing extremism* is generally a reaction against perceived threats to a group's value system, its presumption of superiority, or its sense of specialness. Rightists often try to preserve their value system and special status by aggressively asserting this claimed status. They frequently desire to return to a time of past glory, which, in their belief system, has been lost or usurped by an enemy group or culture. In this sense, right-wing extremism is *nostalgic*.

single-issue terrorism: Terrorism that is motivated by a single grievance.

Sources of International Terrorism in the United States

International terrorism in the United States has historically included anti-Castro movements, Jewish groups opposing the former Soviet Union's emigration policy, Irish Provos (the Provisional Irish Republican Army), and sporadic spillovers from conflicts around the world. Since the collapse of the Soviet Union, most international terrorism in the United States has come from spillovers originating in Middle Eastern conflicts. Incidents such as the September 11, 2001, homeland attacks indicate that practitioners of the New Terrorism have specifically targeted the United States as an enemy interest.

Operatives carrying out Middle East–related attacks inside the United States have often been foreign nationals who attack symbolic targets, specifically intending to kill people. Some of these attacks have been carried out by prepositioned cells. The members of these cells were drawn from groups such as Hamas and Islamic Jihad, which have had operatives and supporters living in the United States. Collaborative efforts by these and other groups illustrate the internationalization of the New Terrorism, its loose organizational structure, and its potential effectiveness inside the United States. Other attacks have been carried out by homegrown jihadists sympathizing with the international movement.

Case in Point: Conspiracy Theories on the American Right

The modern far and fringe right have produced a number of conspiracy theories and rumors. Although they may seem fantastic to nonmembers of rightist movements, many adherents of these theories live their lives as if the theories are an absolute reality. The Patriot movement and related tendencies, in particular, adhere to recent conspiracy theories (further discussion of the Patriot movement is provided later in this chapter). Conspiracy theories reflect the political and social environments that give rise to the theories, and three phases of modern conspiracy beliefs may be identified.

Phase 1 Conspiracies: Communist Invaders During the Cold War

- Rumors "confirmed" that Soviet cavalry units were preparing to invade Alaska across the Bering Strait from Siberia.

- Thousands of Chinese soldiers (perhaps an entire division) had massed in tunnels across the southwestern border of the United States in Mexico.

- Thousands of Vietcong and Mongolian troops had also massed in Mexico across the borders of Texas and California.

Phase 2a Conspiracies: The New World Order Replaces the Communist Menace

- Hostile un-American interests (which may already be in power) include the United Nations, international Jewish bankers, the Illuminati, the Council on Foreign Relations, and the Trilateral Commission.

- Assuming it is Jewish interests who are in power, the U.S. government has secretly become the Zionist Occupation Government (ZOG).

- The government has constructed concentration camps that will be used to intern Patriots and other loyal Americans after their weapons have all been seized (possibly with the assistance of African American street gangs).

- Invasion coordinates for the New World Order have been secretly stuck to the backs of road and highway signs.

- Sinister symbolism and codes have been found in the Universal Product Code (the bar lines on consumer goods), cleaning products, and cereal boxes and on American paper currency (such as the pyramid with the eyeball).

- Sinister technologies exist that will be used when ZOG or the New World Order makes its move. These include devices that can alter the weather and scanners that can read the plastic strips in American paper currency.

- FEMA (the Federal Emergency Management Agency) has built concentration camps for the day when patriotic Americans will be interned after their firearms have been confiscated.

Phase 2b Conspiracies: Formation of "Citizens' Militias"

With these and other conspiracy theories as an ideological foundation, many members of the Patriot and related movements organized themselves into "citizens' militias," and scores of militias began to be formed during the 1990s. At their peak during this growth period, it is estimated that 50,000 Americans were members of more than 800 militias, drawn from 5 to 6 million adherents of the Patriot movement.[4] After the 1995 bombing of the Alfred P. Murrah Federal Building in Oklahoma City, a general decline in Patriot organizations and militias occurred. This decline reversed after the 2008 election of President Barack Obama, when the number of these organizations increased steadily.

Some militia members joined to train as weekend "soldiers," whereas other militias organized themselves as paramilitary survivalists. Survivalism originated during the Cold War's Soviet–U.S. rivalry, when many people believed that a nuclear exchange between the superpowers was inevitable. They moved into the countryside, stocked up on food and weapons, and prepared for the nuclear holocaust. Many militias and members of the Patriot movement adapted this expectation to the New World Order conspiracy theory. Militia members who became survivalists went "off the grid" by refusing to have credit cards, driver's licenses, Social Security numbers, or government records. The purpose of going off the grid was to disappear from the prying eyes of the government and the New World Order or ZOG. Several principles are common to most Patriot organizations and militias:

- The people are sovereign. When necessary, they can resist the encroachment of government into their lives. They can also reject unjust government authority.

- Only an armed citizenry can counterbalance the authority of an oppressive government.

- The U.S. government has become oppressive, so the time is right to organize citizens' militias.

- It is necessary for citizens' militias to train and otherwise prepare for the day when an oppressive government or the New World Order moves in to take away the sovereignty of the people.

- The potential for political violence from some members of the armed, conspiracy-bound Patriot movement has been cited by experts and law enforcement officials as a genuine threat.

Phase 3a Conspiracies: 9/11 "Truther" Conspiracy Theories

A number of conspiracy theories emerged in the aftermath of the September 11, 2001, terrorist attacks, part of the so-called "truther" movement. These include the following theories:

- The U.S. government allowed the attacks to happen.

- Explosives destroyed the Twin Towers in a controlled detonation, as evidenced by the vertical fall of the Towers and debris that was pushed through the windows.

- An American-made cruise missile hit the Pentagon, as evidenced by the small size of two holes in the building.

- World Trade Center Building 7 was brought down by controlled explosions.

Phase 3b Conspiracies: Post–9/11 Conspiracy Theories

Other conspiracy theories gained traction in the years following the September 11 attacks. These include the following theories:

- President Barack Obama was not born in the United States (so-called "birther" conspiracies), is a socialist, and is secretly a Muslim.

- The New World Order is spraying toxic chemicals or biological agents in the atmosphere (the so-called "chemtrail conspiracy"). These may be seen in the contrails (condensation trails) of aircraft. Conspiracy adherents refer to contrails as "chemtrails."

- The Federal Reserve System will be used to create a one-world banking system.

- Military training exercises such as Jade Helm 15 in 2015 are actually preludes for seizing firearms, declaring martial law, and (in the case of Jade Helm 15) invading Texas.

- The QAnon conspiracy theory alleges that a "deep state" was exposed following the election of President Donald Trump. The goal of the deep state was to destabilize the new administration and disenfranchise its political supporters. Among several conspiracies propounded by QAnon is an alleged *coup d'etat* plot by billionaire George Soros, former president Barack Obama, and former senator and secretary of state Hillary Clinton.

Table 9.3 summarizes and contrasts the basic characteristics of contemporary left-wing, right-wing, and international political violence in the United States. This is not an exhaustive profile, but it is instructive for purposes of comparison. Figure 9.1 reports tactics used in terrorist incidents in the United States from 1970 through 2017.

Table 9.3 Attributes of Terrorism in the United States

In the United States, terrorism has typically been conducted by groups and individuals espousing leftist or rightist ideologies or those who engage in international "spillover" conflicts. These interests are motivated by diverse ideologies, operate from different milieus, possess distinctive organizational profiles, and target a variety of interests.

The following table summarizes these profiles.

| Environment | Ideological Profile | Activity Profile | | |
		Bases of Operation	Organizational Profile	Typical Targets
Leftist	Marxist; left-wing nationalist	Urban areas; suburbs	Clandestine groups; movement-based	Symbolic structures; avoidance of human targets
Rightist	Racial supremacist; antigovernment; religious	Rural areas; small towns	Self-isolated groups; cells; lone wolves	Symbolic structures; human targets
"Old" international terrorism	Ethnonationalist	Urban areas	Clandestine groups	Symbols of enemy interest
"New" international terrorism	Religious	Urban areas	Cells	Symbolic structures; human targets

Figure 9.1 Tactics Used in Terrorist Attacks in the United States, 1970–2017

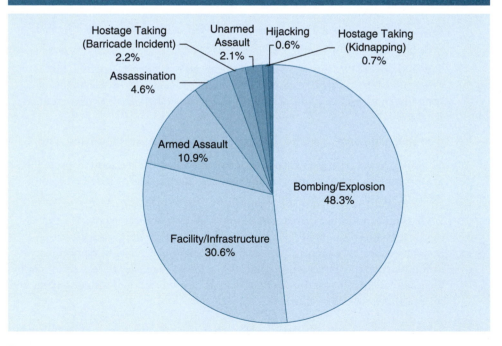

Source: National Consortium for the Study of Terrorism and Responses to Terrorism (2018).

Note: This figure includes up to three tactics per case. We excluded "unknown" (11). The total number of tactics included is 2,878.

LEFT-WING TERRORISM IN THE UNITED STATES

The modern American left is characterized by several movements that grew out of the political fervor of the 1960s. They are fairly interconnected, and understanding their origins provides instructive insight into the basic issues of the left. One should bear in mind that none of these movements were fundamentally violent in nature, and they were not terrorist movements. However, extremist trends within them led to factions that sometimes espoused violent confrontation, and a few engaged in terrorist violence.

Generational Rebellion: New Left Terrorism

The **New Left** was deeply affected by the war in Vietnam, the civil rights movement, and the turmoil in inner-city African American communities. A number of terrorist groups and cells grew out of this environment. Although the most prominent example was the Weatherman group, other groups such as the Symbionese Liberation Army also engaged in terrorist violence. The United Freedom Front proved to be the most enduring of all New Left terrorist groups of the era.

The Weathermen/Weather Underground Organization

The Weatherman group—known as the **Weathermen**—jelled at the June 1969 Students for a Democratic Society (SDS) national convention in Chicago, when SDS splintered into several factions. The Weathermen were mostly young, white, educated members of the middle class. They represented in stark fashion the dynamic ideological tendencies of the era as well as the cultural separation from the older generation. Although they and others were sometimes referred to collectively as the "Crazies," they operated within a supportive cultural and political environment.

From the beginning, the Weathermen were violent and confrontational. In October 1969, they distributed leaflets in Chicago announcing what became known as their **Days of Rage**. The Days of Rage lasted four days and consisted of acts of vandalism and street fights with the Chicago police. In December 1969, the Weathermen held a "war council" in Michigan. Its leadership, calling itself the **Weather Bureau**, advocated bombings, armed resistance, and assassinations. In March 1970, an explosion occurred in a Greenwich Village townhouse in New York City that was being used as a bomb factory. Three Weathermen were killed, several others escaped through the New York subway system, and hundreds of members went underground to wage war.

By the mid-1970s, the Weathermen—renamed the **Weather Underground Organization**—had committed at least 40 bombings, including attacks on the following targets:

- The Pentagon
- Police stations
- National Guard facilities
- Reserve Officers' Training Corps buildings
- The Harvard war research center in Cambridge, Massachusetts
- The Gulf Oil corporate headquarters in Pittsburgh, Pennsylvania

The Weather Underground also freed counterculture guru Timothy Leary from prison, published a manifesto called *Prairie Fire*, and distributed an underground periodical called

New Left: A movement of young leftists during the 1960s who rejected orthodox Marxism and took on the revolutionary theories of Frantz Fanon, Herbert Marcuse, Carlos Marighella, and other new theorists.

Weathermen: A militant faction of Students for a Democratic Society that advocated and engaged in violent confrontation with the authorities. Some Weathermen engaged in terrorist violence.

Days of Rage: Four days of rioting and vandalism committed by the Weathermen in Chicago in October 1969.

Weather Bureau: The designation adopted by the leaders of the Weatherman faction of Students for a Democratic Society.

Weather Underground Organization: The adopted name of the Weathermen after they moved underground.

Symbionese Liberation Army (SLA): A violent terrorist cell that gained notoriety for several high-profile incidents in the mid-1970s.

DeFreeze, Donald: The leader of the Symbionese Liberation Army, a California-based American terrorist group active during the 1970s. DeFreeze adopted the nom de guerre Cinque.

Osawatomie. Their underground network of safe houses and rural safe collectives, which served to hide them and New Left fugitives from the law, was never effectively infiltrated by law enforcement agencies.

By the mid-1970s, members of the Weather Underground had begun to give up their armed struggle and return to aboveground activism, a process they called *inversion*. Those who remained underground (mostly the East Coast wing) committed acts of political violence into the 1980s, and others joined other terrorist organizations.

The Symbionese Liberation Army

A violent terrorist cell known as the **Symbionese Liberation Army (SLA)** gained notoriety for several high-profile incidents in the mid-1970s. The core members were led by **Donald DeFreeze**, who took the nom de guerre Cinque (after the leader of a nineteenth-century rebellion aboard the slave ship *Amistad*). Members trained in the Berkeley Hills of California near San Francisco, rented safe houses, and obtained weapons. In November 1973, the Oakland school superintendent was assassinated by being shot eight times; five of the bullets were cyanide tipped. In a communiqué, the SLA took credit for the attack, using a rhetorical phrase that became its slogan, "Death to the fascist insect that preys upon the people!"

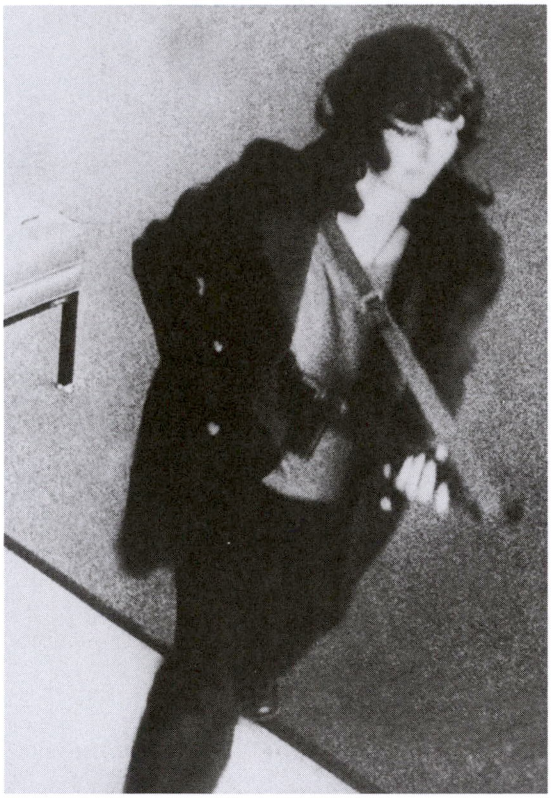

Bettmann/Getty Images

▸ **Photo 9.2** The Symbionese Liberation Army (SLA) in action. A bank camera captures Patricia Hearst exiting a bank after an SLA robbery. Hearst, who joined the group after being kidnapped by them, likely suffered from Stockholm syndrome.

In February 1974, newspaper heiress Patricia Hearst was kidnapped by the cell. She was kept bound and blindfolded in a closet for more than 50 days while under constant physical and psychological pressure, including physical abuse and intensive political indoctrination. She broke down under the pressure, and a tape recording was released in which she stated that she had joined the SLA. In April 1974, Hearst participated in a bank robbery in San Francisco. This was a classic case of Stockholm syndrome.

In May 1974, five of the SLA's core members, including DeFreeze, were killed in a shootout in a house in the Watts neighborhood of Los Angeles.

Patricia Hearst was a fugitive for approximately one year. She was hidden—probably by the Weather Underground—and traveled across the country with compatriots. By 1975, the SLA had a rebirth with new recruits and was responsible for several bank robberies and bombings in California. They referred to themselves as the **New World Liberation Front**. Hearst was captured in September 1975 in San Francisco, along with another underground fugitive. Most of the other members either were captured or disappeared into the underground.

Civil Strife: Ethnonationalist Terrorism on the Left

Ethnonational violence, which is distinguishable from racial supremacist violence, has been rare in the United States. This is primarily because activist environments have not historically supported nationalist terrorism. Exceptions to this general observation grew out of the political environment of the 1960s, when nationalist political violence originated in African American and Puerto Rican activist movements. There have been few nationalist movements outside of these examples.

The following discussion evaluates ethnonational political violence committed by adherents of the Black Liberation and Puerto Rico independence movements. In both examples, the underlying ideological justifications for the violence were Marxist inspired.

The Black Liberation Movement

Racial tensions in the United States were extremely high during the 1960s. African Americans in the South directly confronted Southern racism through collective nonviolence and the burgeoning **Black Power** ideology. In the urban areas of the North and West, cities became centers of confrontation between African Americans, the police, and state National Guards. Many Black Power advocates in the North and West became increasingly militant as the summers became seasons of urban confrontation. During what became known in the 1960s as the "long, hot summers," many cities were social and political powder kegs, and hundreds of riots occurred during the summers from 1964 to 1969. When President Lyndon Johnson and the U.S. Senate organized inquiries into the causes of these disorders, their findings were disturbing. Table 9.4 describes the quality of these findings, which indicate the severity of tensions in urban areas during the mid-1960s.

Within this racially charged environment grew cadres of African American revolutionaries dedicated to using political violence to overthrow what they perceived to be a racist and oppressive system. The most prominent example of African American nationalist terrorism is the **Black Liberation Army (BLA)**, an underground movement whose membership included former members of the **Black Panther Party for Self-Defense** and Vietnam veterans. BLA members were nationalists inspired in part by the 1966 film *The Battle of Algiers*,[5] a semidocumentary of an urban terrorist uprising in the city of Algiers against the French during their colonial war in Algeria. There were at least two cells (or groups of cells) of the BLA—the East Coast and West Coast groups. The BLA is suspected to have committed a number of attacks in New York and California prior to and after these incidents. They are thought to have been responsible for numerous bombings, ambushes of police officers, and bank robberies to "liberate" money to support their cause. Their areas of operation were California and New York City, although some BLA members apparently received training in the South.

The symbolic leader of the BLA was JoAnne Chesimard, a former Black Panther who later changed her name to **Assata Shakur**. In May 1973, a gunfight broke out when she and two other BLA members were stopped on the New Jersey Turnpike by a New Jersey state trooper. The trooper was killed, as was one of the occupants of the automobile. Shakur was captured, tried, and eventually convicted in 1977. She was sentenced to life imprisonment but was freed in 1979 by members of the May 19 Communist Organization and spirited to Cuba. She remains there under the protection of the Cuban government.

Most members of the BLA were eventually captured or killed. Those who were captured were sentenced to long prison terms. Unlike the Weather Underground's network, the BLA network was successfully penetrated and infiltrated by the FBI, using informants. Those who escaped the FBI net re-formed to join other radical organizations.

Puerto Rican Independentistas

Puerto Rico is a commonwealth of the United States, meaning that it is self-governed by a legislature and an executive (a governor) and has a nonvoting delegate to Congress. Opinion about the island's political status is divided among a majority who wish for it to remain a commonwealth, a large number who favor statehood, and a minority who desire national independence. Those who desire independence are nationalists called *independentistas*. Most independentistas use democratic institutions to promote the cause of independence; they

New World Liberation Front: An American terrorist group active during the mid-1970s, organized as a "reborn" manifestation of the Symbionese Liberation Army by former SLA members and new recruits.

Black Power: An African American nationalist ideology developed during the 1960s that stressed self-help, political empowerment, cultural chauvinism, and self-defense.

Black Liberation Army (BLA): An African American terrorist group active during the 1970s. The BLA tended to target police officers and banks.

Black Panther Party for Self-Defense: An African American nationalist organization founded in 1966 in Oakland, California. The Black Panthers eventually became a national movement. It was not a terrorist movement, but some members eventually engaged in terrorist violence.

Shakur, Assata: The symbolic leader of the Black Liberation Army (BLA) in the United States, formerly known as JoAnne Chesimard. A former Black Panther, she was described by admirers as the "heart and soul" of the BLA.

Table 9.4 Racial Conflict in America: The "Long, Hot Summers" of the 1960s

The urban disturbances in the United States during the 1960s caused an unprecedented period of communal discord. The disturbances were widespread and violent and were a culmination of many factors. One factor was the deeply rooted racial polarization in American society. The presidentially appointed National Advisory Commission on Civil Disorders (known as the Kerner Commission) reported in 1968 that "segregation and poverty have created in the racial ghetto a destructive environment totally unknown to most white Americans. What white Americans have never fully understood—but what the Negro can never forget—is that white society is deeply implicated in the ghetto. White institutions created it, white institutions maintain it, and white society condones it."[i]

The following table reports data from a Senate Permanent Subcommittee on Investigations inquiry into urban rioting after the serious disturbances in the summer of 1967.[ii] The table summarizes the environment during three years of civil disturbances.[iii]

Incident Report	Activity Profile		
	1965	1966	1967
Number of urban disturbances	5	21	75
Casualties			
Killed	36	11	83
Injured	1,206	520	1,897
Legal sanctions			
Arrests	10,245	2,298	16,389
Convictions	2,074	1,203	2,157
Costs of damage (in millions of dollars)	$40.1	$10.2	$664.5

Source: U.S. Senate, Permanent Subcommittee on Investigations.

i. *Report of the National Advisory Commission on Civil Disorders* (New York: Bantam Books, 1968), 2.

ii. In 1967, the Senate passed a resolution ordering the Senate Permanent Subcommittee on Investigations to investigate what had caused the 1967 rioting and to recommend solutions.

iii. Senate Permanent Subcommittee on Investigations. Data reported in *Ebony Pictorial History of Black America*, vol. 3 (Chicago: Johnson, 1971), 69.

are activists but are not prone to violence. Many are intellectuals and professionals who are working to build pro-independence sentiment.

Some independentistas are revolutionaries, and a small number have resorted to violence. Puerto Rican nationalist violence on the mainland United States has a history dating to the immediate post–World War II era. For example, on November 1, 1950, two Puerto Rican nationalists attempted to assassinate President Harry Truman at Blair House (the presidential guesthouse) in Washington, D.C. Following the Cuban Revolution, violent Puerto Rican nationalists patterned themselves after Cuban nationalism and viewed the United States as an imperial and colonial power. Cuba did, in fact, provide support for violent independentista groups, especially during the 1980s.

The **Armed Forces for National Liberation (Fuerzas Armadas de Liberación Nacional, or FALN)** was a very active terrorist organization that concentrated its activities on the U.S. mainland, primarily in Chicago and New York City. One important fact stands out about the FALN: It was the most prolific terrorist organization in U.S. history. The group became active in 1974, and from 1975 to 1983, approximately 130 bombings were linked to the FALN or another group known as the **Macheteros** ("Machete-Wielders"), with the vast majority being the responsibility of the FALN. Most attacks by the FALN were symbolically directed against buildings, although some of their attacks were deadly. The group was also responsible for armored car and bank robberies.

In 1977, FALN leader William Morales was captured by the police after being injured in an explosion at an FALN bomb factory in New York City. In 1979, Morales was freed from a hospital in New York by the May 19 Communist Organization, the same group that freed BLA leader Assata Shakur. He escaped to Mexico, where he remained hidden until 1983. In 1983, Morales was captured by Mexican authorities at an international telephone; he was also convicted in absentia of sedition by a federal district court in Chicago for participation in 25 bombings. In 1988, Mexico refused to extradite Morales to the United States, and he was allowed to move to Cuba, where he remains under the protection of Cuban authorities.

In 1980, more than a dozen FALN members were convicted of terrorist-related crimes. Sentences were imposed for seditious conspiracy, possession of unregistered firearms, interstate transportation of a stolen vehicle, interference with interstate commerce by violence, and interstate transportation of firearms with intent to commit a crime. None of these charges were linked to homicides. FALN members' sentences ranged from 15 to 90 years, and they considered themselves to be prisoners of war.

The Revolution Continues: Leftist Hard Cores

The left-wing revolutionary underground re-formed after the decline of groups such as the Weather Underground and the BLA. These new groups were made up of die-hard former members of the Weather Underground and BLA as well as former activists from other organizations, such as the radicalized SDS and the Black Panthers.

The May 19 Communist Organization

The **May 19 Communist Organization (M19CO)** derives its name from the birthdays of Vietnamese leader Ho Chi Minh and Malcolm X. The symbolism of this designation is obvious: It combines domestic and international examples of resistance against self-defined U.S. racism and imperialism. The group was composed of remnants of the Weather Underground, Black Panthers, the BLA, and the Republic of New Africa. M19CO cadres included the founders of the Republic of New Africa and the most violent members of the Weather Underground. Many of the organization's members were people who had disappeared into the revolutionary underground for years.

M19CO was fairly active, engaging in bank and armored car robberies, bombings, and other politically motivated actions. Its more spectacular actions included participation in the October 1981 robbery of a Brink's armored car in suburban Nyack, New York. During the robbery, one security guard was killed. After an automobile chase and shootout at a roadblock, during which two police officers died, four M19CO members were captured. One of them was Kathy Boudin, daughter of prominent attorney Leonard Boudin. She had been one of the survivors of the explosion at the Weatherman group's Greenwich Village townhouse in 1970. Also captured was Donald Weems, a former BLA member.

Armed Forces for National Liberation (Fuerzas Armadas de Liberación Nacional, or FALN): A Puerto Rican independentista terrorist group active during the 1970s and 1980s and responsible for more bombings than any other single terrorist group in American history.

Macheteros: A Puerto Rican independentista terrorist group active during the 1970s and 1980s.

May 19 Communist Organization (M19CO): An American Marxist terrorist group that was active in the late 1970s and early 1980s. It was composed of remnants of the Republic of New Africa, the Black Liberation Army, the Weather Underground, and the Black Panthers. M19CO derived its name from the birthdays of Malcolm X and Vietnamese leader Ho Chi Minh.

M19CO adopted several different names when claiming responsibility for its attacks. These aliases included the Red Guerrilla Resistance, the Revolutionary Fighting Group, and the Armed Resistance Unit. After the Nyack incident, M19CO remained active and engaged in several bombings. The group was finally dissolved when its remaining members were arrested in May 1985.

The United Freedom Front

One case is unique in comparison with other New Left, nationalist, and hard-core groups. Formed in 1975, the **United Freedom Front (UFF)** was underground and active for approximately 10 years. It was a New Left terrorist organization that grew out of a program by former SDS members to educate prison inmates about the political nature of their incarceration.

The UFF is suspected to have committed at least 25 bombings and robberies in New York and New England. The attacks were primarily intended to exhibit anticorporate or antimilitary symbolism. UFF members displayed a great deal of discipline in their activities—for example, taking copious notes at regular meetings that they called "sets." Members went underground in the American suburbs, immersing themselves in the middle class and adopting covers as nondescript residents. The UFF was broken up when its members were arrested in late 1984 and early 1985. Few leftist groups had remained both underground and active for as long as did the UFF.

Single-Issue Violence on the Left

The left has produced violent single-issue groups and individuals who focus on one particular issue to the exclusion of others. To them, their championed issue is the central point—arguably the political crux—for solving many of the world's problems. For example, Ted Kaczynski, also known as the Unabomber, protested the danger of technology by sending and placing bombs that killed three people and injured 22 others during a 17-year campaign.

Typical of leftist single-issue extremism is the fringe environmental movement. Groups such as the **Animal Liberation Front (ALF)** and the **Earth Liberation Front (ELF)** have committed numerous acts of violence, such as arson and vandalism, which they refer to as *ecotage*. Their activity profiles are summarized as follows:

- The ALF favors direct action to protest animal abuse, with the objective of saving as many animals as possible. There is no hierarchy within the movement, and it has operated in small groups.

- The ELF was founded in England by activists who split from the environmentalist group Earth First! because of its decision to abandon criminal activities. It is potentially more radical than the ALF.

The ALF and ELF have coordinated their activities. Several joint claims have been made about property damage and other acts of vandalism, and it is likely that the two groups share the same personnel. For the most part, both the ALF and ELF have been nonviolent toward humans, but they have committed many incidents of property destruction.

ALF and ELF targets include laboratories, facilities where animals are kept, and sport utility vehicles (SUVs). Some of these incidents are vandalism sprees. For example, in 2003, a group of activists apparently affiliated with the ELF went on a firebombing and vandalism

United Freedom Front (UFF): A leftist terrorist group in the United States that was active from the mid-1970s through the mid-1980s.

Animal Liberation Front (ALF): An American-based single-issue movement that protests animal abuse and is responsible for committing acts of violence such as arson and vandalism.

Earth Liberation Front (ELF): A single-issue movement that protests environmental degradation and pollution. A splinter group from the environmentalist group Earth First!, the ELF is potentially more radical than the Animal Liberation Front.

spree in the San Gabriel Valley east of Los Angeles. About 125 SUVs and other vehicles parked at homes and auto dealerships were burned or damaged. The initials *ELF* were also spray-painted. In the latter operation, a doctoral student attending the California Institute of Technology was found guilty of conspiracy and arson.

The FBI estimates that the ELF alone has engaged in 1,200 criminal acts and caused about $100 million in property damage since 1996. In 2001, an ELF firebomb destroyed the University of Washington's Center for Urban Horticulture, which was rebuilt at a cost of $7 million. In one particularly destructive incident in August 2003, the group caused $50 million in damages to a condominium complex under construction in San Diego, California. The ELF has also targeted suburban property developments, as occurred in 2008 when four luxury homes were burned in a suburb north of Seattle, Washington. In September 2009, members of the ELF toppled two radio towers near Seattle, Washington.

RIGHT-WING TERRORISM IN THE UNITED STATES

The modern American right is characterized by several trends that developed from cultural and grassroots sources. Unlike the left, whose characteristics reflect the activism of the 1960s, the right is characterized more by self-defined *value systems*. These value systems have been perceived by many on the right to be under attack and hence in need of protection—often by resorting to activist defense. This tendency is rooted in newly emergent trends such as antigovernment and evangelical religious activism as well as in historical cultural trends such as racial supremacy. Some political controversies, such as illegal immigration, have rallied extremists who promote their own agendas by claiming that such issues justify their extreme beliefs.

The Past as Prologue: The Historical Legacy of the Ku Klux Klan

The **Ku Klux Klan (KKK)** is a racist movement that has no counterpart among international right-wing movements; it is a purely American phenomenon. Its name comes from the Greek word *kuklos*, or "circle." The KKK is best described as an enduring movement that developed the following ideology:

- Racial supremacy

- Protestant Christian supremacy

- American cultural nationalism (also known as **nativism**)

- Violent assertion of Klan racial doctrine

- Ritualistic symbolism, greetings, and fraternal behavior

Klan terminology in many ways is an exercise in racist secret fraternal bonding. Table 9.5 samples the exotic language of the KKK.

KKK terrorism has been characterized by different styles of violence in several historical periods. Not every Klansman has been a terrorist, nor has every Klan faction practiced terrorism. However, the threat of violence and racial confrontation has always been a part of the Klan movement. In order to understand the nature of Klan violence, it is instructive to survey the historical progression of the movement. There have been several manifestations of the KKK, which most experts divide into five eras.

Ku Klux Klan (KKK): A racial supremacist organization founded in 1866 in Pulaski, Tennessee. During its five eras, the KKK was responsible for thousands of acts of terrorism.

kuklos: Literally "circle" in Greek; the insignia of the Ku Klux Klan, consisting of a cross and a teardrop-like symbol enclosed by a circle.

nativism: American cultural nationalism.

Table 9.5   The Fraternal Klan

From its inception in 1866, the Ku Klux Klan has used fraternity-like greetings, symbolism, and rituals. These behaviors promote secrecy and racial bonding within the organization. Examples of Klan language include the following greetings: *Ayak*? (Are you a Klansman?) and *Akia*! (A Klansman I am!). The language used for regional offices is also unique, as indicated in the following examples:

Invisible Empire	National
Realm	State
Klavern	Local

The following table summarizes the activity profiles of official Klan organizational designations.

Klan Official	Duties	Scope of Authority	Symbolic Identification
Imperial wizard	National leader	Invisible Empire	Blue stripes or robe
Grand dragon	State leader	Realm	Green stripes or robe
Exalted cyclops	County leader	Klaverns within county	Orange stripes or robe
Nighthawk	Local security and administration	Klavern	Black robe
Klonsel	General counsel	Invisible Empire	White robe
Citizen	Member	Klan faction	White robe

The First-Era Klan

The KKK was founded in 1866 in the immediate aftermath of the Civil War. Some sources date its origin to Christmas Eve 1865, whereas others cite the year 1866. According to most sources, the KKK was first convened in Pulaski, Tennessee, by a group of Southerners who initially formed the group as a fraternal association. They originally simply wore outlandish outfits and played practical jokes but soon became a full civic organization. Their first *grand wizard*, or national leader, was former Confederate general and slave-trader Nathan Bedford Forrest. Military-style rankings were established, and by 1868, the KKK was a secretive and politically violent underground organization. Its targets included African Americans, Northerners, and Southern collaborators. Northern victims were those who traveled south to help improve the conditions of the former slaves as well as profiteering *carpetbaggers*. Southern victims were collaborators derisively referred to as *scalawags*. The KKK was suppressed by the Union Army and the anti-Klan "Ku Klux laws" passed by Congress. Nathan Bedford Forrest ordered the KKK to be officially disbanded, and their robes and regalia were ceremoniously burned. It has been estimated that the Klan had about 400,000 members during its first incarnation.

The Second-Era Klan

After the Reconstruction era (after the departure of the Union Army from the South and the end of martial law), the KKK re-formed into new secret societies and fraternal groups.

It wielded a great deal of political influence and successfully helped restore racial supremacy and segregation in the South. African Americans lost most political and social rights during this period, ushering in conditions of racial subjugation that did not end until the civil rights movement in the mid-twentieth century. The targets of Klan violence during this period were African Americans, immigrants, Catholics, and Jews.

The Third-Era Klan

During the early part of the twentieth century and continuing into the 1920s, the KKK became a broad-based national movement. In 1915, members gathered at Stone Mountain, Georgia, and formed a movement known as the Invisible Empire. The Klan was glorified in the novel *The Clansman* and in the 1915 film *Birth of a Nation*, which was shown in March 1915 at the White House during the administration of President Woodrow Wilson. During this period, the Invisible Empire had between 3 and 4 million members. In 1925 in Washington, D.C., 45,000 Klansmen and Klanswomen paraded down Pennsylvania Avenue. Also during this period, Klan-inspired violence was widespread. Thousands of people—mostly African Americans—were victimized by the KKK. Many acts of terrorism were ritualistic communal lynchings.

The Fourth-Era Klan

After a decline because of revelations about Third-Era violence and corruption, the Klan was reinvigorated in 1946—once again at Stone Mountain, Georgia. At this gathering, the Invisible Empire disbanded, and new independent Klans were organized at local and regional levels. There was no longer a single national Klan but rather autonomous Klan factions. During the civil rights movement, some Klan factions became extremely violent. The White Knights of Mississippi and the United Klans of America (mostly in Alabama) committed numerous acts of terrorism to try to halt progress toward racial equality in the American South. This era ended following several successful federal prosecutions on criminal civil rights charges, although the Klan itself endured.

The Fifth-Era Klan

Violence during the Fifth Era has been committed by lone wolves rather than organized Klan activity. The modern era of the Ku Klux Klan is characterized by two trends.

The Moderate Klan. Some Klansmen and Klanswomen have tried to moderate their image by adopting more mainstream symbolism and rhetoric. Rather than advocating violence or paramilitary activity, they have projected an image of law-abiding activists working on behalf of white civil rights and good moral values. Those who promote this trend have eschewed the prominent display of Klan regalia and symbols. For example, former neo-Nazi and Klansman David Duke has repeatedly used mainstream political and media institutions to promote his cause of white civil rights. He is the founder of the National Association for the Advancement of White People and the European-American Unity and Rights Organization (EURO).

The Purist Klan. A traditional and "pure" Klan has emerged that hearkens back to the original traditions and ideology of the KKK. This group has held a number of aggressive and vitriolic rallies, many in public at county government buildings. Its rhetoric is unapologetically racist and confrontational. Some factions of the purist trend prohibit the display of Nazi swastikas or other non-Klan racist symbols at KKK gatherings.

KKK membership has ebbed and flowed in the Fifth Era, in part because of changes in the nation's cultural and political environment but also because of competition from other racial supremacist movements such as racist skinhead and neo-Nazi groups. There was also fresh competition beginning in the late 1990s from the neo-Confederate movement.

Racial Mysticism: Neo-Nazi Terrorism

In the modern era, most non-Klan terrorism on the right wing has come from members of the neo-Nazi movement. The American version of Nazism has incorporated mystical beliefs into its underlying ideology of racial supremacy. This mysticism includes **Christian Identity**, **Creativity**, and racist strains of **Ásatrú**. Neo-Nazi terrorism is predicated on varying mixes of religious fanaticism, political violence, and racial supremacy. Proponents' worldview is premised on the superiority of the Aryan race, the inferiority of non-Aryans, and the need to confront an evil global Jewish conspiracy. Another common theme is the belief that a **racial holy war (RaHoWa)** is inevitable. Chapter Perspective 9.3 discusses the Christian Identity creation myth.

The new non-Klan groups came into their own during the 1980s, when Aryan Nations, White Aryan Resistance, and the National Alliance actively disseminated information about supremacist ideology. Members of the new supremacist groups created their own mythologies and conspiracy theories. For example, the novel *The Turner Diaries* is considered by many neo-Nazis to be a blueprint for the Aryan revolution in America. The book inspired the terrorist group The Order (discussed later) in its terrorist campaign as well as Oklahoma City bomber **Timothy McVeigh**. Also on the racist right, the "Fourteen Words" have become

CHAPTER PERSPECTIVE 9.3

Race and the Bible: The Christian Identity Creation Myth

Christian Identity is the Americanized strain of an eighteenth-century quasireligious doctrine called Anglo-Israelism that was developed by Richard Brothers. Believers hold that whites are descended from Adam and are the true Chosen People of God, that Jews are biologically descended from Satan, and that nonwhites are soulless beasts (also called the "Mud People"). Christian Identity adherents have developed two cultish creation stories that are loosely based on the Old Testament. The theories are called One-Seedline Christian Identity and Two-Seedline Christian Identity.

One-Seedline Christian Identity accepts that all humans, regardless of race, are descended from Adam. However, only Aryans (defined as northern Europeans) are the true elect of God. They are the "chosen people" whom God has favored and who are destined to rule over the rest of humanity. In the modern era, those who call themselves the Jews are actually descended from a minor Black Sea ethnic group and, therefore, have no claim to Israel.

Two-Seedline Christian Identity rejects the notion that all humans are descended from Adam. Instead, its focus is on the progeny of Eve. Two-Seedline adherents believe that Eve bore Abel as Adam's son but bore Cain as the son of the Serpent (that is, the devil). Outside of the Garden of Eden lived nonwhite, soulless beasts, who are a separate species from humans. They are the modern nonwhite races of the world and are often referred to by Identity believers as Mud People. When Cain slew Abel, he was cast out of the Garden to live among the soulless beasts. Those who became the descendants of Cain are the modern Jews. They are, thus, biologically descended from the devil and are a demonic people worthy of extermination. There is an international conspiracy by the Jewish "devil race" to rule the world. The modern state of Israel and the Zionist Occupation Government in the United States are part of this conspiracy.

a rallying slogan. Originally coined by David Lane, a convicted member of the terrorist group The Order, the Fourteen Words are as follows: "We must secure the existence of our people and a future for White children." The Fourteen Words have been incorporated into the Aryan Nations' "declaration of independence" for the white race, and the slogan is often represented by simply writing or tattooing *14*.

Although most violence emanating from these beliefs has been expressed as lone-wolf terrorism and hate crimes, several groups have embarked on violent sprees. For example, a neo-Nazi terrorist cell known as **The Order** was active in 1983 and 1984 and was responsible for robberies and murder, primarily in the Pacific Northwest. A neo-Nazi group calling itself the **Aryan Republican Army (ARA)** operated in the Midwest from 1994 to 1996. The ARA was inspired by the example of the Irish Republican Army and robbed 22 banks in seven states before its members were captured. Their purpose had been to finance racial supremacist causes and to hasten the overthrow of the Zionist Occupation Government. Some members also considered themselves Christian Identity fundamentalists called Phineas Priests.

Patriot Threats

The Patriot movement came to prominence during the early 1990s. The movement considers itself to represent the true heirs of the ideals of the framers of the U.S. Constitution. It hearkens back to what it defines as the "true" American ideals: individualism, an armed citizenry, and minimum interference from government. For many Patriots, government in general is not to be trusted, the federal government in particular is to be distrusted, and the United Nations is a dangerous and evil institution. To them, American government no longer reflects the will of the people; it has become dangerously intrusive and violently oppressive. The Patriot movement is not ideologically monolithic, and numerous tendencies have developed, such as the Common Law Courts and Constitutionalists.

Two events from the 1990s served to invigorate paranoid political activism on the Patriot right, giving rise to new conspiracy theories. These events were the tragedies at Ruby Ridge, Idaho, and Waco, Texas.

- *Ruby Ridge.* In August 1992 at Ruby Ridge, Idaho, racial supremacist Randy Weaver and his family, with compatriot Kevin Harris, were besieged by federal agents in response to Weaver's failure to reply to an illegal weapons charge. Weaver's teenage son Sammy and U.S. Deputy Marshal William Degan were killed during a shootout that occurred when Sammy, Randy, and Harris were confronted as they walked along a path. Weaver's wife Vicky was later fatally shot by an FBI sniper as she held her baby in the doorway of the Weaver home. The sniper had previously fired shots at Randy Weaver and Harris. Members of the Patriot movement and other right-wing extremists cite this incident as evidence of a broad government conspiracy to deprive freedom-loving "true" Americans of their right to bear arms and other liberties. Randy Weaver's story has inspired Patriots and other members of the extreme right.

- *Waco.* In early 1993 at Waco, Texas, federal agents besieged the Branch Davidian cult's compound after a failed attempt in February to serve a search warrant for illegal firearms had ended in the deaths of four federal agents and several cult members. On April 19, 1993, during an assault led by the FBI, about 80 Branch Davidians—including more than 20 children—died in a blaze that leveled the compound. Patriots and other rightists consider this tragedy, like the Ruby Ridge incident, to be evidence of government power run amok.

Ásatrú: A mystical belief in the ancient Norse pantheon. Some Ásatrú believers are racial supremacists.

racial holy war (RaHoWa): A term given by racial supremacists to a future race war that they believe will inevitably occur in the United States.

McVeigh, Timothy: A member of the Patriot movement in the United States and probably a racial supremacist. McVeigh was responsible for constructing and detonating an ANFO bomb that destroyed the Alfred P. Murrah Federal Building in Oklahoma City, Oklahoma, on April 19, 1995. One hundred sixty-eight people were killed.

Order, The: An American neo-Nazi terrorist group founded by Robert Jay Mathews in 1983 and centered in the Pacific Northwest. The Order's methods for fighting its war against what it termed the Zionist Occupation Government were counterfeiting, bank robberies, armored car robberies, and murder. The Order had been suppressed by December 1985.

Aryan Republican Army (ARA): A neo-Nazi terrorist group that operated in the midwestern United States from 1994 to 1996. Inspired by the example of the Irish Republican Army, the ARA robbed 22 banks in seven states before its members were captured. Their purpose had been to finance racial supremacist causes and to hasten the overthrow of the "Zionist Occupation Government."

In 1992, former KKK member Louis Beam began to publicly advocate **leaderless resistance** against the U.S. government. Fundamentally a cell-based strategy, leaderless resistance requires the formation of phantom cells to wage war against enemy interests and the government. Dedicated neo-Nazis and Patriots believe that the creation of phantom cells and leaderless resistance will prevent infiltration from federal agencies. The chief threat of violence came from the armed **militias**, which peaked in membership immediately prior to and after the Oklahoma City bombing. After the Oklahoma City bombing, federal authorities broke up at least 25 Patriot terrorist conspiracies. Chapter Perspective 9.4 discusses Timothy McVeigh and the Oklahoma City bombing.

Although the Patriot movement attracted a significant number of adherents during the 1990s and although militias at one point recruited tens of thousands of members, no underground similar to that of the radical left was formed. Relatively few terrorist movements or groups emanated from the Patriot movement, largely because many members were "weekend warriors" who did little more than train and also because law enforcement agencies successfully thwarted a number of true Patriot-initiated plots. Thus, despite many implicit and explicit *threats* of armed violence from Patriots, terrorist conspiracies were rarely carried to completion.

CHAPTER PERSPECTIVE 9.4

The Oklahoma City Bombing

On April 19, 1995, Timothy McVeigh drove a rented Ryder truck to the Alfred P. Murrah Federal Building in Oklahoma City. He deliberately chose April 19 as a symbolic date for the attack; it was the 220th anniversary of the battles of Lexington and Concord and the second anniversary of the law enforcement disaster in Waco, Texas.

McVeigh was a hard-core devotee of the Patriot movement and a believer in New World Order conspiracy theories. He was almost certainly a racial supremacist, having tried to solicit advice from the neo-Nazi National Alliance and the racial separatist Elohim City group about going underground after the bombing. McVeigh had also visited the Branch Davidian site at Waco, Texas,[a] where about 75 members of the Branch Davidian cult died in a fire that was ignited during a paramilitary raid by federal law enforcement officers.

McVeigh had converted the Ryder truck into a powerful mobile ammonium nitrate and fuel oil (ANFO)–based bomb. He used "more than 5,000 pounds of ammonium nitrate fertilizer mixed with about 1,200 pounds of liquid nitromethane, [and] 350 pounds of Tovex."[b] When he detonated the truck bomb at 9:02 a.m., it destroyed most of the federal building and

killed 168 people, including 19 children. More than 500 others were injured.

McVeigh's attack was, in large part, a symbolic act of war against the federal government. He had given careful consideration to achieving a high casualty rate, just as "American bombing raids were designed to take lives, not just destroy buildings."[c]

The deaths of the 19 children were justified in his mind as the unfortunate "collateral damage" against innocent victims common to modern warfare.[d] Timothy McVeigh was tried and convicted, and he was executed in a federal facility in Terre Haute, Indiana, on June 11, 2001. His execution was the first federal execution since 1963.

Notes

a. For a discussion of Timothy McVeigh's immersion in the fringe right—from the perspective of McVeigh himself—see Lou Michel and Dan Herbeck, *American Terrorist: Timothy McVeigh & the Oklahoma City Bombing* (New York: HarperCollins, 2001).

b. Ibid., 164.

c. Ibid., 224.

d. Ibid., 234.

The number of militias declined during the period between the April 1995 Oklahoma City bombing and the American homeland attacks of September 11, 2001. By 2000, the number of Patriot organizations was only one-fourth of the 1996 peak, and this general decline continued after September 11. This occurred for several reasons: First, the 1995 Oklahoma City bombing caused many less-committed members to drift away. Second, the dire predictions of apocalyptic chaos for the new millennium that were embedded in the Patriots' conspiracy theories did not materialize, especially the predicted advent of the New World Order. Third, the September 11, 2001, attacks shifted attention from domestic issues to international threats. Experts noted, however, that the most militant and committed Patriot adherents remained within the movement and that these dedicated members constituted a core of potentially violent true believers. This became evident after the 2008 presidential election, when the number of Patriot organizations and identified militia groups increased markedly. The following trend occurred:

- 1996: 858 Patriot organizations, 370 armed militias

- 2001: 158 Patriot organizations, 73 armed militias

- 2006: 147 Patriot organizations, 52 armed militias

- 2008: 149 Patriot organizations, 42 armed militias

- 2009: 512 Patriot organizations, 127 armed militias

- 2010: 824 Patriot organizations, 330 armed militias

- 2012: 1,360 Patriot organizations, 321 armed militias

- 2013: 1,096 Patriot organizations, 240 armed militias

- 2014: 874 Patriot organizations, about 200 armed militias

- 2017: 689 Patriot organizations, 273 armed militias

- 2018: 612 Patriot organizations, 212 armed militias[6]

CASE IN POINT: MORALIST TERRORISM

Moralist terrorism refers to acts of political violence motivated by a moralistic worldview. Most moralist terrorism in the United States is motivated by an underlying religious doctrine, and this is usually a fringe interpretation of Christianity. Abortion clinics and gay establishments have been targets of moralist violence.

From 1977 to 2017, violence against abortion providers included "11 murders, 26 attempted murders, 42 bombings, [and] 187 arsons."[7] Examples of moralist terrorism and threats against abortion providers include the following incidents:

- June and December 1984: An abortion clinic was bombed twice in Pensacola, Florida.

- March 1993: A physician was shot and killed outside an abortion clinic in Pensacola.

- July 1994: A physician and his bodyguard were killed outside an abortion clinic in Pensacola.

- October 1997: A physician was wounded by shrapnel in Rochester, New York.

- January 1998: A bomb was detonated at an abortion clinic in Montgomery, Alabama, killing a police officer and severely wounding a nurse. Eric Robert Rudolph was convicted for the attack.

- October 1998: A physician was killed in Amherst, New York.

- 1998–2000: Scores of letters with notes claiming to be infected with anthrax bacteria were sent to abortion clinics in at least 16 states.

- 2011–2016: Several cases of arson and at least one bombing occurred at abortion clinics nationwide. Most cases were unsolved.

- November 2015: Robert Lewis Dear killed three people, including a police officer, at a Planned Parenthood clinic in Colorado Springs, Colorado. Dear declared during a court appearance that he was "a warrior for the babies."

- November 2017: An explosive device was deactivated at an abortion clinic in Champaign, Illinois.

Army of God: A shadowy and violent Christian fundamentalist movement in the United States that has attacked moralistic targets, such as abortion providers.

Phineas Priesthood: A shadowy movement of Christian Identity fundamentalists in the United States who believe that they are called by God to purify their race and Christianity. They are opposed to abortion, homosexuality, interracial mixing, and whites who "degrade" white racial supremacy. This is a calling for men only, so no women can become Phineas priests. The name is taken from the Bible at Chapter 25, Verse 6 of the Book of Numbers, which tells the story of a Hebrew man named Phineas who killed an Israelite man and his Midianite wife in the temple.

Examples of violent moralist movements include the **Army of God** and the **Phineas Priesthood**. They are both shadowy movements that apparently have little or no organizational structure, operate as lone wolves or cells, and answer to the "higher power" of their interpretations of God's will. They seem to be belief systems in which like-minded activists engage in similar behavior. The Phineas Priesthood is apparently a "calling" (divine revelation) for Christian Identity fundamentalists, and the Army of God membership is perhaps derived from fringe evangelical Christian fundamentalists. These profiles are speculative, and it is possible that they are simply manifestations of terrorist contagion (copycatting). There has also been speculation that both movements are linked. Nevertheless, it is instructive to review their activity profiles.

The Army of God

The Army of God is a cell-based and lone-wolf movement that opposes abortion and homosexuality. Its ideology is apparently a fringe interpretation of fundamentalist Protestantism, although it has also exhibited racial supremacist tendencies. The methodology of the Army of God has included the use of violence and intimidation, primarily in attacks against abortion providers and gay and lesbian targets. The Army of God posted a website with biblical references and grisly pictures of abortions, and the manifesto disseminated by the group included instructions for manufacturing bombs. The website also pays homage to those the movement considers political prisoners and martyrs in their cause.

The Army of God first appeared in 1982 when an Illinois abortion provider and his wife were kidnapped by members of the group. It has since claimed responsibility for a number of attacks, primarily against abortion providers.

- February 1984: A clinic in Norfolk, Virginia, where abortions were performed was firebombed.

- February 1984: A clinic in Prince George's County, Maryland, where abortions were performed was firebombed.

- July 1994: Paul Hill, an antiabortion activist, shot and killed a physician and his bodyguard, a retired Air Force lieutenant colonel, in Pensacola, Florida. Hill was executed by lethal injection in September 2003. He was the first person to be executed for anti-abortion violence.

- January 1997: A clinic in Atlanta, Georgia, where abortions were performed was bombed.

- February 1997: A nightclub in Atlanta was bombed. Its patrons were largely gays and lesbians.

- January 1998: An abortion clinic in Birmingham, Alabama, was bombed, killing a police officer and severely wounding a nurse.

- October–November 2001: Five hundred fifty letters claiming to be contaminated with anthrax were sent to abortion providers. Notes included with some letters said, "You have chosen a profession, which profits from the senseless murder of millions of innocent children each year . . . we are going to kill you. This is your notice. Stop now or die." Some letters also said, "From the Army of God, Virginia Dare Chapter." Clayton Lee Waagner was convicted of sending the letters. He had also threatened to kill 42 employees of abortion providers.

- May 2009: Physician George Tiller was shot and killed inside his church in Wichita, Kansas, during religious services by an anti-abortion extremist, who confessed to the murder. The killer was accepted by the Army of God as one of its "soldiers."

One apparent affiliate of the Army of God—Eric Robert Rudolph—became a fugitive after he was named a suspect in the Birmingham bombing and the Atlanta bombings. Rudolph was also wanted for questioning for possible involvement in the July 1996 bombing at Centennial Olympic Park in Atlanta during the Summer Olympic Games and was linked to a militia group in North Carolina. He was captured in May 2003 in the mountains of North Carolina. In April 2005, Rudolph pleaded guilty to the Birmingham and Atlanta bombings as well as the Centennial Olympic Park attack. He was also convicted for two other clinic bombings and the bombing of a gay bar.

Regarding the November 2015 attack on a Planned Parenthood clinic in Colorado Springs, Colorado, the following comment was posted on the Army of God Website:

Planned Parenthood Colorado Springs

Robert Lewis Dear aside, Planned Parenthood murders helpless preborn children. These murderous pigs at Planned Parenthood are babykillers and they reap what they sow. In this case, Planned Parenthood selling of aborted baby parts came back to bite them. **Anyone who supports abortion has the blood of babies on their hands.**

The Phineas Priesthood

Phineas Priests were first described in the 1990 book *Vigilantes of Christendom: The History of the Phineas Priesthood.*[8] The book is a fundamentalist interpretation of Christian Identity. In the book, the alleged history of the Phineas Priesthood is traced from biblical times to the modern era. The name is taken from the Bible at Chapter 25, verse 6 of the Book of Numbers, which tells the story of a Hebrew man named Phineas who killed an Israelite man and his Midianite wife in the temple. According to the Book of Numbers, this act stayed the plague from the people of Israel.

Phineas Priests believe that they are called by God to purify their race and Christianity. They are opposed to abortion, homosexuality, interracial mixing, and whites who "degrade"

Phineas actions: Acts of violence committed by individuals who are "called" to become Phineas priests. Adherents believe that Phineas actions will hasten the ascendancy of the Aryan race.

white racial supremacy. Members also believe that acts of violence—called **Phineas actions**—will hasten the ascendancy of the Aryan race. The Phineas Priesthood is a calling for men only, so no women can become Phineas Priests. The calling also requires an absolute and fundamentalist commitment to Christian Identity mysticism. Beginning in the 1990s, acts of political and racial violence have been inspired by this doctrine. Early incidents include the following:

- In 1991, Walter Eliyah Thody was arrested in Oklahoma after a shootout and chase. Thody claimed to be a Phineas Priest and stated that fellow believers would also commit acts of violence against Jews and others.

- In 1993, Timothy McVeigh apparently "made offhand references to the Phineas Priesthood" to his sister.[9]

- From 1994 to 1996, the Aryan Republican Army robbed 22 banks throughout the Midwest. Members of the ARA had been influenced by *Vigilantes of Christendom* and the concept of the Phineas Priesthood.[10]

- In October 1996, three Phineas Priests were charged with bank robberies and bombings in Washington State. They had left political diatribes in notes at the scenes of two of their robberies. The notes included their symbol, 25:6, which denotes Chapter 25, verse 6 of the Book of Numbers.

Typical of more recent incidents is the 2014 lone-wolf attack by Larry Steven McQuilliams in Austin, Texas. On November 28, 2014, McQuilliams fired at a Mexican consulate and tried to set it on fire. He also fired more than 100 shots at a federal building and at a police station. McQuilliams was shot and killed by an Austin police officer. A copy of *Vigilantes of Christendom: The History of the Phineas Priesthood* was found in his residence.

Because the Phineas Priesthood has been a lone-wolf and cell-based phenomenon, it is impossible to estimate its size or even whether it has ever been much more than an example of the contagion effect. Nevertheless, the fact is that a few true believers have considered themselves to be members of the Phineas Priesthood, and the concept of Phineas actions was taken up by some adherents of the moralist and racial supremacist right.

THE NEW TERRORISM IN THE UNITED STATES

International terrorism has been relatively rare in the United States, and the number of international terrorist incidents is much lower than in other countries. During most of the postwar era (prior to the 1990s), international incidents in the United States were spillovers from conflicts in other Western countries and were directed against foreign interests with a domestic presence in the United States. Most of these spillovers ended after a single incident or a few attacks. Some terrorist spillovers were ongoing campaigns. Like the short-term incidents, these campaigns were directed primarily against non-American interests. Examples include the anti-Castroite group **Omega 7** and American suppliers of the Provisional Irish Republican Army. The terrorist environment changed during the 1990s, when American interests began to be directly attacked domestically by international terrorists. A new threat emerged from religious radicals who considered the United States a primary target in their global jihad.

Omega 7: An anticommunist Cuban American terrorist group that targeted Cuban interests.

Jihad in America

The American people and government became acutely aware of the destructive potential of international terrorism from a pattern that emerged during the 1990s and culminated on September 11, 2001. The following incidents were precursors to the modern security environment:

- February 1993: In the first terrorist attack on the World Trade Center, a large vehicular bomb exploded in a basement parking garage. This was a failed attempt to topple one tower into the other. Six people were killed, and more than 1,000 were injured. The mastermind behind the attack was the dedicated international terrorist Ramzi Yousef. His motives were to support the Palestinian people, to punish the United States for its support of Israel, and to promote an Islamic jihad. Several men, all jihadis, were convicted of the attack.

- October 1995: Ten men were convicted in a New York federal court of plotting further terrorist attacks. They had allegedly conspired to attack New York City landmarks, such as tunnels, the United Nations headquarters, and the George Washington Bridge.

These incidents heralded the emergence of a threat to homeland security that had not existed since World War II. The practitioners of the New Terrorism apparently concluded that assaults on the American homeland were desirable and feasible. The key preparatory factors for making these attacks feasible were the following:

- The attacks were carried out by operatives who entered the country for the sole purpose of carrying out the attacks.

- The terrorists had received support from cells or individuals inside the United States. Members of the support group had facilitated the terrorists' ability to perform their tasks with dedication and efficiency.

The support apparatus profile in the United States for these attacks was not entirely unknown; militants had been known to be in the United States since the late 1980s and 1990s. For example, aboveground organizations were established to funnel funds to the Middle East on behalf of Hamas, Hezbollah, and other movements. These organizations—and other social associations—were deliberately established in many major American cities. The fact is that since at least the late 1980s, anti-American jihadi sentiment has existed within the United States among some fundamentalist communities. Significantly, jihad has been overtly advocated by a number of fundamentalist leaders who have taken up residence in the United States.

September 11, 2001

The worst incident of modern international terrorism occurred in the United States on the morning of September 11, 2001. It was carried out by 19 al-Qaeda terrorists who were on a suicidal "martyrdom mission." They committed the attacks to strike at symbols of American (and Western) interests in response to what they perceived to be a continuing process of domination and exploitation of Muslim countries. They were religious terrorists fighting in the name of a holy cause against perceived evil emanating from the West. Their sentiments had been born in the religious, political, and ethnonational ferment that has characterized the politics of the Middle East for much of the modern era.

The symbolism of a damaging attack on homeland targets was momentous because it showed that the American superpower was vulnerable to attack by small groups of determined revolutionaries. The Twin Towers had dominated the New York skyline since the completion of the World Trade Center in 1972. They were a symbol of global trade and prosperity and the pride of the largest city in the United States. The Pentagon, of course, is a unique building that symbolizes American military power, and its location across the river from the nation's capital showed the vulnerability of the seat of government to attack.

The Anthrax Crisis: A Post–9/11 Anomaly

After the September 11 attacks, the activity profile of international terrorism in the United States shifted to cell-based religious terrorist spillovers originating in the Middle East. The threat from the New Terrorism in the United States included the very real possibility of a terrorist campaign using high-yield weapons to maximize civilian casualties.

The potential scale of violence was demonstrated by an anthrax attack immediately after the September 11 attacks when, for the first time in its history, the threat of chemical, biological, and radiological terrorism became a reality in the United States. During October through December 2001, more than 20 people were infected by anthrax-laced letters; five victims died. The attack made use of the U.S. postal system when letters addressed to news organizations and two members of the U.S. Senate were mailed from Princeton, New Jersey. Some of the letters contained references to radical Islam, causing a presumption by authorities and the public that the anthrax incident was part of an ongoing attack against the American homeland.

The crisis led to an extensive manhunt by the FBI, which conducted more than 10,000 interviews on six continents, including intensive investigations of more than 400 people. One person under careful investigation was Dr. Bruce Ivins, a microbiologist and army biodefense scientist. Ivins worked for decades on the army's anthrax vaccination program at the army biodefense laboratory in Maryland. The FBI's investigation involved detailed scrutiny of his behavioral habits, e-mail, his trash, and computer downloads. The FBI's scrutiny included attaching a global positioning satellite device to his automobile. Ivins committed suicide in July 2008 after he learned federal authorities were possibly moving forward with a criminal indictment against him. In February 2010, the FBI released an extensive report that closed its investigation of Ivins. However, debate continued about whether Ivins was responsible for the mailings. In January 2011, the National Academy of Sciences questioned the veracity of the FBI's evidence. In March 2011, a panel of psychiatrists developed a psychological profile of Ivins and concluded that the case against him was persuasive. Nevertheless, prominent scientists and investigative journalists continued to raise serious questions about the FBI's testing procedures and the accuracy of the FBI investigation.

Case in Point: The Threat From Homegrown Jihadists

A significant threat to homeland security in the United States and Europe arose from an unanticipated source: homegrown sympathizers of the international jihadist movement. Domestic security became increasingly challenged in the aftermath of high-casualty terrorist incidents carried out by extremists residing in Western democracies. Such incidents were particularly problematic because many of the perpetrators were seamlessly woven into the fabric of mainstream society.

The Fort Hood Incident

On November 5, 2009, a gunman opened fire in the sprawling military base at Fort Hood, Texas, killing 13 people and wounding 29. The attack occurred inside a Fort Hood medical center, and the victims were four officers, eight enlisted soldiers, and one civilian. The shooter was army major Nidal Malik Hasan, a psychiatrist at the base who treated returning veterans for combat stress.

Hasan is an interesting profile in how someone born and raised in the West can eventually adopt an ideology that advocates violent resistance to Western governments and policies. He was born in Virginia to Palestinian parents. He received an undergraduate degree from Virginia Tech University and eventually graduated from medical school with a specialization in psychiatry. Hasan was a devout Muslim who eventually became outspoken about his opposition to the wars in Afghanistan and Iraq. He also had a history of expressing himself provocatively. For example, at a public health seminar he presented a PowerPoint presentation titled "Why the War on Terrorism Is a War on Islam." At another presentation to medical colleagues, Hasan detailed the torments awaiting non-Muslims in hell. On other occasions, he proselytized his patients on behalf of Islam, argued that

The Boston Globe Exclusive via Getty Images

▶ **Photo 9.3** Dzhokhar Tsarnaev, wearing the reversed white hat on the right, before he and his brother, Tamerlan, detonated two bombs during the 2013 Boston Marathon.

he believed Islamic law (*shari'a*) is superior to the U.S. Constitution, and publicly identified himself as a Muslim first and an army officer second. During his trial in 2013, Hasan represented himself and refused to cross-examine witnesses called by the prosecution, thus essentially refusing to mount a defense on his own behalf. He was found guilty as charged.

The Boston Marathon Bombing

On April 15, 2013, two bombs were detonated at the crowded finish line of the Boston Marathon. Three people were killed, and more than 260 were wounded, many severely. The devices were constructed from pressure cookers and were detonated 13 seconds apart within approximately 210 yards of each other. They were packed with nails, ball bearings, and possibly other metal shards. Emergency response occurred swiftly, in part because medical personnel and emergency vehicles were already on hand to assist runners at the finish line. Law enforcement officers were also present as members of the race's security detail.

Two brothers, Dzokhar and Tamerlan Tsarnaev, were responsible for the attack. The Tsarnaevs were young immigrants from Chechnya who had resided in the United States since about 2002. Tamerlan, the elder brother, became radicalized during a visit to Chechnya when he became a committed Islamist. His and Dzokhar's underlying motive for the attack was to condemn the United States' interventions in the Middle East. It was reported that they downloaded instructions on how to construct pressure-cooker bombs from the online al-Qaeda magazine *Inspire*.

FBI analysis of video and photographic evidence from the scene of the attack eventually focused on images of two men whose behavior and demeanor differed from that of others in the crowd. Images of the men, one wearing a black baseball cap (Tamerlan) and the other wearing a white cap backward (Dzokhar), were disseminated to law enforcement officials, the media, and the public. During the manhunt, the Tsarnaevs shot and killed a Massachusetts Institute of Technology police officer. They also carjacked a vehicle and forced its occupant to withdraw money from an ATM. The victim escaped when the pair stopped at a gas station, ran to another station, and notified the authorities. The victim left his cell phone in the car, which was used by the authorities to track the Tsarnaevs. They were later observed driving a stolen sport utility vehicle and were confronted by the police. An intense gunfight ensued, and Tamerlan Tsarnaev was killed when he was run over by the SUV driven by his brother. Dzokhar Tsarnaev temporarily evaded the police, but he was eventually captured after an intense door-to-door manhunt while hiding in a boat parked in a backyard.

The question of motivation for the Boston Marathon attack is an instructive case study. Young immigrant men from a war-torn country became disaffected and radicalized even though they relocated to a society largely removed from the turmoil in their homeland. This disaffection is not uncommon among some migrants to the West and demonstrates a view of the world that transcends nationality; it represents the adoption of a globalized radical worldview. For disaffected individuals who may be marginalized in their new country of residence, radical ideologies provide a common connection to an international movement.

The San Bernardino Attack

On December 2, 2015, 14 people were killed and 21 injured when two armed assailants—a married couple—attacked the Inland Regional Center in San Bernardino, California. The state-run center assisted people with developmental disabilities. The assailants were Syed Rizwan Farook, who had worked at the regional center for five years, and his wife Tashfeen Malik. Farook was born and raised in the United States, and his wife Malik was born in Pakistan. Farook previously traveled abroad to Pakistan and Saudi Arabia, where he participated in the Muslim *hajj*, the pilgrimage to Mecca. He returned to the United States in July 2014 with Malik, whom he subsequently married.

On the day of the attack, Farook attended a holiday party at the regional center. He left the gathering and went to his home to prepare with Malik for their assault. They left their six-month-old child with Farook's mother, advising her that they were on their way to a medical appointment. Farook and Malik then dressed in paramilitary tactical gear and armed themselves. They returned to the regional center carrying semiautomatic assault rifles and pistols while wearing masks and opened fire on celebrants at the holiday party, killing and wounding at least 35 people. They left the facility and returned home, where the police had posted a stakeout after a tip about the vehicle they were driving. Law enforcement officers identified their vehicle and gave chase when Farook and Malik took to the road. During the chase, Farook and Malik shot at police officers and tossed an inert pipe out of their vehicle, apparently as an attempted ruse that it was a pipe bomb. Both assailants were shot and killed when they halted the vehicle and engaged in an intensive firefight with more than 20 officers.

The incident required extensive prior planning by the couple. Aside from the weapons and tactical gear in their possession during the assault and chase, a search of their home by law enforcement officers uncovered 12 functional pipe bombs, thousands of rounds of ammunition, and material for constructing more bombs. The couple had also placed an improvised explosive device (IED) at the scene of the assault. The IED consisted of three pipe bombs with a remote control detonator that would have been activated by a toy car controller. A law enforcement official reported that an unsuccessful attempt had been made to convert at least one of the semi-automatic assault rifles to fully automatic. Farook and Malik attempted to destroy computer hard drives and other electronic equipment in their home prior to the incident.

The incident also confirmed the reality of a domestic threat environment in the United States that for years had existed in Europe: mass-casualty violence emanating from home-grown terrorists inspired by international terrorist movements. During the attack, Malik posted a message on Facebook, under an alias, pledging allegiance to Abu Bakr al-Baghdadi, leader of Islamic State of Iraq and the Levant (ISIS). Two days later, a pro-ISIS broadcast declared that the couple were supporters of the movement.

Pipe Bomb Clusters in Manhattan and New Jersey

On September 17, 2016, clusters of pipe bombs were rigged to detonate in the Chelsea area of Manhattan; Elizabeth, New Jersey; and Seaside Park, New Jersey. Ahmad Khan Rahimi, an American citizen originally from Afghanistan, was arrested and prosecuted for planting the devices. Rahimi had been in the United States since 1995 and became a citizen in 2011. At an unknown date, he began to consider himself as a soldier in the Islamist war against the United States.

On September 17, 2016, in Seaside, New Jersey, a cluster of bombs placed in a trash can partially detonated near the starting line of the Seaside Semper Five road race. There were no casualties from the explosion. Two additional devices were placed in Chelsea, one of which detonated, injuring 31 people. Another cluster of five pipe bombs was found in Elizabeth, New Jersey, in a backpack placed in a trash can. The Elizabeth cluster did not detonate.

Rahimi was convicted in October 2017 in federal court for the Chelsea bombs. In February 2018, Rahimi was sentenced in federal court to two life sentences for the Chelsea incident. Although also suspected of responsibility for the Elizabeth and Seaside Park incidents, he was not definitively tied to these events during the Chelsea-related trial.

Case in Point: The Orlando Mass Shooting—Jihadist Terrorism and Hate Crime

On June 12, 2016, gunman Omar Mir Seddique Mateen shot 102 people at the Pulse nightclub in Orlando, Florida, with an assault rifle and semi-automatic handgun, killing 49 of his victims and wounding 53. Pulse was a popular nightclub frequented by members of the LGBT (lesbian, gay, bisexual, transgender) community and was hosting a "Latin night" music and dance theme on the day of the attack. The attack was the most lethal mass shooting by one individual in U.S. history. As discussed in Chapter Perspective 9.2, some cases of political violence may be classified as both acts of terrorism and hate crimes. The mass shooting in Orlando is a case in point of this nexus between terrorist events and hate crimes, in this case bias-motivated violence directed toward a protected group (the LGBT community) by an Islamist-inspired extremist.

Omar Mateen was a first-generation Afghan American, born in Queens, New York City, and raised in Port St. Lucie, Florida. He had an extensive history of behavioral challenges dating from elementary school. He was described in school records and by school officials as an aggressive and confrontational student and classmate, and received discipline on dozens of occasions. Significantly, classmates reported that 14-year-old Mateen imitated an exploding airplane on his school bus soon after the September 11, 2001, terrorist attack. As he matured, Mateen became a dedicated body builder, attended prayers at local mosques, and attempted to pursue a career in law enforcement. His career goal was cut short when he was terminated from a corrections department trainee program because he joked about bringing a firearm to class, poor attendance, and sleeping in class. He was eventually hired as a security guard by a private firm. Mateen's personal life was turbulent, and his first wife divorced him after less than one year of marriage because of repeated physical abuse. He also allegedly stalked

Pulse Nightclub Shooting
Orlando, Florida

OMAR MIR SEDDIQUE MATEEN - DECEASED

U.S. Federal Bureau of Investigation

▶ **Photo 9.4** The Federal Bureau of Investigation's bulletin seeking information about Omar Mateen, perpetrator of the Pulse nightclub mass shooting in Orlando, Florida.

a woman he met via an online dating service while he was married to a second wife.

Mateen attracted the attention of the FBI in 2013 when the security company he was employed with removed him from his post at the St. Lucie County Courthouse when he commented on his alleged ties to Lebanon's Shi'a Hezbollah movement and the Sunni al-Qaeda network—groups that are rivals, not allies. The FBI made inquiries and concluded that not enough evidence existed to continue investigating Mateen. In 2014, the FBI again made inquiries after Moner Mohammad Abu-Salha, who attended the same mosque as Mateen, carried out a suicide bombing in Syria on behalf of an al-Qaeda-affiliated group. The FBI concluded Mateen and Abu-Salha were only minimally acquainted. In June 2016, Mateen legally purchased a SIG Sauer MCX assault rifle and a Glock 9mm handgun, the weapons he used during the Pulse nightclub attack. He had unsuccessfully attempted to purchase body armor.

Omar Mateen deliberately selected an LGBT site to carry out his attack. Mateen's first wife reported that he exhibited homophobic tendencies, and his father reported Mateen was angered when he saw two men kissing. Ironically, patrons at Pulse nightclub reported Mateen had visited Pulse on numerous occasions, appearing to enjoy himself at the nightclub. He again visited Pulse on the evening of the attack, and returned later with his firearms. Mateen opened fire as he entered the nightclub, shooting patrons and exchanging gunfire with an off-duty police officer. He continued firing, retreating to a restroom when police officers began arriving on the scene. Mateen shot a number of patrons who tried to take refuge in the restroom. While in the restroom, he dialed the local 911 emergency service and professed his allegiance to ISIS. He also participated in three conversations with a crisis negotiation team, during which Mateen claimed he was an "Islamic soldier" demanding an end to American intervention in Iraq and Syria. Other claims were made that he had a suicide vest, bombs had been planted outside the nightclub, and he had associates who were planning additional attacks. Police responders attempted to blast a hole in the restroom's wall, and when this failed used an armored vehicle to breach the wall. They engaged Mateen, who died during the ensuing firefight.

Omar Mateen's declaration of allegiance to ISIS, his selection of an LGBT target, and his stated opposition to U.S. foreign policy strongly indicate that the Orlando attack was both an act of terrorism and a hate crime. It led to significant partisan political division within the United States on the questions of domestic security, counterterrorism, and the Second Amendment right to keep and bear arms. A debate also ensued on the media's reporting of this and other similar incidents, in particular on whether such publicity could result in copycat incidents.

LONE-WOLF TERRORISM IN THE UNITED STATES

As reported from cases presented in this chapter, many incidents of terrorist violence have been committed by individual extremists who act alone without clearly identifiable associations with terrorist organizations or networks. Such individuals certainly profess an intellectual or ideological identification with extremist causes, but they are lone operators who act on their own initiative or are sent on lone missions by extremist organizations. This phenomenon—the lone-wolf model—occurs with regularity in the United States,

Europe, Israel, and elsewhere. The United States, in particular, has a high incidence of lone-wolf terrorism, with an estimated 40 percent of incidents occurring in the United States during the past four decades.[11]

It is a very difficult model to combat because, conceptually, terrorist cells can be as small as a single individual who is unknown to law enforcement or intelligence services. Historically, most lone-wolf attacks in the United States have been racially motivated, anti-government, or religion-related attacks. Religion-related attacks tend to be motivated by either moralist or jihadist beliefs. Plausible threat scenarios concerning radicalization from international extremists also pose an increasing risk because of the prevalence of individual access to the Internet, social networking media, and other digital technologies.

Supremacist and Antigovernment Lone Wolves

In the modern era, most violence emanating from supremacist sentiment in the United States has been conducted as lone-wolf terrorism and hate crimes by perpetrators professing overtly racist or anti-Semitic motivations. In comparison, antigovernment lone wolves tend to attack individual officials or government offices because they symbolize their dissatisfaction with government policies. Some lone wolves combine antigovernment sentiment with racial supremacy or anti-Semitism. For example, on April 4, 2009, in Pittsburgh, Pennsylvania, Richard Poplawski opened fire on police officers inside his mother's home with a shotgun, .357 Magnum handgun, and an AK-47 assault rifle, killing three and wounding three others. Poplawski had lain in wait for the officers after summoning them to the house, ambushing them when they appeared. He was motivated by antigovernment and anti-Zionist sentiment.

The following cases further illustrate the type of violence emanating from supremacist and antigovernment lone wolves.

Richard Baumhammers

A typical example of neo-Nazi lone-wolf violence is the case of Richard Baumhammers. Baumhammers was a racist immigration attorney influenced by neo-Nazi ideology who murdered five people and wounded one more on April 28, 2000, near Pittsburgh, Pennsylvania. He methodically shot his victims during a 20-mile trek. The victims were a Jewish woman, two Indian men, two Asian men, and an African American man. The sequence of Baumhammers's assaults occurred as follows:

- Baumhammers went to his Jewish neighbor's house and fatally shot her. He then set a fire inside her home.

- He next shot two Indian men at an Indian grocery store. One man was killed, and the other was paralyzed by a .357 slug that hit his upper spine.

- Baumhammers shot at a synagogue, painted two swastikas on the building, and wrote the word *Jew* on one of the front doors.

- He then drove to a second synagogue, where he fired shots at it.

- Baumhammers shot two young Asian men at a Chinese restaurant, killing them both.

- Finally, Baumhammers went to a karate school, pointed his revolver at a white man inside the school, and then shot to death an African American man who was a student at the school.

Richard Baumhammers was convicted in May 2001 and received the death penalty.

James Wenneker von Brunn

On June 10, 2009, a gunman opened fire inside the entrance to the U.S. Holocaust Memorial Museum in Washington, D.C. An African American security guard who opened the door for him was shot with a .22 caliber rifle and later died of his wounds. Other security guards returned fire, wounding the assailant. The attacker was James Wenneker von Brunn, a known racial supremacist and Holocaust denier—he believed that the Nazi-led genocide during World War II never occurred. Von Brunn was a known extremist and had an arrest record from an incident in 1981 when he entered a federal building armed with weapons and attempted to place the Federal Reserve Board under "citizen's arrest."

The police later found a notebook containing a list of other sites in Washington, D.C., and the following entry:

> You want my weapons—this is how you'll get them. The Holocaust is a lie. Obama was created by Jews. Obama does what his Jew owners tell him to do. Jews captured America's money. Jews control the mass media. The 1st Amendment is abrogated—henceforth.[12]

Von Brunn died in January 2010 before he could be brought to trial on charges of murder and firearms violations.

Frazier Glenn Cross

On April 13, 2014, Frazier Glenn Cross shot to death a 14-year-old Eagle Scout and the boy's grandfather in the parking lot of a Jewish community center in the suburban community of Overland Park, Kansas, near Kansas City. He then went to a nearby Jewish retirement home and killed another victim. It was reported that Cross shouted, "Heil Hitler!" several times as the police took him into custody. The 73-year-old Cross had a long history of activity in the American racial supremacist movement, including leadership in a group originally affiliated with the Ku Klux Klan that eventually reformed as the White Patriot Party, a Christian Identity organization.

Cross was sentenced to death in November 2014. He shouted, "Heil Hitler!" several times as the judge read his sentence.

Dylann Roof

On June 17, 2015, Dylann Storm Roof shot twelve people attending a Bible study meeting at the Emmanuel African Methodist Episcopal Church in Charleston, South Carolina. All victims were African Americans, and nine died during the assault. Roof was an avowed racial supremacist who carried out the attack after being welcomed by the Bible study participants and sitting with them for approximately one hour. He confessed to the crimes and stated he sought to set an example by his actions, which he intended to be a "spark" to ignite a race war.

Prior to the shootings, Dylann Roof posted a website titled *The Last Rhodesian* that was a discourse on what he considered to be the plight of the white race at the hands of non-whites and Jews. Using racist expletives and perspectives, he concluded several times that the white race is naturally superior and must reestablish its hegemony over nonwhite races and Jews. Several photographs were posted on the website of Roof posing with the Confederate, Rhodesian, and apartheid-era South African flags as symbols of racial supremacy. He is also shown posing as he burned and spat on the American flag.

Roof was charged with nine counts of murder and three counts of attempted murder as well as possession of a firearm during the commission of a felony.

Cesar Sayoc

During the week of October 22, 2018, thirteen pipe bombs were discovered in packages addressed to prominent politicians and other public individuals. The first device was found on October 22, 2018, in the mailbox of billionaire and Democratic Party supporter George Soros. On October 23 and 24, the Secret Service intercepted devices addressed to Bill and Hillary Clinton and former president Barack Obama. Additional package bombs were intercepted in the following locations:

- The New York City mailroom of CNN, addressed to former CIA director John Brennan

- The Florida office of Representative Debbie Wasserman Shultz, addressed to former attorney General Eric Holder but with Schultz's return address. The address for Holder was incorrect.

- Two devices addressed to Representative Maxine Waters, one in Los Angeles and another in Washington, D.C.

- Actor Robert DeNiro's film company in New York City

- Two mail facilities in Delaware, addressed to Senator Joe Biden

- A device intercepted by the FBI in Florida, addressed to New Jersey Senator Cory Booker

- A mail sorting facility in New York addressed to CNN, with the addressee entered as former Director of National Intelligence James Clapper

- A Sacramento mail facility, addressed to Senator Kamala Harris

- In California, a device addressed to billionaire Tom Steyer, a Democratic Party supporter

Florida resident Cesar Sayoc was arrested by the FBI using fingerprint and DNA evidence taken from devices, as well as tracking his mobile telephone and Twitter account. Sayoc had posted angry partisan political statements on the Internet and covered his van with stickers supportive of President Donald Trump. Also on his van were additional stickers of Democratic leaders with bull's eyes drawn over their images. Sayoc selected his targets because of their political affiliations with the Democratic Party.

Robert Bowers

On October 27, 2018, Robert Bowers shot 17 people attending religious services at the Tree of Life synagogue in Pittsburgh, Pennsylvania. All victims were Jewish, and 11 died during the assault. Four of the injured were police officers.

Bowers entered the synagogue during morning worship services armed with an AR-15 assault rifle and three Glock semi-automatic handguns. He shouted "all Jews must die!" and began shooting attendees. Police responded quickly, and SWAT team members subdued Bowers after wounding him in an exchange of gunfire.

Bowers had posted numerous anti-Semitic messages on the Gab online social network website. The Gab website billed itself as a free speech forum but was in fact a platform used

▶ **Photo 9.5** Richard Reid, popularly known as the "shoe bomber," attempted to detonate plastic explosives hidden in his shoe aboard a trans-Atlantic flight in December 2001.

U.S. Army

by many racial supremacists, white nationalists, and anti-Semites. Bowers expressed particular animosity toward the Hebrew Immigrant Aid Society refugee aid organization. His final post, immediately before entering the synagogue, was "screw your politics, I'm going in."

Jihadist Lone Wolves

The United States and Europe experience terrorist violence from individuals and small cells who are motivated by international jihadist ideologies. Jihadist movements such as ISIS and al-Qaeda have specifically encouraged lone-wolf and small-cell attacks on Western nations. Messages broadcast by these groups on the Internet and other technologies are easily received by potential sympathizers. For example, on June 1, 2009, Carlos Bledsoe conducted a drive-by shooting at an army recruiting center in Little Rock, Arkansas, killing one soldier and wounding another. Bledsoe was a convert to Islam who was radicalized in a Yemeni prison where he attempted to join al-Qaeda. He was also inspired by U.S.-born jihadist cleric Anwar al-Awlaki. He returned to the United States and carried out his lone-wolf attack.

The following cases further illustrate the type of violence emanating from jihadist lone wolves who apparently received inspiration and, in some cases, training from jihadist movements.

The "Shoe Bomber"

An instructive example of a single-member Islamist cell is the case of Richard C. Reid, a British resident who converted to Islam in prison in the UK and who became known as the "shoe bomber." Reid was detected by an alert flight attendant and overpowered by passengers on December 22, 2001, when he attempted to ignite plastic explosives in his shoe on American Airlines Flight 63, a Boeing 767 carrying 198 passengers and crew from Paris to Miami.[13] In Reid's case, he was a self-professed follower of al-Qaeda and may have been trained by the organization in Afghanistan. He was sentenced to life imprisonment in a super-maximum-security prison after pleading guilty before a federal court in Boston.

The "Underwear Bomber"

On December 25, 2009, Nigerian national Umar Farouk Abdulmutallab ignited explosive chemicals aboard Northwest Airlines Flight 253 with approximately 290 people aboard as it approached Detroit, Michigan, in the United States. According to a federal criminal complaint and FBI affidavit, Abdulmutallab attempted to detonate an improvised explosive containing PETN (pentaerythritol tetranitrate), which had been attached to his leg. He used a syringe to detonate the PETN, but fortunately, the device merely caught fire and did not fully detonate. Passengers reported that immediately prior to the incident, Abdulmutallab had been in the restroom for approximately 20 minutes. He pulled a blanket over himself, and passengers heard cracking sounds comparable to firecrackers, sensed an odor, and observed Abdulmutallab's pants leg and the airplane wall on fire. Abdulmutallab was subdued by passengers and members of the crew, who also extinguished the fire. He was calm throughout the incident and replied, "Explosive device," when asked by a flight attendant what he had in his pocket.

Abdulmutallab had recently associated with religious militants in Yemen and visited there from August to December 2009. He said to officials that he had been trained in Yemen to make explosives; he also claimed that Yemenis had given him the chemicals used on Flight 253. His name had been listed in a U.S. terrorism database during November 2009 after his father reported to the U.S. embassy in Nigeria that his son had been radicalized and was associating with religious extremists. However, Abdulmutallab was not placed on an airlines watch list for flights entering the United States because American authorities concluded they had insufficient information to do so. In fact, Abdulmutallab possessed a two-year tourist visa, which he received from the U.S. embassy in London in June 2008, and he had traveled to the United States on at least two occasions. In October 2011, Abdulmutallab pleaded guilty to eight counts of terrorism-related criminal charges.

GLOBAL PERSPECTIVE: LONE-WOLF TERROR IN NORWAY

Extreme right-wing activism in Europe usually involves the formation of nationalist political parties, grassroots populist movements, and skinhead subcultures. Occasional outbreaks of violence are directed against immigrant populations and ideological opponents. This chapter's Global Perspective discusses right-wing extremist Anders Breivik's lone-wolf spree shooting in the vicinity of Oslo, Norway.

GLOBAL PERSPECTIVE

LONE-WOLF TERROR IN NORWAY

On July 22, 2011, Anders Breivik, a self-professed right-wing ideologue, detonated a car bomb in the government district of Oslo and methodically shot to death nearly 80 people at a Norwegian Labor Party youth summer camp on the island of Utøya. His victims were government workers, bystanders, and teenage residents of the camp. The sequence of Breivik's assault occurred as follows:

- Breivik detonated a car bomb in Oslo's government district using ANFO explosives. The blast killed eight people and wounded at least a dozen more.

- He next drove nearly two hours to a youth summer camp on the island of Utøya. The camp was sponsored by the youth organization of the ruling Norwegian Labor Party, and hundreds of youths were in attendance. Breivik was disguised as a policeman.

- When Breivik arrived on the island, he announced that he was a police officer who was

following up on the bombing in Oslo. As people gathered around him, he drew his weapons and began shooting.

- Using a carbine and semiautomatic handgun, Breivik methodically shot scores of attendees on Utøya, most of them teenagers. The attack lasted for approximately 90 minutes and ended when police landed on the island and accepted Breivik's surrender.

In August 2012, Breivik was convicted of murdering 77 people and received Norway's maximum sentence of 21 years imprisonment. Under Norwegian law, his incarceration may be extended indefinitely if he is deemed to be a risk to society.

The Breivik case illustrates how the lone-wolf scenario involves an individual who believes in a certain ideology but who is not acting on behalf of an organized group.

CHAPTER SUMMARY

This chapter introduced domestic terrorism in the United States, discussing threats from local and international sources. The purpose of this discussion was to identify and define several sources of domestic terrorism, to differentiate terrorism emanating from foreign and domestic sources, and to provide cases in point for these concepts.

DISCUSSION BOX

Domestic Terrorism in the American Context

This chapter's Discussion Box is intended to stimulate critical debate about the idiosyncratic nature of domestic terrorism in the United States.

The subject of domestic terrorism in the United States is arguably a study in idiosyncratic political violence. Indigenous terrorist groups reflected the American political and social environments during historical periods when extremists chose to engage in political violence.

In the modern era, left-wing and right-wing political violence grew from very different circumstances. Leftist violence evolved from a uniquely American social environment that produced the civil rights, Black Power, and New Left movements. Rightist violence grew out of a combination of historical racial and nativist animosity combined with modern applications of religious and antigovernment ideologies.

In the early years of the new millennium, threats continued to emanate from right-wing antigovernment and racial supremacist extremists. Potential violence from leftist extremists remained low in comparison with the right. The September 11, 2001, attacks created a new security environment with an international dimension. The question of terrorism originating from domestic sources inspired by international events and ideologies became very plausible.

Discussion Questions

1. Assume that a nascent anarchist movement continues in its opposition to globalism. How should the modern leftist movement be described? What is the potential for violence originating from modern extremists on the left?

2. Keeping in mind the many conspiratorial and mystical beliefs of the American right, what is the potential for violence from adherents of these theories to the modern American environment?

3. As a matter of policy, how closely should hate and antigovernment groups be monitored? What restrictions should be imposed on their activities? Why?

4. Is the American activity profile truly an *idiosyncratic* profile, or can it be compared with other nations' environments? If so, how? If not, why not?

5. What are the factors that may give rise to a resurgence of a rightist movement on the scale of the 1990s Patriot movement? What trends indicate that it *will* occur? What trends indicate that it *will not* occur?

KEY TERMS AND CONCEPTS

The following topics were discussed in this chapter and can be found in the glossary:

ON YOUR OWN

Get the tools you need to sharpen your study skills. SAGE edge offers a robust online environment featuring an impressive array of free tools and resources.

Access practice quizzes, eFlashcards, video, and multimedia at **edge.sagepub.com/martinhs3e**

RECOMMENDED WEBSITES

The following websites provide information about domestic extremism and terrorism in the United States. Also included are links to dissident organizations and movements.

Anarchist Cookbook: www.anarchistcookbook.com

Animal Liberation Front: www.animalliberation front.com

Anti-Defamation League: www.adl.org

Army of God: www.armyofgod.com

Christian Exodus: christianexodus.org

Council of Conservative Citizens: www.cofcc.org

Earth First! Journal: www.earthfirstjournal.org

Earth Liberation Front: www.earth-liberation-front.com

Hate Directory: hatedirectory.com

Revolutionary Communist Party, USA: www.revcom.us

Southern Poverty Law Center: www.splcenter.org

WEB EXERCISE

Using this chapter's recommended websites, conduct an online investigation of terrorism in the United States.

1. How would you describe the typologies of groups that predominate in the United States?

2. Conduct a Web search of American monitoring organizations, read their mission statements, and assess their services. Which organizations do you think provide the most useful data? Why?

3. If you were an American dissident extremist (leftist or rightist), how would you design your website?

To conduct an online search on terrorism in the United States, activate the search engine on your Web browser and enter the following keywords:

"American jihad"

"Domestic terrorism"

RECOMMENDED READINGS

The following publications discuss the nature of terrorism in the United States and the root causes of political violence in American society.

Bader, Eleanor J. and Patricia Baird-Windle. 2001. *Targets of Hatred: Anti-Abortion Terrorism.* New York: Palgrave.

Dunbar, David and Brad Reagan, eds. 2006. *Debunking 9/11 Myths: Why Conspiracy Theories Can't Stand Up to the Facts.* New York: Hearst Books.

Emerson, Steven. 2002. *American Jihad: The Terrorists Living Among Us.* New York: Free Press.

Emerson, Steven. 2006. *Jihad Incorporated: A Guide to Militant Islam in the US*. Amherst, NY: Prometheus.

Gartenstein-Ross, Daveed and Laura Grossman. 2009. *Homegrown Terrorists in the U.S. and U.K.* Washington, DC: FDD Press.

George, John and Laird Wilcox. 1996. *American Extremists: Militias, Supremacists, Klansmen, Communists, and Others*. Amherst, NY: Prometheus.

German, Mike. 2007. *Thinking Like a Terrorist: Insights of a Former FBI Undercover Agent*. Washington, DC: Potomac Books.

Gerstenfeld, Phyllis B. 2013. *Hate Crimes: Causes, Controls, and Controversies*. Thousand Oaks, CA: Sage.

Graebmer, William. 2008. *Patty's Got a Gun: Patricia Hearst in 1970s America*. Chicago: University of Chicago Press.

Kaczynski, Theodore. 2010. *Technological Slavery: The Collected Writings of Theodore J. Kaczynski, a.k.a. "The Unabomber."* Port Townsend, WA: Feral House.

Kurst-Swanger, Karl. 2008. *Worship and Sin: An Exploration of Religion-Related Crime in the United States*. New York: Peter Lang.

MacDonald, Andrew. 1978. *The Turner Diaries*. New York: Barricade.

McCann, Joseph T. 2006. *Terrorism on American Soil: A Concise History of Plots and Perpetrators From the Famous to the Forgotten*. Boulder, CO: Sentient Publications.

McCarthy, Timothy Patrick and John McMillian. 2003. *The Radical Reader: A Documentary History of the American Radical Tradition*. New York: New Press.

Michel, Lou and Dan Herbeck. 2001. *American Terrorist: Timothy McVeigh & the Oklahoma City Bombing*. New York: Regan Books.

Ogbar, Jeffrey O. G. 2004. *Black Power: Radical Politics and African American Identity*. Baltimore, MD: Johns Hopkins University Press.

Perkins, Samuel. 2011. *Homegrown Terror and American Jihadists: Assessing the Threat*. Hauppauge, NY: Nova.

Ridgeway, James. 1995. *Blood in the Face: The Ku Klux Klan, Aryan Nations, Nazi Skinheads, and the Rise of a New White Culture*. 2nd ed. New York: Thunder's Mouth.

Ronczkowski, Michael. 2006. *Terrorism and Organized Hate Crime: Intelligence Gathering, Analysis, and Investigations*. 2nd ed. Boca Raton, FL: CRC Press.

Sargent, Lyman Tower, ed. 1995. *Extremism in America: A Reader*. New York: New York University Press.

Simon, Jeffrey D. 2013. *Lone Wolf Terrorism: Understanding the Growing Threat*. New York: Prometheus Books.

Smith, Brent L. 1994. *Terrorism in America: Pipe Bombs and Pipe Dreams*. Albany: State University of New York Press.

Stern, Kenneth S. 1997. *A Force Upon the Plain: The American Militia Movement and the Politics of Hate*. New York: Simon & Schuster.

Wilkerson, Cathy. 2007. *Flying Close to the Sun: My Life and Times as a Weatherman*. New York: Seven Stories Press.

Zakin, Susan. 2002. *Coyotes and Town Dogs: Earth First! and the Environmental Movement*. Tucson: University of Arizona Press.

Zeskind, Leonard. 2009. *Blood and Politics: The History of the White Nationalist Movement From the Margins to the Mainstream*. New York: Farrar, Straus and Giroux.

PREPAREDNESS AND RESILIENCE

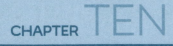

CHAPTER TEN

POROUS NODES
Specific Vulnerabilities

Chapter Learning Objectives

This chapter will enable readers to do the following:

1. Evaluate policy challenges inherent in border protection, aviation security, and port security

2. Analyze aviation security policies, procedures, and challenges

3. Explain the complexities of securing the nation's borders

4. Discuss policy options for border control

5. Analyze the unique characteristics of port and maritime infrastructure

Opening Viewpoint: Aviation Terrorism and Manipulation of the Media—The Case of TWA Flight 847

Lebanon's Hezbollah has demonstrated its skill at conducting extraordinary strikes, some of which ultimately affected the foreign policies of France, Israel, and the United States. It regularly markets itself to the media by disseminating grievances as press releases, filming and photographing moving images of its struggle, compiling "human interest" backgrounds of Hezbollah fighters and Shi'a victims, and packaging its attacks as valiant assaults against Western and Israeli invaders and their proxies. This has been done overtly and publicly, and incidents are manipulated to generate maximum publicity and media exposure.

On June 14, 1985, three Lebanese Shi'a terrorists hijacked TWA Flight 847 as it flew from Athens to Rome. It was diverted to Beirut, Lebanon, and then to Algiers, Algeria. The airliner was flown back to Beirut, made a second flight to Algiers, and then flew back to Beirut. During the odyssey, the terrorists released women, children, and non-Americans until 39 American men remained on board the aircraft. At the final stop in Beirut, the American hostages were offloaded and dispersed throughout the city.

As the hijacking unfolded, the media devoted an extraordinary amount of airtime to the incident. The television networks ABC, CBS, and NBC broadcast almost 500 news reports, or 28.8 per day, and devoted two-thirds of their evening news programs to the crisis.[i] "During the 16 days of the hijacking, CBS devoted 68% of its nightly news broadcasts to the event while the corresponding figures at ABC and NBC were 62% and 63% respectively."[ii]

The hijackers masterfully manipulated the world's media. They granted carefully orchestrated interviews, held press conferences, and selected the information they permitted the news outlets to broadcast.

It was reported later that the terrorists had offered to arrange tours of the airliner for the networks for a $1,000 fee and an interview with the hostages for $12,500.[iii] After the hostages were dispersed in Beirut, Nabih Berri, the leader of Lebanon's Syrian-backed Shi'a Amal movement (an ally and occasional rival of the Shi'a Hezbollah movement), was interviewed by news networks as part of the negotiations to trade the hostages for concessions. In the end, the terrorists' media-oriented tactics were quite effective. They successfully broadcast their grievances and demands to the world community and achieved their objectives. "[The] media exposure of the hostages generated enough pressure for the American president to make concessions."[iv]

The hostages were released on June 30, 1985.

As a postscript—which was sometimes forgotten during the episode—a U.S. Navy diver had been severely beaten, shot, and thrown down to the Beirut airport's tarmac by the terrorists. The murder occurred during the second stopover in Beirut. The leader of the terrorist unit, Imad Mughniyah, and three others were later indicted by U.S. prosecutors for the killing. One hijacker, Mohammed Ali Hamadi, was convicted in Germany of the Navy diver's murder and sentenced to life in prison. Imad Mughniyeh was assassinated in Damascus, Syria, in February 2008 by a car bomb.

Notes

i. Bruce Hoffman, *Inside Terrorism* (New York: Columbia University Press, 1998), 132.

ii. Gabriel Weimann and Conrad Winn, *The Theater of Terror: Mass Media and International Terrorism* (New York: Longman, 1994), 1.

iii. Steven E. Barkan and Lynne L. Snowden, *Collective Violence* (Boston: Allyn & Bacon, 2000), 84.

iv. Lawrence Howard, ed., *Terrorism: Roots, Impact, Responses* (New York: Praeger, 1992), 102.

This chapter explores several centers, or *nodes*, of specific homeland security vulnerability. These areas are sources of possible danger to homeland security, primarily from threats originating outside the United States. The characteristics of these centers of vulnerability lend themselves to terrorist threat scenarios emanating from terrorist groups, individual extremists, and aggressive governments wishing to carry out attacks inside the United States. Accepted policy priorities require that special attention be given to securing these points of entry, although it is understood that complete security cannot be absolutely guaranteed under all scenarios.

An important concept in this area of homeland security policy is **critical infrastructure/key resources (CIKR)**. All of the security nodes discussed in this chapter—borders, aviation, and ports—are subsumed under CIKR. As framed by the Department of Homeland Security (DHS), CIKR refers to the networks, assets, and systems, whether virtual or physical, so vital to the United States that their destruction or incapacitation would have a debilitating effect on national public health or safety, national economic security, security, or any

critical infrastructure/ key resources (CIKR): The assets, systems, and networks, whether physical or virtual, so vital to the United States that their incapacitation or destruction would have a debilitating effect on security, national economic security, national public health or safety, or any combination thereof.

combination thereof.[1] Table 10.1 reports the CIKR responsibilities of the sector-specific agencies previously discussed in Chapter 5.

American entry points are extensive, and because of this, they are vulnerable in several respects. The sheer number of entry points and departure locations is daunting from a domestic security perspective. Hundreds of seaports require extensive monitoring. Hundreds of airports require safety-related security protocols. The northern and southern borders are so extensive that border security personnel and resources are severely extended.

Table 10.1	Sector-Specific Agencies for Critical Infrastructure and Key Resources

The following list includes those federal departments and agencies identified in Homeland Security Presidential Directive 7 as responsible for CIKR protection activities in specified CIKR sectors.

Sector-Specific Agency	Critical Infrastructure and Key Resources Sector
Department of Agriculture **Department of Health and Human Services**	Agriculture and Food
Department of Defense	Defense Industrial Base
Department of Energy	Energy
Department of Health and Human Services	Public Health and Healthcare
Department of the Interior	National Monuments and Icons
Department of the Treasury	Banking and Finance
Environmental Protection Agency	Drinking Water and Water Treatment Systems
Department of Homeland Security *Office of Infrastructure Protection*	Chemical Commercial Facilities Dams Emergency Services Nuclear Reactors, Materials, and Waste
Office of Cyber Security and Communications	Information Technology Communications
Transportation Security Administration	Postal and Shipping
Transportation Security Administration/U.S. Coast Guard	Transportation Systems
Immigration and Customs Enforcement/ Federal Protective Service	Government Facilities

Source: U.S. Federal Emergency Management Agency.

The myriad challenges inherent in securing vulnerable nodes against credible dangers truly highlight the complexities of coordinating the work of individual agencies tasked with preserving domestic security. Implementing policies and procedures is an expensive and vast bureaucratic undertaking. Nevertheless, sustainable and viable homeland security policies are dependent upon the efficient implementation of these policies by each agency participating in the overall domestic security mission. Failure to do so may result in dire consequences.

The discussion in this chapter will review the following:

- Aviation security
- Border security
- Border control: Securing national entry points
- Port security

AVIATION SECURITY

Assaults against aviation targets have been a favored tactic among violent extremists. The reasons for selecting such targets are logical: Aviation assets possess significant symbolic value, and they are legitimate hubs of commerce with a great deal of economic value. Because airports and aircraft are integral to sustaining the economies of nations and regions, attacks can have a substantial effect on these locations. In the modern era of mass-casualty terrorism, crowded and busy transportation hubs also provide opportunities for concentrated death and destruction.

Symbolic Value: Airline Attacks and Maximum Propaganda Effect

Attacks against aviation targets and other modes of transportation are conducted for maximum propaganda effect. Modern terrorists learned during the late twentieth century that airline hijackings, attacks at airports, in-flight bombings, and similar assaults garnered immediate global attention. The selection of international passenger carriers as targets by terrorists creates unambiguous international implications. They are relatively "soft" targets that easily garner international media attention when attacked. Passengers on these carriers are considered to be legitimate symbolic targets, so that terrorizing or killing them is justifiable in the minds of terrorists.

The following weapons pose special threats to aircraft, airline terminals, and other aviation targets:

- *Rocket-propelled grenades (RPGs).* As discussed in Chapter 2, RPGs are light, self-propelled munitions and are common features of modern infantry units.

- *Precision-guided munitions (PGMs).* As discussed in Chapter 2, PGMs are weapons that can be guided to their targets by using infrared or other tracking technologies.

- *Plastic explosives.* As discussed in Chapter 2, plastic explosives are putty-like explosive compounds that can be easily molded.

- *Pipe bombs.* These devices are easily constructed from common pipes, which are filled with explosives (usually gunpowder) and then capped on both ends. Nuts, bolts, screws, nails, and other shrapnel are usually taped or otherwise attached to pipe bombs. Many hundreds of pipe bombs have been used by terrorists.

- *Vehicular bombs.* Ground vehicles that have been wired with explosives are a frequent weapon in the terrorist arsenal. Vehicular bombs can include car bombs and truck bombs; they are mobile, are covert in the sense that they are not readily identifiable, are able to transport large amounts of explosives, and are rather easily constructed. They have been used on scores of occasions throughout the world.

- *Barometric bombs.* These bombs use triggers that are activated by changes in atmospheric pressure. An altitude meter can be rigged to become a triggering device when a specific change in pressure is detected. Thus, an airliner can be blown up in midair as the cabin pressure changes. These are sophisticated devices.

Because terrorist attacks against aviation targets create an immediate international profile, such operations are frequently conducted in conjunction with the seizure of hostages. The following case in point illustrates this reality. It recounts activities of the Popular Front for the Liberation of Palestine and the origin of modern terrorism against aviation targets.

Case in Point: The PFLP and the Dawn of Modern Aviation Terrorism

Beginning in the late 1960s, terrorists began attacking international ports of call used by travelers, especially aviation nodes. International passenger carriers—primarily airliners—became favorite targets of terrorists. In the beginning, hijackings were often the acts of extremists, criminals, or otherwise desperate people trying to escape their home countries to find asylum in a friendly country. This profile changed, however, when the Popular Front for the Liberation of Palestine (PFLP) staged a series of aircraft hijackings as a way to publicize the cause of the Palestinians before the world community. The first successful high-profile PFLP hijacking was conducted by Leila Khaled in August 1969. The PFLP struck again in September 1970, when it attempted to hijack five airliners, succeeding in four of the attempts. These incidents certainly directed the world's attention to the Palestinian cause, but they also precipitated the Jordanian army's Black September assault against the Palestinians. Nevertheless, passenger carriers have since been frequent targets of international terrorists.

Since the Black September hijackings, many airline attacks clearly have been executed by varied extremist movements and individuals for maximum media effect. More recently, some hijacking attempts have become outright suicide operations intending to kill as many victims as possible. The post–9/11 cases presented in Chapter 9—Richard Reid and Umar Farouk Abdulmutallab—are instructive examples of this trend.

The Complexity of Aviation Security

Comprehensive aviation security is difficult to achieve because of the size and complexity of the nation's aviation systems. The *National Strategy for Aviation Security*, published by DHS, acknowledges this complexity:

> The Aviation Transportation System comprises a broad spectrum of private and public sector elements, including: aircraft and airport operators; over 19,800 private and public use airports; the aviation sector; and a dynamic system of facilities, equipment, services, and airspace. The Aviation Transportation System continues to grow rapidly, as more and more passengers regularly choose

to fly. On a daily basis, thousands of carrier flights arrive, depart, or overfly the continental United States, while each year millions of tons of freight and thousands of tons of mail are transported by air in the United States.[2]

Within the context of homeland security demands, such a large and complicated system necessitates careful policy consideration by federal and local governments.

Lessons From the New Terrorism Environment

As reported in 2003 by *The National Strategy for the Physical Protection of Critical Infrastructures and Key Assets*, "the September 11 attacks offered undeniable proof that our critical infrastructures and key assets represent high-value targets for terrorism."[3] The 2003 *National Strategy* cogently identified the following three deleterious effects of a terrorist attack:

- *Direct infrastructure effects*: Cascading disruption or arrest of the functions of critical infrastructures or key assets through direct attacks on a critical node, system, or function. The immediate damage to facilities and disruption of services that resulted from the attacks on the World Trade Center towers, which housed critical assets of the financial services sector, are examples of direct infrastructure effects.

- *Indirect infrastructure effects*: Cascading disruption and financial consequences for government, society, and economy through public- and private-sector reactions to an attack. Public disengagement from air travel and other facets of the economy as a result of the September 11 attacks exemplifies this effect. Mitigating the potential consequences from these types of attacks will require careful assessment of policy and regulatory responses, understanding the psychology of their impacts, and appropriately weighing the costs and benefits of specific actions in response to small-scale attacks.

- *Exploitation of infrastructure*: Exploitation of elements of a particular infrastructure to disrupt or destroy another target. On September 11, terrorists exploited elements of the aviation infrastructure to attack the World Trade Center and the Pentagon, which represented seats of U.S. economic and military power. Determining the potential cascading and cross-sector consequences of this type of attack is extremely difficult.[4]

Because of these operational effects, the 2003 *National Strategy* further focused on what were termed the "unique protection challenges for aviation," which were identified as follows:

- *Volume*: U.S. air carriers transport millions of passengers every day and at least twice as many bags and other cargo.

- *Limited capabilities and available space*: Current detection equipment and methods are limited in number, capability, and ease of use.

- *Time-sensitive cargo*: "Just-in-time" delivery of valuable cargo is essential for many businesses—any significant time delay in processing and transporting such cargo would negatively affect the U.S. economy.

- *Security versus convenience*: Maintaining security while limiting congestion and delays complicates the task of security and has important financial implications.

- *Accessibility*: Most airports are open to the public; their facilities are close to public roadways for convenience and to streamline access for vehicles delivering passengers to terminals.[5]

It is important to observe that the foregoing findings and policy priorities reveal a post–9/11 emphasis on proactive planning rather than reactive policymaking.

Implementing Aviation Security Priorities

An extensive reexamination of aviation security priorities occurred in the wake of the September 11 attacks. The focus of the reexamination was uncomplicated: to prevent bombs and other potential weapons from entering aviation facilities and to prevent hijackings. The 2003 *National Strategy* emphasized the following collaborative aviation protection initiatives:

- *Identify vulnerabilities, interdependencies, and remediation requirements.* DHS and DoT will work with representatives from state and local governments and industry to implement or facilitate risk assessments to identify vulnerabilities, interdependencies, and remediation requirements for operations and coordination-center facilities and systems, such as the need for redundant telecommunications for air traffic command and control centers.

▶ **Photo 10.1** A Transportation Security Administration officer stands guard at an entryway to an airport terminal. TSA officers routinely search luggage and screen travelers.

- *Identify potential threats to passengers.* DHS and DoT will work with airline and airport security executives to develop or facilitate new methods for identifying likely human threats while respecting constitutional freedoms and privacy.

- *Improve security at key points of access.* DHS and DoT will work with airline and airport security executives to tighten security or facilitate increased security at restricted access points within airport terminal areas, as well as the perimeter of airports and associated facilities, including operations and coordination centers.

- *Increase cargo screening capabilities.* DHS and DoT will work with airline and airport security officials to identify and implement or facilitate technologies and processes to enhance airport baggage-screening capacities.

- *Identify and improve detection technologies.* DHS and DoT will work with airline and airport security executives to implement or facilitate enhanced technologies for detecting explosives. Such devices will mitigate the impact of increased security on passenger check-in efficiency and convenience, and also provide a more effective and efficient means of assuring vital aviation security.[6]

Aviation and Transportation Security Act of 2001 (ATSA): Passed in November 2001, the Aviation and Transportation Security Act required the following: (a) creation of the Transportation Security Administration, (b) replacement of the previous screening system of contractors with professional TSA screeners by November 2002, and (c) screening of all baggage through an explosives detection system by December 2002.

This analysis led to an important redirection in national policies and procedures. Tangible changes occurred rapidly after the attack, including a transformation of the screening system into the Transportation Security Administration (TSA), creating an unprecedented national system of trained airport screeners. The previous system had relied on poorly trained and paid private contractors. In less than a year, TSA transformed the screening system into a new system of 60,000 federal employees. This administrative transformation was the result of requirements incorporated in the **Aviation and Transportation Security Act of 2001 (ATSA)**. Passed in November of that year, ATSA mandated the following bureaucratic and policy changes: (a) creation of the Transportation Security Administration,

Joe Raedle/Getty Images

(b) by November 2002, replacement of the previous screening system of contractors with professional TSA screeners, and (c) by December 2002, screening of all baggage through an explosives detection system.

Another immediate procedural response was the nationwide restriction of entry to airport terminals and gates. The pre–9/11 access of nontravelers to departure gates was halted unless nontravelers received special authorization. Entry to airport terminals was carefully restricted to reduce accessibility to departure areas. Screenings of passengers became increasingly more stringent as new screening technologies became available and in response to lessons learned as new terrorist plots were carried out or thwarted. Inside airport terminals, bomb-sniffing dog patrols became routine, and passengers were instructed during routine security-related public announcements to report suspicious bags and parcels. Onboard security measures included posting air marshals on aircraft and installing reinforced locked cockpit doors.

Nevertheless, the creativity and resiliency of dedicated extremists became manifest as new attacks were planned and attempted. The TSA and other agencies have responded accordingly—for example, the "shoe bomber" case discussed in Chapter 9 led to implementation of the now-routine practice of mandating the removal of passengers' shoes during security screening at U.S. airports. Another instructive case in point is the procedural outcome from the *Chemical Plot*.

Case in Point: The Chemical Plot

Passengers are required by the TSA to comply with the **3-1-1 rule** for carrying liquids on airliners—no more than three ounces of liquid placed in one clear quart-sized plastic bag in one carry-on bag. The reason for this requirement is a thwarted attempt by Islamists to detonate liquid explosives aboard transatlantic flights in late 2006, a conspiracy that has been termed the **Chemical Plot**. Several trials in the United Kingdom resulted in the eventual convictions of ten men. Prosecutors alleged during an initial trial of eight coconspirators that the suspects intended to board flights from London bound for several destination cities in the United States and Canada. Liquid explosives were to be disguised as soft drinks in innocuous containers and detonated using household ingredients and an electronic source such as a camera.

BORDER SECURITY

Protecting the nation's borders is primarily a federal responsibility, with states providing supportive services as needed. In the present era, DHS has been delegated principal responsibility for border security. The most important DHS agencies tasked with securing the border are the U.S. Coast Guard, Customs and Border Protection, and Immigration and Customs Enforcement. The TSA is concurrently tasked with securing international-air ports of entry.

Border security, drug smuggling, and illegal immigration are socially and politically charged issues that, within the context of homeland security, are linked by plausible domestic security threat scenarios. For example, some experts warn of extremist operatives entering the United States via one of the illicit networks specializing in transporting illicit goods (mostly drugs) and unauthorized immigrants across the border. The flows of illicit drugs and unauthorized individuals are unresolved social and political concerns, despite regular proposals to "fortify" the border by constructing border fences and increasing border patrols. Because of this, it is very conceivable that violent extremists could take advantage of illicit

3-1-1 rule: The Transportation Security Administration requires passengers to comply with the 3-1-1 rule for carrying liquids on airliners—no more than three ounces of liquid placed in one clear quart-sized plastic bag in one carry-on bag. The reason for this requirement is a thwarted plot by Islamists to detonate liquid explosives aboard transatlantic flights in late 2006.

Chemical Plot: A thwarted plot by Islamists to detonate liquid explosives aboard transatlantic flights in late 2006. The suspects intended to board flights from London to several destination cities in the United States and Canada. Liquid explosives were to be disguised as soft drinks in innocuous containers and detonated using household ingredients and an electronic source such as a camera.

networks to gain entry into the country. Many officials agree that massive federal allocations of economic and human resources have yet to result in acceptable border security.

There has been some success in securing points of entry. Improved aviation and maritime interdiction is arguably one reason for the shift of illicit activity to overland routes along the southern border region. Drug- and human-trafficking organizations continue to use aviation and maritime points for illicit smuggling, but overland routes on the southern border are the primary points of entry.

▶ **Photo 10.2** A barrier fence on the U.S.–Mexico border. Such barriers hinder, but do not stem, the migration of unauthorized immigrants into the United States.

Unauthorized Immigration and Homeland Security

The social and political challenge of unauthorized immigration into the United States encompasses a significant homeland security dimension. This dimension is the subject of extensive discussion and debate, and it is instructive to understand the homeland security implications within the overall context of unauthorized immigration.

Because the United States is historically and culturally a nation grown from strong immigrant roots, the modern social and political debate has spawned a multiplicity of strongly held and fervently expressed opinions. These opinions represent a number of countervailing positions, including the following:

- The borders should be sealed to the greatest extent possible to stem the flow of unauthorized immigrants, with the use of military assets if necessary. An opposing position is that the United States must establish moderate but effective procedures to focus only on illicit immigration and immigration from regions that potentially give refuge to terrorists.

- Unauthorized immigrants residing in the United States should be identified, taken into custody, and deported to their home countries. In contrast, many people hold that the United States must establish reasonable amnesty procedures to integrate into mainstream society unauthorized immigrants residing in the United States as of a specified cutoff date. There were approximately 10.7 million unauthorized immigrants residing in the United States in 2016.[7]

These points of view frame much of the political debate about immigration and are often discussed within the context of homeland security. For example, some commentators who favor *de facto* fortification of the southern border argue that only in this way can entry by terrorists be thwarted. Others counter this assertion by observing that most unauthorized immigrants simply seek a new life in the United States and to settle into jobs and communities—in essence, that there is a minimal homeland security threat from unauthorized residents.

Research has shown that, in the past, bona fide terrorists tended to enter the United States using legal documents, such as temporary student or visitor visas, and then simply violated the terms of their entry documents.[8] Stated another way, violent extremists did not avail themselves of cross-border human smuggling networks to enter the United States to a significant extent. Research shows that the immigration-related dimension of homeland security is also an immigration *enforcement* problem and not exclusively a border security problem. Nevertheless, plausible scenarios project that determined extremists can conceivably enter the United States with the assistance of transnational organized crime's extensive and efficient human-trafficking networks. It is arguably simply a matter of time before extremists use human-trafficking and drug-smuggling networks to infiltrate people and matériel into the United States.

Plausible Scenarios: Crime and Insecurity on the Southern Border

There exists an unprecedented convergence of transnational organized crime and extensive smuggling networks in some regions of Latin America. This convergence has created potential opportunities for dedicated extremists wishing to strike the United States from within. It has also created significant challenges for homeland security professionals. Two crime-related security environments are of particular concern. These are the Tri-Border Area of South America and the Mexican drug war.

John Moore/Getty Images

▶ **Photo 10.3**
Unauthorized immigrants on the journey north to the U.S. border. Professional human traffickers known as "coyotes" are often retained by immigrants for transportation to the border and across.

Case in Point: The Tri-Border Area of South America

The Tri-Border Area, also known as the Triple Frontier, is a region in central South America straddling the borders of Brazil, Paraguay, and Argentina. It is a remote area where government authority is weak, and it has become home to an illicit economy specializing in drug trafficking, money laundering, and the transfer of financial resources to the Middle East, including to extremist groups. Much of the smuggling network is coordinated by Lebanon's Hezbollah, an operation that is possible because the Tri-Border Area is home to a diaspora of approximately 25,000 Arab residents whose ancestral homes are largely from the Levant (eastern Mediterranean region) of Lebanon.

The region is known for its thriving illegal smuggling and financial criminal activities. Smugglers regularly cross international borders, and Hezbollah is quite adept at raising funds and laundering them to extremist causes. Narcotic trafficking alone generates billions of dollars in profit, and other contraband goods (including cash) add to the lucrative illicit economy.

Ready access to three countries friendly to the United States poses a plausible security risk to the region because extremists could pose as travelers and enter the United States through neighboring countries. Motivated extremists could also travel to Mexico and easily cross the border into the United States. Other countries are also vulnerable to attack, as evidenced by Hezbollah's 1992 bombing of the Israeli embassy in Buenos Aires and the July 1994 bombing of the Argentine Israelite Mutual Association in Buenos Aires.

The Tri-Border Area's nexus of weak government control; an organized, vibrant criminal economy; and political extremism pose a significant security challenge to the region and potentially to the United States.

Case in Point: The Mexican Drug War

The drug trade has become particularly prominent in the financing of some extremist movements, and many terrorists and extremist movements have become adept drug traffickers. This is a result of the enormous profits derived from the global underground drug market, to which American drug users contribute $64 billion each year.[9] There exists a very fluid and

▶ **Photo 10.4** Mexican soldiers secure a drug smuggling tunnel beneath the border between Tijuana and San Diego. Such tunnels are used to transport tons of illicit drugs into the United States and could theoretically be used by violent extremists.

intricate web that links profit-motivated traditional criminal enterprises to ideologically motivated criminal political enterprises.

Criminal gangs in Mexico have historically been involved in banditry and traditional organized criminal activity, such as extortion and prostitution. With the rise of the cocaine trade in the 1970s and 1980s, Colombian drug cartels hired Mexican gangs to transship cocaine overland to the United States. The gangs were subordinate to the cartels and were initially paid in cash. As Mexican gangs became proficient smugglers, they began to demand marijuana and cocaine as payment, which they then sold to their own customers. The gangs eventually became independent and equal partners with the Colombians, growing into criminal cartels. They also became adept at using narco-terrorism to defend their enterprises.

Several large and lucrative criminal enterprises were organized. By 1999, the most dominant of these were the Sinaloa cartel, the Carillo Fuentes organization in Ciudad Juárez, the Caro-Quintero organization in Sonora, and the Arellano-Félix organization in Mexicali and Tijuana. Newer organizations eventually arose, including Los Zetas, Cartel Pacifico Sur, the Gulf Cartel, Jalisco New Generation, and La Familia Michoacana. They have prospered as drug traffickers, and the Mexican trade in marijuana and cocaine became a multibillion-dollar industry.

Mexican criminal enterprises tend to not cooperate with each other. Feuding is common, and the demise of one cartel leads to exceptional drug-related violence over its former turf. At the same time, narco-terrorist activities of the groups continue to include attacks against government officials, journalists, and other critics. Also, from the time when Mexican president Felipe Calderon launched a crackdown on drug traffickers in December 2006, the toll from drug-related violence has been high. From 2006 to 2018, approximately 150,000 Mexicans were killed by the cartels and approximately 34,000 were missing.[10] There were approximately 12,000 deaths in 2011, approximately 11,000 deaths in 2012, and "[i]n 2017, government statistics from the National Public Security System indicate there were more than 29,000 intentional homicides—a new record that exceeded the previous high in 2011."[11]

Despite the sheer intensity of the Mexican drug war, the flow of cocaine and other drugs across the border to the United States continues unabated. In fact, profits for traffickers often reach record highs. Because of the cartels' success in transporting drugs to the United States despite anarchic violence, it is quite conceivable that dedicated extremists could retain the services of smugglers for their own purposes.

BORDER CONTROL: SECURING NATIONAL ENTRY POINTS

The nation's *border* conceptually includes continental and maritime points of entry to the United States, and securing these points of entry is a massive undertaking. Border control is a major domestic security concern for the simple reason that borders grant access to the nation. The continental border of the United States extends 1,933 miles with Mexico and 5,525 miles with Canada.[12] The maritime border extends along 12,383 miles of coastline, composing 2,069 miles on the Atlantic coastline, 7,623 miles on the Pacific coastline, and 1,621 miles on the Gulf of Mexico coastline.[13]

Security Considerations and Patrolling the Border

Policies and procedures for border control and security evolved and changed markedly during the period from the early twentieth century to the early twenty-first century. Many policies and procedures were far-reaching and intrusive, primarily because each new permutation of border control reflected the predominant political and security priorities of the time, many of which were considered to be critical to domestic security. Border control and domestic security are currently well recognized as homeland security priorities. In particular, the smuggling of illicit goods and unauthorized persons have been classified as plausible homeland security risks.

As a practical matter, there are no enforceable restrictions on travel within the borders of the United States, so entry into the country after crossing the border affords virtually unlimited access to cities, critical infrastructure, and population centers. Border control is essential for homeland security because both legal and illegal penetration pose potentially serious risks to domestic security. Risks include illegal entry of violent extremists as well as smuggled CBRNE (chemical, biological, radiological, nuclear, and explosives) agents and weapons, conventional weapons, and dangerous substances. CBRNE threat scenarios are discussed in Chapter 12.

Historical Context: The Evolution of the U.S. Border Patrol

The U.S. Border Patrol was founded in 1924. From the early twentieth century until the founding of the Border Patrol, the Mounted Guards and later the Mounted Inspectors were organized to halt the unauthorized immigration of Chinese people across the U.S.–Mexico border. In 1932, the Border Patrol's primary mission was shifted to halting the bootlegging of illegal alcohol across the U.S. border with Canada. In 1933, the Immigration and Naturalization Service (INS) was created by the administration of President Franklin D. Roosevelt, initially housed in the U.S. Department of Labor and then in the U.S. Department of Justice beginning in 1940. The first Border Patrol Academy was established in 1934.

During the Second World War, the Border Patrol's mission was changed to collaboration with Coast Guard patrols to find saboteurs, protect diplomats, and guard detention camps. During the 1950s, the Border Patrol participated in implementing a massive deportation policy aimed at rounding up and deporting unauthorized Mexican immigrants to Mexico. The Border Patrol was granted national jurisdiction to locate unauthorized persons. Tens of thousands of unauthorized immigrants were deported to Mexico during this period. During the 1980s and 1990s, the Border Patrol's manpower and resources were increased in response to significant increases in the number of unauthorized immigrants entering the United States. Border Patrol personnel were repositioned and concentrated in key entry points.

After the September 11 attacks, the Border Patrol's renewed mission emphasized the homeland security function assigned to many federal law enforcement agencies. In 2003, the Border Patrol's and INS's functions were transferred to the U.S. Department of Homeland Security. INS was disbanded and the agency's mission was distributed among three DHS agencies.

Hardening the Border: Three Agencies

Three agencies in DHS are now responsible for border control and security. They are the **Citizenship and Immigration Services (USCIS), Immigration and Customs Enforcement (ICE)**, and **Customs and Border Protection (CBP)**, all introduced in Chapter 5. Within the context of homeland security, in many respects, they are responsible for "hardening the border" against plausible security threats. It is within this context of hardening the border that the following discussion summarizes their missions.

Citizenship and Immigration Services (USCIS): An agency within the Department of Homeland Security responsible for safeguarding the procedures and integrity of the U.S. immigration system, particularly by supervising the conferral of citizenship and immigration status on legally eligible individuals.

Immigration and Customs Enforcement (ICE): An agency within the Department of Homeland Security that serves as the investigative and enforcement arm for federal criminal and civil laws overseeing immigration, border control, customs, and trade. ICE is the second largest federal law enforcement investigative agency.

Customs and Border Protection (CBP): An agency within the Department of Homeland Security that serves as the primary border security arm of the federal government.

Citizenship and Immigration Services

The border control and domestic security function of USCIS is distinguished by the agency's supervision of legal immigration to the United States. USCIS is responsible for safeguarding the integrity and procedures of the U.S. immigration system, particularly by supervising the conferral of immigration status and citizenship on legally eligible individuals. Because of the characteristics of the modern homeland security environment and plausible threats to domestic security, the implementation of immigration laws, regulations, and procedures is bureaucratic and time-consuming, necessitating careful attention to detail. Among the sweeping duties of the USCIS are the following:

- Implementation of immigration laws, regulations, and procedures

- Approval or denial of citizenship status

- Approval and processing of employment status for foreign nationals

- Verification of foreign nationals' legal right to work in the United States

- Administration of procedures for the immigration of family members of U.S. citizens and permanent residents

Violent extremists are demonstrably interested in manipulating the immigration system to attain admittance to the United States. Because of this reality, USCIS is necessarily tasked with examining literally millions of cases to ascertain the propriety of approving entry to the United States. Therefore, USCIS is a large agency employing approximately 18,000 people based at approximately 250 sites worldwide. As indicated in Figure 10.1, it is organized into task-centered program offices and directorates to accomplish its mission.

Immigration and Customs Enforcement

The border control and domestic security function of ICE is distinguished by its mission as the investigative and enforcement arm for federal criminal and civil laws pertaining to immigration, border control, customs, and trade. As the second-largest federal law enforcement investigative agency, ICE deploys approximately 20,000 personnel in 50 states and 47 countries.[14] ICE is the principal border control agency within the nation's homeland security enterprise. As indicated in Figure 10.2, it is organized into task-centered program offices and directorates to accomplish its mission.

Managing the ICE Mandate: Two Directorates. Two of ICE's assigned missions are administered by two distinct directorates: **Enforcement and Removal Operations (ERO)** and **Homeland Security Investigations (HSI)**.[15]

The ERO directorate "identifies and apprehends removable aliens, detains these individuals when necessary and removes illegal aliens from the United States."[16] Priority is given to individuals who pose a potential threat to national security, criminals, fugitives, and recent entrants to the United States.

The HSI directorate is an important homeland security directorate, responsible for investigating "immigration crime, human rights violations and human smuggling, smuggling of narcotics, weapons and other types of contraband, financial crimes, cybercrime and export enforcement issues."[17] HSI employs approximately 10,000 special agents and other personnel in 200 cities and 47 countries.

Managing the ICE Mandate: Task Forces and Special Units. Within the context of the missions assigned to the ERO and HSI directorates, ICE engages in several other critical

Enforcement and Removal Operations (ERO): A directorate within Immigration and Customs Enforcement that identifies and apprehends removable aliens, detains these individuals when necessary, and removes illegal aliens from the United States.

Homeland Security Investigations (HSI): A directorate within Immigration and Customs Enforcement responsible for investigating "immigration crime, human rights violations and human smuggling, smuggling of narcotics, weapons and other types of contraband, financial crimes, cybercrime and export enforcement issues."

Figure 10.1   Organization Chart of U.S. Citizenship and Immigration Services (USCIS)

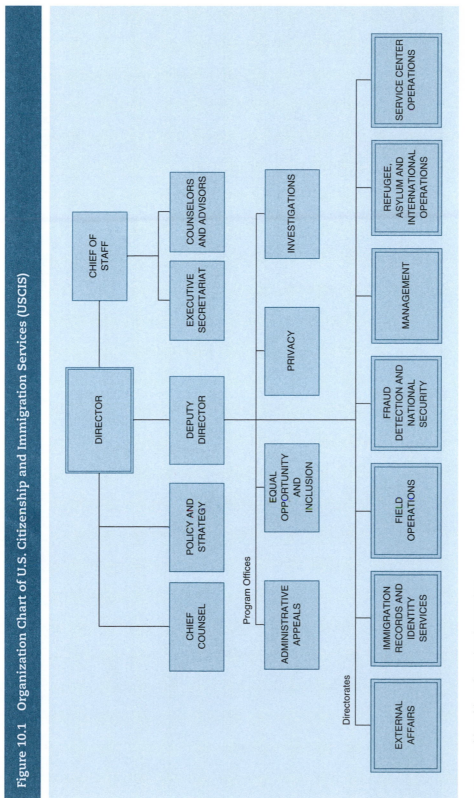

Source: U.S. Citizenship and Immigration Services.

Figure 10.2 Organization Chart of U.S. Immigration and Customs Enforcement (ICE)

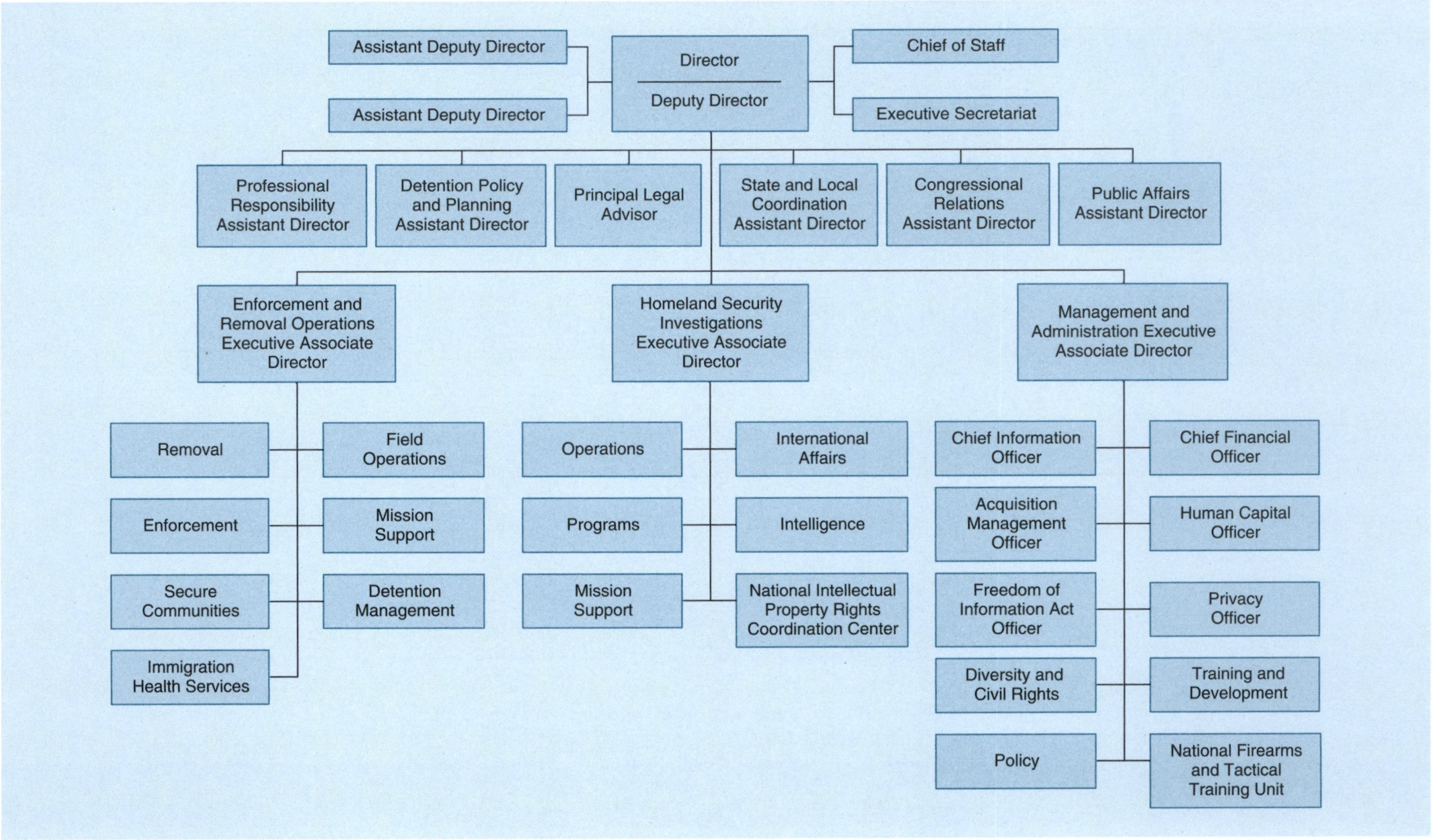

Source: U.S. Department of Homeland Security.

domestic security initiatives that place the agency and its personnel as key participants in the homeland security enterprise. ICE initiatives include the following:

- The *Border Enforcement Security Task Force* (BEST) was formed in 2005 by HSI as a collaborative national and transnational initiative comprising U.S. Customs and Border Protection; federal, state, and local agencies; and international law enforcement agencies. In December 2012, the Jaime Zapata Border Enforcement Security Task Force (BEST) Act was signed into law. It authorized the secretary of DHS to establish BEST teams, initially resulting in the formation of approximately 35 teams in 17 states and territories, representing 100 law enforcement agencies. These BEST teams have "jointly committed to investigate transnational criminal activity along the Southwest and Northern Borders and at our nation's major seaports."[18]

- The *National Security Investigations Division* (NSID) is a component of ICE's HSI directorate that "enhances national security through criminal investigations; prevents acts of terrorism by targeting people, money and materials that support terrorist and criminal activities; and identifies and eliminates vulnerabilities in the nation's border, economic, transportation and infrastructure security."[19] Included within its administrative purview are the Counter-Proliferation Investigations Program, the National Security Program, and the Student and Exchange Visitor Program.

- The *Counter-Proliferation Investigations Unit* "prevents sensitive U.S. technologies and weapons from reaching the hands of adversaries."[20] The unit combats trafficking in weapons of mass destruction, conventional weapons, and explosives; business transactions by embargoed countries and organizations; and controlled dual-use commodities and technology.

- The *National Security Unit* oversees initiatives that constitute the National Security Program. The unit broadly "integrates the agency's national security investigations and counter-terrorism responsibilities into a single unit."[21]

- The *Counterterrorism and Criminal Exploitation Unit* "prevents terrorists and other criminals from exploiting the nation's immigration system through fraud."[22] The unit comprises three sections: the National Security Threat Task Force, the SEVIS Exploitation Section, and the Terrorist Tracking and Pursuit Group.

- The *Joint Terrorism Task Force* (JTTF) is a multijurisdictional collaborative initiative bringing together law enforcement agencies at every level of government. As discussed in detail in Chapter 7, JTTFs are established in more than 100 cities nationwide, and ICE contributes the largest federal presence in the JTTF efforts nationally. As its underlying mission, the JTTF "investigates, detects, interdicts, prosecutes and removes terrorists and dismantles terrorist organizations."[23]

Customs and Border Protection (CBP)

The border control and domestic security function of CBP is distinguished by its front-line mission of securing the borders of the United States at the port of entry. Founded in 2003, the agency is the federal government's primary border security arm, and its mission is described as follows: "CBP is one of the Department of Homeland Security's largest and most complex components, with a priority mission of keeping terrorists and their weapons

Figure 10.3 Organization Chart of U.S. Customs and Border Protection (CBP)

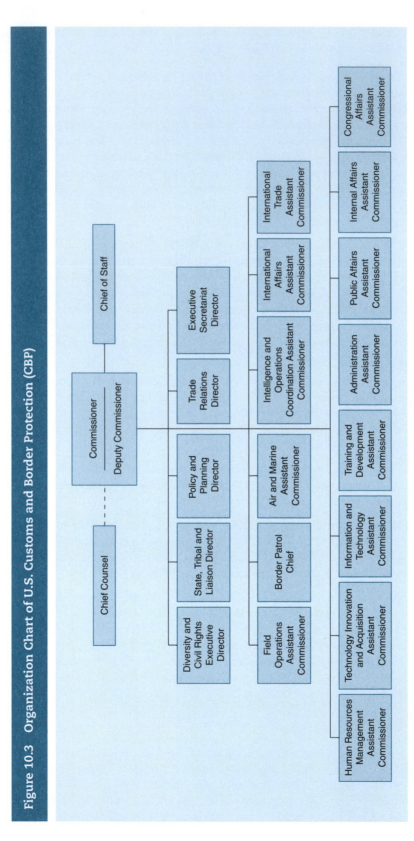

Source: U.S. Department of Homeland Security.

out of the U.S. It also has a responsibility for securing the border and facilitating lawful international trade and travel while enforcing hundreds of U.S. laws and regulations, including immigration and drug laws."[24]

CBP's principal mission is homeland security. The agency's broad domestic security mission necessitates a rather complex organizational structure. CBP's organization is divided into task-focused directorates and commissions, as indicated in Figure 10.3.

The Border Patrol, discussed previously, is a central component of CBP. In addition to the Border Patrol, the Office of Air and Marine (OAM) "is the world's largest aviation and maritime law enforcement organization, and is a critical component of CBP's layered enforcement strategy for border security."[25] OAM employs approximately 1,200 agents, 276 aircraft, and 289 marine vessels and operates from 84 locations in the United States.[26] CBP's field operations and port security operations include cargo examinations, agricultural inspection, and immigration inspection. Another ongoing campaign of CBP is its participation in the overall effort to end human trafficking.

PORT SECURITY

Secure ports of entry are critical to the economy of the United States because of the immense volume of maritime commerce passing through its 361 seaports.[27] Should terrorists successfully strike an important seaport or other maritime target, the consequences to the U.S. economy could be significant because most commerce and trade travel through maritime conveyances.

Balancing port security and economic imperatives is a delicate and serious task because protecting the flow of international commerce through maritime infrastructure requires efficient cooperation between agencies tasked with patrolling and managing maritime entry. DHS agencies tasked with securing the nation's ports include the U.S. Coast Guard and Customs and Border Protection. The Department of Transportation also plays a central role in enforcing regulations and laws concerning port security.

Understanding the Terrorist Threat to Ports and Maritime Targets

Threat analysis involving terrorist operations against maritime infrastructure is uncomplicated, and scenarios are readily apparent. For example, terrorists could conceivably convert cargo containers into weapons of mass destruction by transporting CBRNE agents. Also, terrorists adept at identifying sensitive junctures in containerized trade could conceivably disrupt such trade with carefully focused attacks, thus disrupting maritime trade and thereby possibly affecting national economies.

Terrorist Strategic Priorities and Maritime Targeting

Although genuine threat scenarios are quite disturbing, historically there has not been a high probability that maritime targets are a priority for terrorists. Unlike the documented violence terrorists have prolifically directed against aviation infrastructure, assaults against ports and maritime targets have not been common practice among violent extremists. In fact, terrorists have rarely exhibited an inclination to systematically attack maritime infrastructure. There are several possible explanations for the relatively few attacks directed against maritime targets vis-à-vis aviation infrastructure and other targets.

Terrorist organizations have not historically attempted to penetrate the international trade in cargo containers. The likely reason for this is a logical calculation that other means are available for carrying out more effective terrorist operations. From the terrorists' perspective, cargo containers are capable of transporting operatives and tools of the terrorist trade. For example, groups of terrorist operatives could be secreted inside containers and, thereby, infiltrated into the United States. Similarly, weapons could be smuggled in via cargo containers, or containers could simply be converted into large explosive devices. The rarity of these tactics indicates it is likely that terrorist organizations have so far determined that other means of infiltration are more efficient and also that there is not yet a need to convert containers into explosive weapons.

Another consideration is that maritime infrastructure does not usually offer the opportunity for mass-casualty attacks as easily as aviation infrastructure and other targets. Aside from assets such as passenger cruise ships, few maritime targets provide the readily available concentrations of people found in aviation hubs and other venues.

There is an important caveat regarding past patterns of terrorist behavior. Should an organization or cell successfully carry out a spectacular attack against maritime infrastructure, it is very likely that other violent extremists will attempt to replicate their success. It is simply a matter of clearly identifying vulnerabilities in port security and then exploiting these vulnerabilities. This process, referred to previously in Chapter 8 as the contagion effect, theoretically occurs when terrorists receive widespread media attention for an incident, and other terrorists replicate the perceived success.

Case in Point: Maritime Cargo Containers

More than 200 million cargo containers are transported internationally annually, composing about 90 percent of all seaborne cargo. Fewer than 2 percent of cargo containers are inspected to verify their content. The combination of this huge volume of containerized trade and minimal supervision lends itself to exploitation for illicit purposes. For example, transnational organized crime often uses cargo containers to smuggle illicit drugs, people, and other goods.

It is not implausible to presume that terrorist organizations would be able to exploit this system for their own purposes; all that is required is imagination and the will to do so.

Transnational organized crime and violent extremists are adept at operating clandestinely, and there is evidence of some collaboration between members of these groups. Maritime ports of entry offer many of the same opportunities for exploitation to terrorists as they do for other illicit organizations. Thus, to appreciate why policymakers select certain initiatives and priorities for port security, the terrorist threat to maritime targets must first be analyzed.

Port Security Initiatives and Priorities

Port security requires protecting a multiplicity of infrastructure components. These include the ports themselves, offloaded cargo, ships, waterways, and transportation systems that pass through ports. The synergy between offloaded cargo and transportation systems is of particular relevance for promoting homeland security because ports are gateways to the American heartland. Absent vigilance at the nation's maritime hubs, maritime cargo could conceivably pose significant homeland security vulnerabilities. Thus, the sheer complexity and size of the nation's maritime infrastructure create significant challenges for building a viable system of port security.

▶ **Photo 10.5** Shipping containers stored at the Port of Miami awaiting transport. Many thousands of containers are offloaded in the United States.

Mark Elias/ Bloomberg via Getty Images

The federal government, led by the Department of Transportation, conducted an extensive risk assessment of seaports in the aftermath of the September 11, 2001, attacks. Security practices, vessel notification requirements, and other procedures were reviewed. In particular, container security was specifically addressed with the implementation of the 2002 **Container Security Initiative (CSI)**. The objective of the CSI is to secure cargo entering U.S. seaports and is summarized as follows: "CSI is a program intended to help increase security for maritime containerized cargo shipped to the United States from around the world. CSI addresses the threat to border security and global trade posed by the potential for terrorist use of a maritime container to deliver a weapon."[28]

Initiatives such as CSI have successfully integrated seaport security into a nationally coordinated homeland security system. As noted in the 2003 *National Strategy*, "These assessments have helped refine critical infrastructure and key asset designations, assess vulnerabilities, guide the development of mitigation strategies, and illuminate best practices."[29] However, securing the nation's ports is an ongoing process requiring homeland security specialists to focus policies and resources on specific initiatives. The 2003 *National Strategy* summarized these initiatives as follows:

- *Identify vulnerabilities, interdependencies, best practices, and remediation requirements.* DHS and DoT will undertake or facilitate additional security assessments to identify vulnerabilities and interdependencies, . . . share best practices, and issue guidance or recommendations on appropriate mitigation strategies.

- *Develop a plan for implementing security measures corresponding to varying threat levels.* DHS and DoT will work closely with other appropriate Federal departments and agencies, port security committees, and private-sector owners and operators to develop or facilitate the establishment of security plans to minimize security risks to ports, vessels, and other critical maritime facilities.

- *Develop processes to enhance maritime domain awareness and gain international cooperation.* DHS and DoT will work closely with other appropriate Federal departments and agencies, port security committees, and port owners and operators, foreign governments, international organizations, and commercial firms to establish a means for identifying potential threats at ports of embarkation and monitor identified vessels, cargo, and passengers en route to the U.S.

- *Develop a template for improving physical and operational port security.* DHS and DoT will collaborate with appropriate Federal departments and agencies and port owners and operators to develop a template for improving physical and operational port security. A list of possible guidelines will include workforce identification measures, enhanced port-facility designs, vessel hardening plans, standards for international container seals, guidance for the research and development of noninvasive security and monitoring systems for cargo and ships, real-time and trace-back capability information for containers, prescreening processes for high-risk containers, and recovery plans. Activities will include reviewing the best practices of other countries.

- *Develop security and protection guidelines and technologies for cargo and passenger ships.* DHS and DoT will work with international maritime organizations and industry to study and develop appropriate guidelines and technology requirements for the security of cargo and passenger ships.

- *Improve waterway security.* DHS and DoT, working with state and local government owners and operators, will develop guidelines and identify needed support for improving security of waterways, such as developing electronic monitoring systems for waterway traffic; modeling shipping systems to identify and protect critical components; and identifying requirements and procedures for periodic waterway patrols.[30]

The post–9/11 increase in maritime security assets such as personnel and screening procedures have had the effect of promoting a proactive security environment. This is a daunting undertaking but a necessary one. Inspection of every item of every shipment entering the maritime jurisdiction of the United States is not feasible. Nevertheless, controls and policies attempt to mitigate the likelihood of terrorist exploitation of maritime infrastructure.

CHAPTER SUMMARY

This chapter introduced plausible homeland security threats emanating from terrorists who exploit entry points to the United States. The nation's borders, aviation infrastructure, and ports are all *porous nodes* that are gateways into the homeland for violent extremists and the deadly tools of their trade. The demonstrated and historical ability of transnational organized criminals to successfully transport people and illicit goods across borders, through the air, and in seaborne cargo containers indicates that terrorist organizations are likewise capable of replicating criminal methods of entry. Doing so is simply a question of will and tactical know-how.

DISCUSSION BOX

Threat Scenario: WMDs Hidden in Cargo Containers

This chapter's Discussion Box is intended to stimulate critical debate about the transporting of weapons of mass destruction via containerized cargo.

Many millions of cargo containers are transported each year on seaborne carriers. These containers are unloaded at ports around the world. The terrorist threat scenario is clear and uncomplicated: Weapons of mass destruction and other explosives could conceivably be hidden within seaborne cargo containers. The plausibility of this threat suggests that the use of cargo containers as conveyances for weapons of mass destruction may simply be a matter of time.

Discussion Questions

1. Why haven't terrorist organizations more aggressively attempted to infiltrate the container trade?

2. Under what conditions will terrorist organizations begin utilizing cargo containers as weapons?

3. What counterterrorist measures can be implemented to deny terrorists access to the container trade?

4. What is the worst-case scenario for the aftermath of a terrorist attack using cargo containers?

5. What is the future of maritime security?

KEY TERMS AND CONCEPTS

The following topics were discussed in this chapter and can be found in the glossary:

Aviation and Transportation Security Act of 2001 (ATSA)   222

Chemical Plot   223

Citizenship and Immigration Services (USCIS)   227

Container Security Initiative (CSI)   235

critical infrastructure/key resources (CIKR)   217

Customs and Border Protection (CBP)   227

ON YOUR OWN

Get the tools you need to sharpen your study skills. SAGE edge offers a robust online environment featuring an impressive array of free tools and resources.

Access practice quizzes, eFlashcards, video, and multimedia at **edge.sagepub.com/martinhs3e**

RECOMMENDED WEBSITES

The following websites provide information on issues pertaining to border security, aviation security, and port security:

   Aviation Secure USA: www.aviationsecureusa.com

   FEMA Port Security Grant Program: www.fema.gov/port-security-grant-program

House Committee on Homeland Security: homeland.house.gov/issue/border-security

Port of Long Beach: www.polb.com/about/security

Port of New York and New Jersey: www.panynj.gov/port/port-security.html

WEB EXERCISE

Using this chapter's recommended websites, conduct an online investigation of critical infrastructure security.

1. Are there certain potential infrastructure targets that appear to be more vulnerable than others?

2. Read the mission statements of agencies responsible for infrastructure security. Do they reflect objective and professionally credible approaches to securing critical infrastructure?

3. In your opinion, how effective are these agencies?

To conduct an online search on state terrorism, activate the search engine on your Web browser and enter the following keywords:

   "Critical infrastructure security"

   "Airline and maritime security"

RECOMMENDED READINGS

The following publications provide discussions on border security, aviation security, and port security:

Alden, Edward. 2008. *The Closing of the American Border: Terrorism, Immigration, and Security Since 9/11*. New York: HarperCollins.

Carr, Jeffrey. 2011. *Inside Cyber Warfare: Mapping the Cyber Underworld*. 2nd ed. Sebastopol, CA: O'Reilly Media.

Clarke, Richard A. and Robert K. Knake. 2010. *Cyber War: The Next Threat to National Security and What to Do About It*. New York: HarperCollins.

Davis, Anthony. 2008. *Terrorism and the Maritime Transportation System*. Livermore, CA: WingSpan Press.

Elias, Bartholomew. 2010. *Airport and Aviation Security: U.S. Policy and Strategy in the Age of Global Terrorism*. Boca Raton, FL: Auerbach/Taylor & Francis.

Janczewski, Lech J. and Andrew M. Colarik, eds. 2008. *Cyber Warfare and Cyber Terrorism*. Hershey, PA: Information Science Reference.

McNicholas, Michael. 2007. *Maritime Security: An Introduction*. Waltham, MA: Butterworth-Heinemann.

Price, Jeffrey and Jeffrey Forrest. 2013. *Practical Aviation Security: Predicting and Preventing Future Events*. 2nd ed. Waltham, MA: Butterworth-Heinemann.

Winterdyk, John A. and Kelly W. Sundberg, eds. 2009. *Border Security in the Al-Qaeda Era*. New York: CRC Press.

CHAPTER ELEVEN

ALWAYS VIGILANT
Hardening the Target

Chapter Learning Objectives

This chapter will enable readers to do the following:

1. Explain the concept of cybersecurity and the vulnerabilities of the information system

2. Describe the importance of securing information infrastructure

3. Evaluate options for target hardening

4. Analyze options for infrastructure security

5. Apply target-hardening concepts to the elements of transportation security

Opening Viewpoint: The Long Reach of Homeland Security

The Capture of Mir Aimal Kansi

The capture of Mir Aimal Kansi is an instructive case study of the international reach of federal agencies. Law enforcement and intelligence assets effectuated his capture in collaboration with diplomatic liaisons with the Pakistani government.

Prior to the September 11, 2001, terrorist attack, domestic security was challenged by international terrorism operating in the United States. During the 1990s, many of these threats were precursors to the 9/11 attack. Federal agencies were not only tasked to conduct domestic investigations to bring alleged perpetrators to justice but were also tasked to conduct international manhunts to capture suspects.

On the morning of January 25, 1993, a man armed with an AK-47 assault rifle began firing on employees of the Central Intelligence Agency who were waiting in their cars to enter the CIA's headquarters in Langley, Virginia. Two people were killed, and three were wounded.

The person responsible was Mir Aimal Kansi, a Pakistani who had been a resident of the United States since 1991. After the shootings, he immediately fled for sanctuary in Pakistan and Afghanistan. He apparently traveled between the two countries, though he found refuge among relatives and local Pakistanis in Quetta.

The United States posted a $2 million reward for Kansi's capture and distributed wanted posters throughout the region. His photograph was distributed on thousands of matchbooks, printed in newspapers, and placed on posters. The hunt was successful, because still-unidentified individuals contacted U.S. authorities in Pakistan and arranged Kansi's capture. In June 1997, Kansi was arrested by a paramilitary FBI team and, with the permission of the Pakistani government, flown to the United States to stand trial in a Virginia state court.

At his trial, prosecutors argued that Kansi had committed the attack in retaliation for U.S. bombings of Iraq. He was convicted of murder on November 10, 1997. Kansi was executed by lethal injection at the Greensville Correctional Center in Virginia on November 14, 2002.

▶ **Photo 11.1** The future of warfare has begun—cyberattacks and cyberterrorism. Cyberattacks committed by nations and extremists have occurred and remain a plausible warfighting and terrorist scenario.

iStock.com /TommL

This chapter explores the characteristics of "hardening" several levels of domestic targets. Efforts to make potential targets more difficult to attack include an antiterrorist measure termed **target hardening**. This is a key component of antiterrorism, which attempts to deter or prevent terrorist attacks. The homeland security enterprise is deeply invested in hardening several categories of potential targets. These include the nation's critical infrastructure, international borders and ports of entry, the transportation system, and information technologies. Protection of these categories is an extremely complicated proposition because each category includes subclassifications that are separately difficult to harden. Although protecting these potential targets is difficult, and they pose ever-evolving challenges, domestic security cannot be achieved without contingency plans for each category.

The discussion in this chapter will review the following:

- Cybersecurity
- Hardening cyberspace: Cyberwar and information security
- Hardening critical infrastructure
- Transportation security

target hardening: An antiterrorist measure that makes potential targets more difficult to attack. Target hardening includes increased airport security, the visible deployment of security personnel, and the erection of crash barriers at entrances to parking garages beneath important buildings.

CYBERSECURITY

The Internet is a universal component of modern enterprise, and it provides opportunities for commercial, private, and political interests to spread their message and communicate with outsiders. The Internet is also a resource available to political extremists, and the use of the Internet by extremists has become a common feature of the modern era. Information technologies are being invented and refined constantly and are used extensively as a feature of the New Terrorism. New technologies designed for the Internet and social networking media facilitate networking and communication between groups and cells and permit propaganda to be spread widely and efficiently. The Internet, e-mail, and social networking media are used to send instructions about overall goals, specific tactics, new bomb-making techniques, and other facets of the terrorist trade. Both overt and covert information networks permit widely dispersed cells to exist and communicate covertly. Readily available encryption technologies facilitate communication and networking among extremists by "hiding the trail" of participants.

Expansion of the Internet consistently results in the increasing assignment of fundamental responsibilities to computer systems and networks. The result is that government and industry become increasingly dependent on information technologies. Because of this, targeted attacks on computer systems and networks can conceivably result in significant social, political, and economic disruption. Motivated extremists who understand this reality are likely to attempt to disrupt these systems when opportunities present themselves.

Understanding Threats to Cybersecurity

There are several categories of cyberattacks that must be differentiated. Although the threat from potential cyberterrorism is often at the forefront of the discussion on homeland cybersecurity planning, it is only one possible type of cyberattack. The cybersecurity discussion must also consider the following delineations of the sources of and motivations behind cyberattacks:[1]

- **Cyberattack** is "a deliberate computer-to-computer attack that disrupts, disables, destroys, or takes over a computer system, or damages or steals the information it contains."[2]

- **Cyberwar** "refers to offensive computer assaults that seek to damage or destroy networks and infrastructures or deter them from waging cyberattacks of their own." It "is largely, but not exclusively, the domain of states."[3]

- **Hacktivism** "is a form of 'contentious politics' carried out by non-state actors in support of a variety of political, social or religious causes, frequently in opposition to government policy."[4] An instructive example of hacktivism is the loose anarchist collective calling itself Anonymous.

- **Cyberterrorism** "refers to computer-generated attacks that target other computers in cyberspace or the information they contain . . . [i]t is, in this sense, the 'convergence of terrorism and cyberspace,' with computer technology serving as both weapon and target."[5]

Malicious Use of Cyber Technology: Examples

Cyber technologies can be wielded as virtual weapons by extremists and governments. In this regard, a significant number of cyberattacks have occurred, and it is quite conceivable that emergent technologies will be used by terrorists to destroy information and communications systems when they become available.

Three early incidents illustrate the potential damage that can be wrought by motivated activists and extremists. The following cyberattacks occurred in 1998 and are among the first confirmed examples of the destructive use of cyber technologies by political activists and extremists:

- Members of the Liberation Tigers of Tamil Ealam (also known as the Tamil Tigers) inundated Sri Lankan embassies with 800 daily e-mails during a two-week period. The e-mail messages read, "We are the Internet Black Tigers and we're doing this

cyberwar: As an offensive mode of warfare, terrorists use new technologies to destroy information and communications systems. As a counterterrorist option, cyberwar involves the targeting of terrorists' electronic activities by counterterrorist agencies. Bank accounts, personal records, and other data stored in digital databases can theoretically be intercepted and compromised.

cyberterrorism: The use of technology by terrorists to disrupt information systems.

to disrupt your communications." This attack was the first known cyberattack by a terrorist organization against an enemy country's computer grid.

- Animal liberation activists dropped an "e-mail bomb" on the server of Sweden's Smittskyddinstitutet. Its entire database crashed when 2,000 e-mail messages were sent on one day, followed on a second day by 3,000 messages. The institute was targeted because of its use of monkeys in medical experiments.

- A three-week e-mail campaign targeted approximately 100 Israeli Internet sites, resulting in the destruction of data. The campaign was launched by Lebanese Americans living in Texas.

Recent incidents are much more intensive and intrusive than the first incidents, and they demonstrate the potentially disastrous scale of destruction from cyberattacks when unleashed by determined adversaries.

- In 2008, immediately prior to the Russian invasion of neighboring Georgia, the government of Georgia was the subject of numerous cyberattacks.

- In 2009–2010, Iran's nuclear facility in Natanz was infected by the Stuxnet worm, severely damaging the uranium enrichment program at the facility. The United States and Israel are suspected to have embedded the worm, but both governments deny knowledge of the incident.

- In August and September 2012, cyberattacks were directed at U.S. financial institutions by hackers calling themselves the Izz ad-Din al-Qassam Cyber Fighters.

- In November 2014, a sophisticated cyberattack against Sony Pictures accessed and released a large amount of confidential data. A group calling itself the Guardians of Peace demanded that Sony Pictures cancel its release of the comedy film *The Interview*, which depicted a plot to assassinate North Korean dictator Kim Jong-un. The government of North Korea is suspected to have hacked Sony Pictures, but the North Koreans vigorously denied the allegation.

- During the 2016 election season in the United States, Russian hackers launched cyberattacks on a variety of systems affiliated with national and local elections. Using malicious software technologies, the hackers created false online profiles and personalities such as Guccifer 2.0 and DCLeaks. In July 2018, the U.S. Department of Justice indicted twelve Russian intelligence officers for allegedly targeting and hacking Democratic Party officials during the 2016 elections. They were also accused of allegedly conspiring to hack computers affiliated with state-level election agencies and officials. All named defendants were officers of the GRU (Main Directorate of the General Staff of the Armed Forces of the Russian Federation), the foreign military intelligence agency of the Russian intelligence services.

Because of incidents such as these, the possibility of state- and nonstate-initiated cyber-terrorism and cyberwar has become a central threat scenario of the modern terrorist environment. As one plausible scenario suggests, "a variation on [the] theme of terrorism as an asymmetric strategy goes further to suggest that unconventional modes of conflict will stem . . . from a shift in the nature of conflict itself. In this paradigm, unconventional terrorist attacks on the sinews of modern, information-intensive societies will become the norm, replacing conventional conflicts over the control of territory or people."[6]

Cyberattacks: Potential Targets

Private corporations, telecommunications companies, and government agencies (particularly defense centers) are logical targets for politically motivated cyberattacks. For example, the disruption of telecommunications grids could severely affect urban centers and regions. Attacks against government targets could plausibly include both disruption (such as destruction of data) and obtainment of sensitive information. This is particularly concerning for members of the private defense industry and the Department of Defense.

Cyberterror: Feasibility and Likelihood

The *feasibility* of terrorists' acquisition and use of emerging technologies must be calculated by addressing the availability of these technologies. Acquiring new and exotic technologies frequently requires specialized access and knowledge, and hence, not all technologies are readily available. Having said this, the old adage that "where there is a will there is a way" is very relevant because some practitioners of the New Terrorism (such as al-Qaeda and ISIS and their affiliates) have exhibited great patience and resourcefulness. They have also proven to be meticulous in their planning.

Technical instructions for manipulating new information technologies are readily available. In fact, a great deal of useful information is available for terrorists on the Internet, including instructions on bomb assembly, poisoning, weapons construction, and mixing lethal chemicals. Extremists who wish to use computer and Internet technologies to attack political adversaries can also obtain the technical knowledge to do so. For example, information about how to engage in computer hacking is easy to acquire; instructions have been published in print and posted on the Internet. There is also an underground of people who create computer viruses for reasons that range from personal entertainment to anarchistic sentiment.

The *likelihood* that new technologies will be acquired and used is perhaps a moot consideration because modern terrorists have already acquired and used many of these technologies. For example, apparently apolitical and anarchistic hackers—some of them teenagers—have vandalized information and communications systems, thus demonstrating that cyberwar is no longer an abstract concept. There is little reason to presume that this trend will diminish and many good reasons to presume that it will increase. The increasing availability of new technologies, when combined with the motivations and morality of the New Terrorism, suggests very strongly that technology is an increasingly potent weapon in the arsenals of terrorists.

HARDENING CYBERSPACE: CYBERWAR AND INFORMATION SECURITY

Hardening information infrastructure has become a domestic security priority. The reason is simple and disturbing: New technologies allow terrorists to communicate efficiently, broaden their message, and wield unconventional weapons in unexpected ways. This is a central characteristic of asymmetrical warfare. Because of incremental improvements in communications and computer technologies, it is highly plausible that the trend among terrorists and their supporters will be to use them extensively.

Hardening cyberspace is a central concern for government authorities, who are designing protocols and assigning security duties to specified command centers. For example, the U.S. military has actively responded to the threat of cyberterrorism by forming joint centers of cybersecurity. Also, the United States Strategic Command has been tasked with designing countermeasures to counter the threat of cyberterrorism. A Joint Task Force–Global Network Operations was established to protect and service the military's Global Information Grid. The U.S. Air Force established its own Cyber Command in 2006, which was upgraded to the Twenty-Fourth Air Force in 2009 as a component of the multiservice United States Cyber Command, which became active in 2010.

Cyberwar as an Antiterrorist Option

In the digital age, information is recorded and transferred in forms that can be intercepted, altered, and destroyed. Bank accounts, personal records, and other data are no longer stored on paper but instead in digital databases. Terrorist movements that maintain or send electronic financial and personal information run the risk of having that information intercepted and compromised. Thus, new technologies have become imaginative counterterrorist weapons. They have also become a new mode of warfare, with a hidden and potentially potent array of cyber weaponry and defense systems.

In June 2011, U.S. president Barack Obama signed executive orders approving guidelines for military applications of computer-initiated actions against adversaries. The guidelines governed a range of options, including espionage and aggressive cyberattacks. The overall objective of the executive orders was to embed cyber technology into American warfighting capabilities—in essence, to link cyberwarfare to traditional modes of warfare. Examples of using weaponized cyber technologies include uploading destructive computer viruses, hacking secure sites, and carrying out massive attacks to neutralize communications systems, defense networks, and power grids. A natural corollary to the wielding of weaponized cyber technology is the necessity for creating new cyber defenses to protect friendly computers, networks, and grids against attacks from terrorists or hostile nations.

Modern Surveillance Technologies

Electronic surveillance has moved far beyond the days when law enforcement officers literally tapped the telephone wires of criminal suspects. In the modern era, digital technologies, fiber optics, and satellite communications have moved state security agencies into the realm of technology-based surveillance. Surveillance can be conducted quite remotely, literally from facilities on other continents.

Some technologies are visible and taken for granted by residents of major cities. For example, remote cameras have become common features on London streets and at Los Angeles intersections. Other technologies are neither visible nor well known. For example, **biometric technology** allows the faces of wanted suspects to be matched against digital photographs of faces; such technology is especially useful for antiterrorist screens at ports of entry, such as border crossings and airports. Interestingly, biometrics was used at American football's 2001 Super Bowl championship, when cameras scanned the faces of fans as they entered the stadium and matched their digital images against those of criminal fugitives and terrorists. The game became derisively known as the "Snooperbowl."

David L. Ryan/The Boston Globe via Getty Images

▶ **Photo 11.2** A second bomb detonates in Boston's business district during the running of the 2013 Boston Marathon. Privately owned infrastructure more often bears the brunt of terrorist attacks.

biometric technology: Technology that allows digital photographs of faces to be matched against those of wanted suspects; such technology is especially useful for antiterrorist screens at ports of entry, such as airports and border crossings.

Surveillance technologies are central components of counterterrorist systems. It is technologically feasible to access private electronic transactions, including telephone records and conversations, computer transactions and communications (such as e-mail), social networking media, and credit card records. Digital fingerprinting and facial imaging permit security agencies to access records virtually instantaneously. The FBI has used biometric technologies to collect and analyze unique human traits. These technologies permit the storage of data such as iris scans, facial recognition, fingerprints, hand geometry, speech verification, and vascular recognition. Chapter Perspective 11.1 explores the utility of monitoring private social networking media by homeland security and emergency response authorities.

Because such technologies are inherently intrusive, they have been questioned by political leaders and civil libertarian organizations. As discussed in Chapter 4, the application of technologies in efforts such as the National Security Agency's (NSA's) PRISM and XKeyscore data-mining operations has been criticized by civil libertarians as overly broad and intrusive. Nevertheless, surveillance technologies are considered to be invaluable counterterrorist instruments.

CHAPTER PERSPECTIVE 11.1

The Utility of Monitoring Social Networking Media

Modern social networking media technologies allow users to upload and post information electronically on provider websites. Popular providers include Twitter, MySpace, Facebook, and dating enterprises, and users typically use provider websites as social networking resources. Users upload information about themselves that may be accessed by other users and the public. Information typically includes photographs, videos, statements of interest, and personal information.

Social networking media have proven to be very useful systems for disseminating information when natural or intentional disasters occur. Real-time information about the effect of hurricanes, tornadoes, and other natural events has assisted the media and emergency responders in assessing the magnitude and geographic location of critical incidents. Similarly, social media have alerted the public to unfolding terrorist events, such as in the aftermath of the 2013 Boston Marathon bombing.

Unfortunately, criminals also utilize social media technologies. For example, confidence artists and child sexual predators have been known to target unknowing potential victims.

Social media provide intelligence and law enforcement officials with resources to monitor individuals and investigate crimes. Law enforcement agencies have found social media information to be a useful forensic tool for pursuing active investigations and for framing "sting" operations to take criminals into custody. For example, undercover law enforcement officers have posed as children online and have successfully captured adult predators. Similarly, intelligence analysts are able to examine the use of Web-based social media by extremist individuals and organizations. This can be useful for projecting the intentions of violent extremists. For example, increased communications chatter could suggest an increased likelihood of actual activity by extremists, especially when combined with increased activity on social networking websites.

Discussion Questions

1. How should the examination of social media be regulated?

2. Within which scenarios should homeland security authorities be given broad authority to examine social networking media?

3. What kind of civil liberties issues arise when social networking media are monitored by homeland security authorities?

Case in Point: The Echelon Satellite Surveillance Network

The National Security Agency manages a satellite surveillance network called **Echelon**. The NSA's network is apparently managed in cooperation with its counterparts in Australia, Canada, Great Britain, and New Zealand. Its purpose is to monitor voice and data communications. Echelon is a kind of global wiretap. It filters communications using satellites, antennae, and other technologies. Types of communications that can reportedly be intercepted include Internet transfers, telephone conversations, and data transmissions. It is not publicly known how much communications traffic can be intercepted or how it is done, but the network is apparently very capable. How the traffic is tapped is unknown, but it is likely done with technologies that can pinpoint keywords and interesting websites. It can also be done the old-fashioned way: In 1982, an American listening device was reportedly found on a deep-sea communications cable; it was never reported whether this was an Echelon-style operation.

HARDENING CRITICAL INFRASTRUCTURE

Critical infrastructure must be secured against terrorist threats, and this objective represents a central tenet of successful domestic security. Should critical infrastructure come under attack, the fundamental stability of the nation may be affected. The purpose of target hardening is to minimize the likelihood that terrorists will succeed in damaging critical infrastructure targets.

Securing Critical Infrastructure: Background and Context

To appreciate the importance of hardening potential critical infrastructure targets, it is necessary to understand the contextual background of these concepts.

Defining Critical Infrastructure

Critical infrastructure is defined by the Department of Homeland Security (DHS) as "the assets, systems, and networks, whether physical or virtual, so vital to the United States that their incapacitation or destruction would have a debilitating effect on security, national economic security, national public health or safety, or any combination thereof."[7] It comprises many core priorities necessary for domestic security and stability. The DHS Office of Infrastructure Protection oversees designing and implementing national policies and programs for the protection of critical infrastructure.

Since 1998, the protection of critical infrastructure has been an essential element of domestic security policymaking. In 1998, President Bill Clinton released **Presidential Decision Directive/NSC-63** following terrorist attacks upon U.S. installations abroad. A comprehensive policy directive, it was designed to consolidate efforts to protect critical infrastructure. In 2002, the Homeland Security Act tasked DHS with coordinating comprehensive protection for critical infrastructure. Also in 2002, **Homeland Security Presidential Directive 7 (HSPD-7)** promoted collaborative efforts among different levels of government by specifically establishing a systematic approach toward accomplishing such collaboration. HSPD-7 classified 17 sectors as critical infrastructure sectors and assigned responsibility for these sectors to specified federal sector-specific agencies.

Echelon: A satellite surveillance network maintained by the U.S. National Security Agency. It is a kind of global wiretap that filters communications using antennae, satellites, and other technologies. Internet transfers, telephone conversations, and data transmissions are among the types of communications that can reportedly be intercepted.

Presidential Decision Directive/NSC-63: A 1998 directive released by President Bill Clinton following terrorist attacks upon U.S. installations abroad. It was a comprehensive policy directive designed to consolidate efforts to protect critical infrastructure.

Homeland Security Presidential Directive 7 (HSPD-7): A directive by President George W. Bush that specifically created a systematic approach to collaborative efforts among different levels of government. HSPD-7 classified 17 sectors as critical infrastructure sectors and assigned responsibility for these sectors to specified federal sector-specific agencies. It was superseded in 2013 by Presidential Policy Directive 21.

Presidential Policy Directive 21 (PPD-21): A directive released in 2013 by President Barack Obama, superseding HSPD-7. Titled *Critical Infrastructure Security and Resilience*, PPD-21 identified 16 critical infrastructure sectors, designated specified federal agencies as sector-specific agencies, and tasked those agencies with implementing policies and programs for the protection of their assigned sectors.

In 2013, HSPD-7 was superseded by **Presidential Policy Directive 21 (PPD-21)**, titled *Critical Infrastructure Security and Resilience*. PPD-21 identified 16 critical infrastructure sectors,[8] designated specified federal agencies as sector-specific agencies, and tasked those agencies with implementing policies and programs for the protection of their assigned sectors.

Target Hardening: The Concept

Enhanced security is intended to deter would-be terrorists from selecting hardened facilities as targets. These measures are not long-term solutions for ending terrorist environments, but they serve to provide immediate protection for specific sites. Target hardening includes the erection of crash barriers at entrances to parking garages beneath important buildings, the visible deployment of security personnel, and increased airport security. In the United States, the digital screening of fingerprints and other physical features is one technological enhancement at ports of entry.

A typical example of target hardening occurred when vehicular traffic was permanently blocked on Pennsylvania Avenue in front of the White House because of the threat from high-yield vehicular bombs in the aftermath of the 1993 World Trade Center and 1995 Oklahoma City bombings. The affected area on Pennsylvania Avenue became a pedestrian mall.

Target Hardening: The Public–Private Context

As will be discussed further in Chapter 12, the National Infrastructure Protection Plan (NIPP) was enacted to delineate the public–private relationship between DHS and the private sector. The NIPP protocols institutionalized collaborative efforts for the protection of private critical infrastructure. Sixteen critical infrastructure sectors were identified to form the core of the NIPP planning protocols, each of which were assigned a sector-specific plan (SSP). These critical infrastructure sectors are

- Chemical
- Commercial facilities
- Communications
- Critical manufacturing
- Dams
- Defense industrial base
- Education facilities
- Emergency services
- Energy
- Financial services
- Food and agriculture

John Moore / Getty Images

▶ **Photo 11.3** Members of the Arizona National Guard on border patrol duty. Federal agencies are authorized to enforce border security, but state agencies and units can provide backup resources.

- Health care and public health

- National monuments and icons

- Nuclear reactors, materials, and waste

- Transportation systems

- Water and wastewater systems

A terrorist incident directed against any of these critical infrastructure sectors would have significant ramifications for the private sector. Thus, hardening such targets is a priority from the perspectives of both the national homeland security enterprise and local private sector interests. Figure 11.1 represents the NIPP Sector Partnership Model proposed by the NIPP for public–private collaboration. The Sector Partnership Model proposes "organizational structures and partnerships committed to sharing and protecting the information needed to achieve the NIPP goal and supporting objectives."[9]

Federal Guidance: The DHS Building and Infrastructure Protection Series

The U.S. Department of Homeland Security's Science and Technology Directorate reported the latest recommendations on how to protect buildings and infrastructure in the United States by releasing a series of manuals. Known as the **Building and Infrastructure Protection Series**, or **BIPS**, the series is published by the Directorate's Infrastructure Protection and Disaster Management Division. The main objective of BIPS is to "bring to the design community, researchers, and building owners and operators, state-of-the art [sic] information and tools . . . these publications and tools are devoted to protect[ing] the built environment and expand[ing] the understanding of how lives can be saved, and buildings and structures remain operational in the aftermath of a disaster event."[10] Although BIPS is published as an all-hazards series, it includes several domestic security manuals:

- BIPS 05: *Preventing Structures From Collapsing to Limit Damage to Adjacent Structures and Additional Loss of Life When Explosives Devices Impact Highly Populated Urban Centers.*[11]

- BIPS 06/FEMA 426: *Reference Manual to Mitigate Potential Terrorist Attacks Against Buildings, Edition 2.*[12]

- BIPS 07/FEMA 428: *Primer to Design Safe School Projects in Case of School Shooting and Terrorist Attacks, Edition 2.*[13]

BIPS publications are written as practice manuals for federal, state, and local agencies tasked with domestic security duties.

Case in Point: BIPS 06/FEMA 426

As an installment in its BIPS publications, DHS published the second edition of the FEMA 426 reference manual on hardening targets against terrorist attacks. At more than 500 pages in length, the ***Reference Manual to Mitigate Potential Terrorist Attacks Against Buildings, Edition 2*** (**BIPS 06/FEMA 426**) is a widely distributed guide for government

Building and Infrastructure Protection Series (BIPS): A series of manuals released by the U.S. Department of Homeland Security's Science and Technology Directorate that report the latest recommendations on how to protect buildings and infrastructure in the United States.

***Reference Manual to Mitigate Potential Terrorist Attacks Against Buildings, Edition 2* (BIPS 06/FEMA 426):** A widely distributed guide for government and private agencies and organizations that reports a considerable number of infrastructure security and target-hardening recommendations.

Figure 11.1 Public–Private Partnership Councils

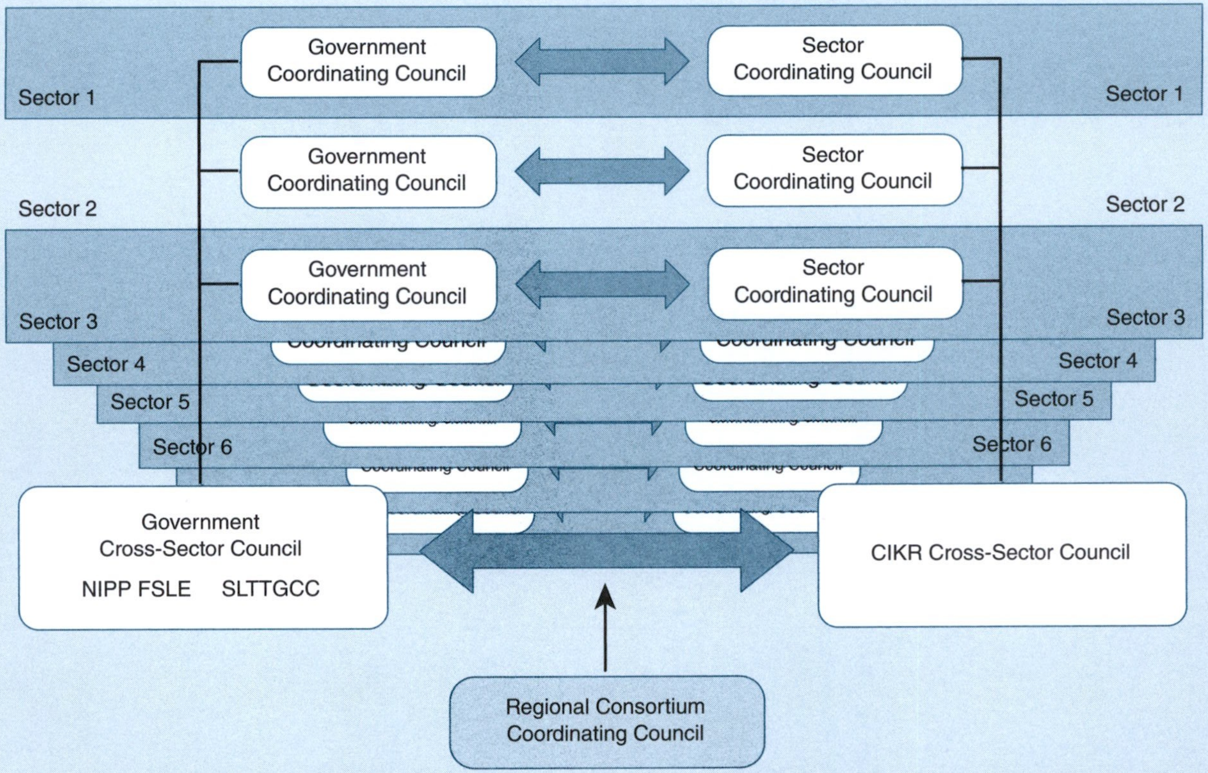

The Critical Infrastructure Partnership Advisory Council (CIPAC) supports a legal framework for collaboration between public sector and private sector partners.

Source: U.S. Department of Homeland Security.

and private agencies and organizations. BIPS 06/FEMA 426 reports a considerable number of infrastructure security and target-hardening recommendations, including the following procedures:

- Design buildings with an underlying goal of increasing security. For example, dense building clusters allow planners to concentrate and simplify security options. Dispersed building clusters spread out and complicate security options.

- Install recommended *mitigation features* in new building designs to reduce the effects of explosions. In existing buildings, the installation of new mitigation features is recommended to reduce explosive effects.

- Create distance between an infrastructure target and a possible blast. This refers to creating "standoff" distance between a target and a terrorist threat. For example, place buildings back and away from where traffic passes and at a distance from where terrorists may launch an assault.

- Install building designs in anticipation of terrorist attacks that are also effective against nondomestic security incidents. For example, design ventilation systems that expel intentional and accidental releases of chemical, biological, and radiological hazards. Also, install windows that resist flying debris from explosions as well as from natural events such as hurricanes.

- Road access to buildings and parking facilities should be designed with the purpose of minimizing velocity and "calming" traffic. This can be accomplished using uncomplicated measures such as speed bumps, winding roads, and barriers.

These and other options harden structural targets against terrorist attacks. They cannot completely *prevent* attacks, but they can deter and minimize possible attacks.

Options for Critical Infrastructure Security and Target Hardening

There exist a range of additional options, some quite innovative, for securing critical infrastructure. These innovations involve the application of readily adaptable engineering and technological models to the hardening of potential targets. For example, creative adaptations of existing engineering and technological models could include the following designs for securing critical infrastructure:

- Create perimeter security around potential targets. Options include perimeter patrols, secure gates and fences, and restricted entry points.

- Use urban design features and landscaping to harden potential targets. Urban design features could include hardened streetlights and other lighting, walls and fences, and bollards (sturdy concrete posts). Landscaping could include hillocks, secured planters, and shrubbery.

- Make use of glazing techniques for installing glass during building construction. For example, thermally tempered glass is significantly stronger than plate glass. When shattered, tempered glass forms cubed fragments rather than shards. Such glass reduces shrapnel effects when shattered by blasts or other terrorist events.

- Install architectural barriers such as antiramming obstructions, reinforced doors and gates, blast walls, and urban design features.

- Employ trained security personnel who engage in stationary security and roving patrols.

- Install electronic warning and detection technologies. These security measures include alarms (silent or audible), sensors, closed-circuit television, and cameras.

- Install electronic access controls at entryways and doors. Control technologies include identification badge readers, pass codes, and biometric sensors.

FEMA's Building Science Branch offers expert guidance on building construction, focusing on construction designs that mitigate the effects of potential disasters. Figure 11.2 illustrates the recommendation and information feedback loop of services offered by the Building Science Branch.

TRANSPORTATION SECURITY

The terrorist attacks on September 11, 2001, used advanced technology from the nation's transportation system as terrorist weaponry. Aircraft routinely used as commercial passenger conveyances were hijacked and converted to ballistic missiles. The attacks occurred within the context of a national transportation system that relied on relatively diffuse and privatized security systems. Following the attack, the nation's transportation system was overhauled to comprehensively enhance transportation security.

In the modern security environment, transportation is a complex globalized system of conveyances that moves literally millions of people and goods each day, frequently

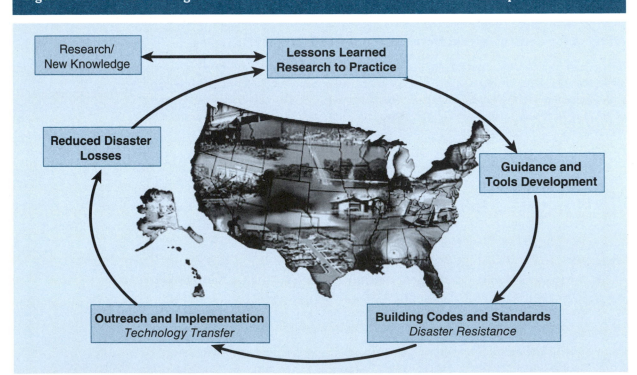

Figure 11.2   FEMA Building Science Branch Recommendation and Feedback Loop

Research/New Knowledge

Lessons Learned
Research to Practice

Reduced Disaster Losses

Guidance and Tools Development

Outreach and Implementation
Technology Transfer

Building Codes and Standards
Disaster Resistance

across international borders. Because of the system's complexity, security challenges are equally complex. The U.S. Department of Homeland Security's **Transportation Security Administration (TSA)** provides leadership in securing transportation in collaboration with the U.S. Department of Transportation and the U.S. Coast Guard.

A Complex Arrangement: Components of the Transportation System

The transportation system moves people and goods through a variety of systems using conveyances such as aircraft, motor transportation, and railways. The nation's commerce and economy depend on seamless transportation networks and would be significantly affected should terrorists disrupt these networks. Components of the transportation system include the following:

Aviation

The aviation component comprises commercial passenger transportation and air freight. Nationally, there are approximately 20,000 general aviation airports (primarily for recreational flying), including about 500 commercially certified airports (for transportation and freight). The U.S. **Federal Aviation Administration (FAA)** administers civil aviation, issuing regulations and other requirements to control air traffic and promote safety. For example, certifications are issued for personnel, airports, airline companies, and aircraft. Prior to the September 11 attacks, airport security was a relatively decentralized system managed by private contractors, states, and private owners. After the attacks, these duties were transferred to the TSA, discussed in the next section.

Hundreds of millions of passengers are served annually by the aviation component, flying from commercially certified airports. At the same time, approximately 10 million tons of freight is transported nationally and internationally by aircraft from the United States. Air freight is delivered within a much faster time frame than that delivered by other methods of transportation, which is critical for the safe transportation of perishable goods. Aircraft used to ship freight include both commercial passenger and exclusive freight-only aircraft.

Because of the foregoing and other considerations, aviation security planning is a priority mission for federal agencies. To that end, National Security Presidential Directive-47/ Homeland Security Presidential Directive-16 (NSPD-47/ HSPD-16) "details a strategic vision for aviation security while recognizing ongoing efforts, and directs the production of a National Strategy for Aviation Security and supporting plans. The supporting plans address the following areas: aviation transportation system security; aviation operational threat response; aviation transportation system recovery; air domain surveillance and intelligence integration; domestic outreach; and international outreach."[14]

Motor Transportation Networks

Motor transportation networks include commercial and private vehicles served by roadways and highways. More than 250 million registered vehicles use this system, including

Transportation Security Administration (TSA): Responsibility for civil aviation security was transferred from the Federal Aviation Administration to the Transportation Security Administration in 2002. The agency promulgates and manages the implementation of transportation security regulations to support its aviation security mission.

Federal Aviation Administration (FAA): The Federal Aviation Administration administers civil aviation, issuing regulations and other requirements to control air traffic and promote safety.

▶ **Photo 11.4** National Guard soldier during a community service rescue mission. State National Guard units perform diversified duties, including disaster relief, domestic order maintenance, and federal activation for deployment.

U.S. Navy; Jobsatnationalguard.com

approximately 135 million automobiles, 35 million trucks, 7 million motorcycles, and 830,000 buses. There are approximately 150,000 miles in the U.S. highway system, nearly 600,000 bridges, and approximately 3.8 million miles of other roadways. A great deal of commerce is conveyed via motor transportation. The infrastructure of this system and the vehicles that move through it are vital to the national economy, so disruption of these networks could be disastrous.

A significant security consideration is the protection of transportation choke points. These are nodes such as border crossings, bridges, terminals, and tunnels. Disruption of such choke points could plausibly create massive traffic blockages, thus exposing passengers to attack.

Railways

U.S. railroads move a large quantity of freight between cities, ports, and distribution centers along approximately 140,000 miles of railroad track. Commodities and raw materials represent the bulk of freight transported via the nation's railways. Security for the railroad system is a complex consideration because there exist more than 500 railway companies that must collaborate with each other when moving freight on railways.

Passenger railways include commuter trains between cities and within local urban areas. Ridership on these conveyances is very active, with approximately 45 million commuters annually riding subways and trains managed by urban transit agencies and approximately 20 million additional commuters riding intercity trains.

▶ **Photo 11.5** The National Terrorism Advisory System is designed to disseminate information about contemporary threat environments.

Federal Air Marshal Service (FAMS): The federal "sky marshals" service. FAMS is a relatively clandestine law enforcement agency whose size, number of marshals, and presence on aircraft are classified. In 2005, FAMS was reassigned to the Transportation Security Administration from U.S. Immigration and Customs Enforcement.

The Transportation Security Administration

Prior to the September 11 terrorist attacks, airport security was largely the responsibility of state and local authorities and employees of private security firms retained as contractors. The quality of screening, searches, and inspections often varied because of a lack of centralized guidelines and regulations between state and local governments and among the private contractors. Following the September 11 attacks, the Aviation and Transportation Security Act of 2001 (ATSA) established the Transportation Security Administration. Originally assigned to the U.S. Department of Transportation, TSA came under the auspices of the U.S. Department of Homeland Security in 2003 and is a separate agency within DHS.

TSA has been responsible for civil aviation security since 2002, when responsibility for civil aviation security was transferred to TSA from the Federal Aviation Administration. To support its aviation security mission, the agency manages the implementation of transportation security regulations and promulgates them as required. In essence, the former decentralized and privatized screening and security system has been replaced by a nationally federalized central model. All airport screening and inspections are now conducted by TSA personnel. In 2005, the **Federal Air Marshal Service (FAMS)** was reassigned to TSA from U.S. Immigration and Customs Enforcement. FAMS is a relatively clandestine law enforcement agency whose size, number of marshals, and presence on aircraft are classified.

TSA administers an extensive and complex security system by appointing more than 30,000 personnel who serve as screeners, law enforcement agents, and airport security staff, among other critical positions. Security measures include installing and operating explosives detection equipment, overseeing canine explosives detection teams, screening air cargo, and seizing millions of prohibited items before travelers board aircraft.

GLOBAL PERSPECTIVE: INTERNATIONAL LAW ENFORCEMENT CONSORTIA

The United States collaborates with international law enforcement consortia that are designed to share information on transnational investigations and other crime-related inquiries. This chapter's Global Perspective discusses collaborative international law enforcement organizations.

GLOBAL PERSPECTIVE

THE ROLE OF THE "INTERNATIONAL POLICE"

Internationally, there is no enforcement mechanism for violations of international criminal law, other than treaties and other voluntary agreements between nations. Nevertheless, many nations have become members of the International Criminal Police Organization, more commonly known as INTERPOL.

INTERPOL is an international association of more than 140 nations that agree to share intelligence and provide assistance in the effort to suppress international crime. The association is based in Lyon, France, and each member nation has a bureau that serves as a liaison with INTERPOL. INTERPOL is more of an investigative consortium than a law enforcement agency. Its value lies in the cooperative sharing of information between members as well as the coordination of counterterrorist and criminal investigations.

Similarly, the **European Police Office (EUROPOL)** is a cooperative investigative consortium of members of the European Union. The association is based in The Hague, Netherlands. EUROPOL has been designated as the European Union's law enforcement agency, and it assists members in combatting terrorism and serious crimes. For example, EUROPOL assisted French authorities in the aftermath of the November 13, 2015, terrorist assault on Paris.

Discussion Questions

1. Are international police organizations viable counterterrorist institutions?

2. How should international police organizations be regulated?

3. Why are such organizations necessary?

CHAPTER SUMMARY

This chapter introduced policy issues regarding homeland security target hardening.

Options for successful hardening of critical infrastructure targets include implementation of recommendations included in the Building and Infrastructure Protection Series published by the U.S. Department of Homeland Security. BIPS represents an ongoing effort to anticipate threats and plausible contingencies as they arise. Border control requires a substantial investment of resources and massive deployment of personnel. Agencies tasked with border control have the daunting mission of successfully patrolling and inspecting long stretches of international crossings, ports of entry, and people entering the country.

Transportation security is a multifaceted category that requires creative measures to harden potential targets. Because it is such a multifaceted classification, transportation systems pose an enormously complex challenge for homeland security agencies. Information security is a new front in the war on terrorism, and cyberwar is an ever-expanding reality. It requires the deployment of personnel with cutting-edge skills and technologies to prevent terrorists from penetrating and damaging private and governmental information systems.

DISCUSSION BOX

This chapter's Discussion Box is intended to stimulate critical thinking about the problem of securing the homeland against threats from international terrorists.

Homeland Security and Responding to the Stateless Revolutionaries

Some terrorist movements operate exclusively on an international scale and have little or no domestic presence in a home country. There are different reasons for this strategy: Some groups espouse a global ideological agenda that requires them to fight on behalf of a vague concept of *the oppressed* of the world. Other groups operate within an environment that mandates as a matter of practicality that they operate internationally. They strike from operational havens across state borders and often move around from country to country.

These movements are essentially *stateless*, in the sense that they have no particular home country that they seek to liberate, there is no homeland to use as a base, or their group has been uprooted from the land that they are fighting for. Among these stateless extremist movements are secular ideological revolutionaries, sectarian radicals fighting on behalf of a faith, and representatives of stateless ethnonational groups.

Discussion Questions

1. How should domestic agencies with homeland security duties prepare for the possibility of incidents from stateless revolutionaries?

2. When a domestic incident occurs, what responses by federal domestic security agencies should be implemented?

3. What measures should be taken to ensure collaboration and coordination between national defense and homeland security institutions?

4. How should the United States deploy its military to counter the threat from stateless revolutionaries?

5. Which agencies should have priority authority in designing policies to counter the threat from stateless revolutionaries?

KEY TERMS AND CONCEPTS

The following topics were discussed in this chapter and can be found in the glossary:

biometric technology   243
Building and Infrastructure
    Protection Series (BIPS)   247
cyberterrorism   240
cyberwar   240
Echelon   245
Federal Air Marshal Service
    (FAMS)   252

Federal Aviation Administration
    (FAA)   251
Homeland Security Presidential
    Directive 7 (HSPD-7)   245
Presidential Decision Directive/
    NSC-63   245
Presidential Policy Directive 21
    (PPD-21)   246

*Reference Manual to Mitigate Potential
    Terrorist Attacks Against Buildings,
    Edition 2* (BIPS 06/FEMA
    426)   247
target hardening   239
Transportation Security
    Administration (TSA)   251

ON YOUR OWN

Get the tools you need to sharpen your study skills. SAGE edge offers a robust online environment featuring an impressive array of free tools and resources.

Access practice quizzes, eFlashcards, video, and multimedia at **edge.sagepub.com/martinhs3e**

RECOMMENDED WEBSITES

The following Web sites provide information on counter-terrorism and infrastructure security.

Air University Library: www.maxwell.af.mil/au/aul/bibs/bib97.htm

Center for Defense Information: www.cdi.org

Combating Terrorism Center at West Point: www.ctc.usma.edu

Counterterrorism Office, U.S. Department of State: www.state.gov/s/ct

Diplomatic Security Service, Rewards for Justice: www.rewardsforjustice.net

International Policy Institute for Counter-Terrorism: www.ict.org.il

Jaffe Center for Strategic Studies (Tel Aviv): www.tau.ac.il/jcss/lmas.html#Terror

Memorial Institute for the Prevention of Terrorism: www.mipt.org

Middle East Media Research Institute: memri.org

Naval Postgraduate School Center on Terrorism & Irregular Warfare: www.nps.edu/Academics/Centers/CTIW/index.html

Patterns of Global Terrorism (State Department): www.state.gov/s/ct/rls

Terrorism Research Center: www.terrorism.com

WEB EXERCISE

Using this chapter's recommended websites, conduct an online investigation of target hardening.

1. What are the primary missions of U.S. border and transportation security agencies?

2. How effective are modern measures in hardening critical infrastructure and other potential targets?

3. In your opinion, what is the most vulnerable category of potential targets?

To conduct an online search on intelligence and target hardening, activate the search engine on your Web browser and enter the following keywords:

"Cybersecurity"

"Critical infrastructure"

RECOMMENDED READINGS

The following publications provide discussions on target hardening, response, and recovery:

Alden, Edward. 2008. *The Closing of the American Border: Terrorism, Immigration, and Security Since 9/11*. New York: HarperCollins.

Carr, Jeffrey. 2011. *Inside Cyber Warfare: Mapping the Cyber Underworld*. 2nd ed. Sebastopol, CA: O'Reilly Media.

Clarke, Richard A. and Robert K. Knake. 2010. *Cyber War: The Next Threat to National Security and What to Do About It*. New York: HarperCollins.

Davis, Anthony. 2008. *Terrorism and the Maritime Transportation System*. Livermore, CA: WingSpan Press.

Elias, Bartholomew. 2010. *Airport and Aviation Security: U.S. Policy and Strategy in the Age of Global Terrorism*. Boca Raton, FL: Auerbach/Taylor & Francis.

Janczewski, Lech J. and Andrew M. Colarik, eds. 2008. *Cyber Warfare and Cyber Terrorism*. Hershey, PA: Information Science Reference.

McNicholas, Michael. 2007. *Maritime Security: An Introduction*. Waltham, MA: Butterworth-Heinemann.

Price, Jeffrey and Jeffrey Forrest. 2013. *Practical Aviation Security: Predicting and Preventing Future Events*. 2nd ed. Waltham, MA: Butterworth-Heinemann.

Winterdyk, John A. and Kelly W. Sundberg, eds. 2009. *Border Security in the Al-Qaeda Era*. New York: CRC Press.

CHAPTER TWELVE

CRITICAL RESOURCES
Resilience and Planning

Chapter Learning Objectives

This chapter will enable readers to do the following:

1. Apply the concept of resilience

2. Critically assess the role of proper preparedness and planning activities

3. Analyze issues that may arise when planning responses to plausible hazard-related scenarios

4. Discuss prevention and mitigation planning issues

5. Explain the political and other considerations that influence the preparedness and planning process

Opening Viewpoint: Planning and the Necessity of Multi-Tiered Collaboration

The national homeland security enterprise requires efficient and effective teamwork among federal, state, and local agencies. Federal agencies, led primarily by the Department of Homeland Security (DHS), are tasked with communicating and collaborating with each other as well as with state and local agencies. At the same time, private sector collaboration has increasingly become a priority for national homeland security planning.

Federal outreach to state and local law enforcement and emergency management agencies is particularly crucial to ensuring that proper planning occurs. These initiatives represent recognition that state and local agencies serve as first responders in the planning, response, and prevention process. For example, the new homeland security mission for law enforcement agencies is a significant augmentation of the pre–September 11 law enforcement mission. Because of this reality, multi-tiered collaboration led by the federal homeland security establishment is a critical component for the successful administration of a viable homeland security enterprise.

This chapter explores homeland security preparedness and planning, within the overarching context of resilience for recovering from the effects of an emergency event. An underlying theme of this discussion is the need for cooperation in preparing and implementing policies and procedures for a range of potential hazards. Designing effective responses to incidents involving weaponized hazards is a necessary policy priority for establishing a secure domestic environment. The commitment of financial and human resources to preparedness and planning is an investment that will, if properly undertaken, reap immeasurable benefits in the security of local communities and the nation at large. However, political and financial priorities often influence which policies and procedures are selected as well as their preparation and implementation. Nevertheless, the ultimate goal of establishing a secure environment necessitates ongoing coordination of preparedness and planning activity—in order to sustain the ability to achieve resilience.

The discussion in this chapter will review the following:

- Understanding resilience

- The role of proper planning

- Prevention and mitigation planning

- Chemical and biological hazard planning

- Radiological and nuclear hazard planning

UNDERSTANDING RESILIENCE

Resilience refers to the ability to recover from emergencies as rapidly as possible. It is a doctrine that is embedded in the homeland security enterprise. The DHS included resilience as an instrumental goal when it published the *Quadrennial Homeland Security Review* (QHSR) in 2014 (previously introduced and discussed in Chapter 1). The QHSR "established a series of goals and objectives in the areas of critical infrastructure, global movement, and supply chain systems, and cyberspace."[1] Furthermore, the document reported five missions, one of which—Mission 5—is titled "Strengthening National Preparedness and Resilience." Similarly, the *National Security Strategy* of 2017, published through the U.S. White House, states,

Resilience includes the ability to withstand and recover rapidly from deliberate attacks, accidents, natural disasters, as well as unconventional stresses, shocks and threats to our economy and democratic system.[2]

The QHSR and *National Security Strategy* of 2017 are illustrative of foundational concepts that frame the overall approach toward achieving resilience in the United States. To that end, both DHS and the Federal Emergency Management Agency (FEMA) report that their efforts for achieving resilience are intended "to build a culture of preparedness through insurance, mitigation, preparedness, continuity and grant programs"[3] within the homeland security enterprise. Thus, resilience is intended to be a convergence of preparedness policies and a strong preparedness culture.

THE ROLE OF PROPER PLANNING

Federal, state, local, and private institutions each have a vested interest in planning for challenges to homeland security. Each level of authority is subject to victimization by violent extremists, and without effective planning, the consequences are potentially quite severe. For this reason, concerted efforts have been undertaken to establish planning protocols and systems to secure multijurisdictional and multilevel collaboration. Such collaboration is especially required between the public and private sectors because although violent extremists often direct their grievances against government interests, in the era of the New Terrorism, entire societies and populations are deemed legitimate targets.

Foundational Concept: The National Incident Management System

On February 28, 2003, President George W. Bush directed the secretary of homeland security to construct a national Incident Command System (ICS). The president's directive, detailed in **Homeland Security Presidential Directive 5 (HSPD-5)**, sought to secure the development of coordinated incident response procedures by institutionalizing new protocols to do so. The HSPD-5 protocols addressed the Incident Command Systems of federal, state, and local authorities and were intended to synchronize these systems in the event of a terrorist incident. Titled *Management of Domestic Incidents*, HSPD-5's stated purpose was "to enhance the ability of the United States to manage domestic incidents by establishing a single, comprehensive national incident management system."[4]

The directive further stated that, as a matter of overarching national policy, "the United States Government shall establish a single, comprehensive approach to domestic incident management."[5] Thus, HSPD-5 specifically sought to create an efficient national Incident Command System to respond to the unprecedented threat scenarios posed by the New Terrorism. To accomplish this goal, the secretary of homeland security was tasked with creating a **National Incident Management System (NIMS)** that would

> provide a consistent nationwide approach for Federal, State, and local governments to work effectively and efficiently together to prepare for, respond to, and recover from domestic incidents, regardless of cause, size, or complexity. To provide for interoperability and compatibility among Federal, State, and local capabilities, the NIMS [was to] include a core set of concepts, principles, terminology, and technologies covering the incident command system; multi-agency coordination systems; unified command; training; identification and management of resources.[6]

In December 2008, DHS published *National Incident Management System*, thus fully implementing HSPD-5 and disseminating the newly created protocols to federal, state, and local homeland security authorities.[7] From its inception, NIMS was designed to serve as "the template for the management of incidents"[8] on a national scale. NIMS was established as a "systematic, proactive approach"[9] for homeland security authorities at all levels of government. Significantly, NIMS is an integrated system for implementing the National Response Framework, which "provides the structure and mechanisms for national-level policy for incident management."[10] Full consideration of the National Response Framework is presented in Chapter 13.

Local Planning: The Whole Community Approach

In 2011, FEMA published a document that framed an overarching conceptual framework for managing emergencies. Titled *A Whole Community Approach to Emergency Management: Principles, Themes, and Pathways for Action*,[11] its principles became conceptually embedded in the homeland security enterprise.

Popularly referred to as the Whole Community approach, the philosophical underpinning of this concept is that the homeland security enterprise cannot rely solely on government intervention to manage and achieve resiliency and recovery during emergencies. Rather,

> Whole Community is a means by which residents, emergency management practitioners, organizational and community leaders, and government officials can

collectively understand and assess the needs of their respective communities and determine the best ways to organize and strengthen their assets, capacities, and interests. By doing so, a more effective path to societal security and resilience is built. In a sense, Whole Community is a philosophical approach on how to think about conducting emergency management.[12]

The Whole Community approach relies on collaboration by all sectors and communities in society for creating a holistic approach to successful preparedness and planning. This is theoretically preferable because

the benefits of Whole Community include a more informed, shared understanding of community risks, needs, and capabilities; an increase in resources through the empowerment of community members; and, in the end, more resilient communities. A more sophisticated understanding of a community's needs and capabilities also leads to a more efficient use of existing resources regardless of the size of the incident or community constraints.[13]

In this way, resilience and planning become the duty of all sectors of society. All members of the community are stakeholders and, therefore, have vested interests in creating holistic policies that theoretically promote sound responses to emergency events.

Local Planning: The Private Sector

The private sector is arguably the sector of society most affected when a terrorist incident occurs. The reason for this is uncomplicated: Most terrorist attacks are directed against or occur within close proximity to private property and businesses. The private sector consists of private citizens and private property and, within this context, may also include hospitals, education institutions, religious institutions, and businesses.

Most economic damage is experienced by private interests, and most casualties are suffered among civilians. The motivation by extremists for selecting these targets is their appreciation of the fact that although many symbolic values can be created to justify the legitimacy of most targets, many civilian institutions are "soft" from a security perspective and, therefore, more easily assaulted. In fact, the most defining terrorist attack in modern history—the September 11 homeland attacks—was, to a considerable extent, committed against symbolic civilian infrastructure. The attack used civilian aircraft to destroy major civilian infrastructure in New York, and it killed approximately 3,000 people (mostly civilians), with the consequence that the enormous economic damage was primarily suffered by the private sector.

Table 12.1 reports worldwide targets of terrorist attacks in 2017, most of which involved targeting of the private sector by violent extremists. Analysis of these incidents clearly poses a significant challenge for policy planners due to the sheer number of plausible scenarios involving private sector targets.

U.S. Federal Emergency Management Agency

▶ **Photo 12.1** Recovery workers survey hazardous debris in the aftermath of the September 11, 2001, terrorist attacks on the World Trade Center in New York City. The disaster site contained massive amounts of debris, much of it unstable and otherwise hazardous.

Table 12.1   Targets of Terrorist Attacks Worldwide, 2017

Although all terrorist attacks are directed against what they perceive to be symbolic targets, there is a pattern of defining this symbolism within the context of private sector people and infrastructure. For this reason, private citizens and property, businesses, and private institutions are the most frequently attacked targets. Terrorist incidents can reasonably be characterized as primarily political violence directed against "soft" civilian targets.

Target Type	Number of Targets
Private Citizens & Property	3,422
Police	1,596
Government (General)	919
Business	797
Military	457
Religious Figures/Institutions	228
Educational Institution	162
Utilities	159
Terrorists/Non-State Militia	158
Transportation	149
Violent Political Party	149
Journalists & Media	123
Government (Diplomatic)	92
Other	84
Non-Governmental Organizations	55
Telecommunication	33
Tourists	14
Airports & Airlines	13
Maritime	12
Food or Water Supply	10
Abortion Related	1
Total	**8,633**

Source: U.S. Department of State.

Promoting Homeland Security and Public–Private Collaboration

The homeland security enterprise has responded by emphasizing collaboration between national institutions, such as DHS and FEMA, and local private interests. These public–private collaborative efforts have become institutionalized aspects of the national homeland security system. For example, DHS established a Private Sector Office as an important outreach and coordinating resource for members of the private sector. Within its stated mission, the DHS Private Sector Office carries out the following tasks:

- Advises the Secretary on the impact of the Department's policies, regulations, processes, and actions on the private sector.

- Creates and fosters strategic communications with the private sector to enhance the primary mission of the Department to protect the American homeland.

- Interfaces with other relevant Federal agencies with homeland security missions to assess the impact of these agencies on the private sector.

- Creates and manages private sector advisory councils composed of representatives of industries and associations designated by the Secretary.

- Works with Federal laboratories, federally funded research and development centers, other federally funded organizations, academia, and the private sector to develop innovative approaches to address homeland security challenges to produce and deploy the best available technologies for homeland security missions.

- Assists in the development and promotion of private sector best practices to secure critical infrastructure.

- Coordinates industry efforts regarding Department functions to identify private sector resources that could be effective in supplementing government efforts to prevent or respond to a terrorist attack.[14]

Similarly, in 2007, FEMA established a Private Sector Division within its Office of External Affairs. The purpose of the Private Sector Division is to "communicate, cultivate and advocate for collaboration between the U.S. private sector and FEMA, to support FEMA's capabilities and to enhance national preparedness, protection, response, recovery, and mitigation of all hazards."[15] Its vision statement projects future collaboration via "establish[ing] and maintain[ing] a national reputation for effective support to our private sector stakeholders through credible, reliable and meaningful two-way communication."[16]

Private Sector Preparedness

In 2009, DHS published and promulgated the second edition of the **National Infrastructure Protection Plan (NIPP).** Conceptually, "the overarching goal of the NIPP is to build a safer, more secure, and more resilient America by preventing, deterring, neutralizing, or mitigating the effects of a terrorist attack or natural disaster, and to strengthen national preparedness, response, and recovery in the event of an emergency."[17] The NIPP identifies 16 critical infrastructure sectors to accomplish this; within these sectors, collaborative initiatives will be undertaken between the public and private sectors. The collaborative initiatives will be done in concert with the sector-specific plans assigned to each critical infrastructure sector. The ultimate goal of the NIPP public–private system is to establish mutually collaborative interests for seamlessly coordinating interaction between public and private institutions. In essence, ensuring efficient private sector participation in homeland security planning and preparedness initiatives is the primary purpose of NIPP.

In addition to the NIPP process, a series of information centers have been established by private stakeholders of critical infrastructure as a way to secure their infrastructure interests, all done in collaboration with DHS initiatives. Assistance in planning for the 16 identified critical infrastructure sectors is conceptually provided by centers known as

National Infrastructure Protection Plan (NIPP): The National Infrastructure Protection Plan identifies 16 critical infrastructure sectors, within which collaborative initiatives will be undertaken between the public and private sectors. This will be done in concert with sector-specific plans assigned to each critical infrastructure sector. The NIPP's purpose is to ensure efficient private sector participation in homeland security planning and preparedness initiatives.

Information Sharing and Analysis Centers (ISACs). Each ISAC is assigned to a specific critical infrastructure sector to accomplish this task. The NIPP and critical infrastructure sectors were discussed in Chapter 11.

Local Planning: Emerging Roles at the Local Government Level

Through NIMS, the National Response Framework, the NIPP, and other planning protocols, DHS acknowledges that collaborative planning with state and local agencies and officials is central to its overall mission. State and local planners likewise acknowledge that their own effectiveness is governed by strong collaboration with federal institutions. It is at the state and local levels that the day-by-day prevention of and first responses to terrorist events occur. State and local authorities also assemble local resources, such as medical care, the private sector, law enforcement, and community organizations. This expanded local homeland security role necessitates federal guidance because, from the federal perspective, it is necessary for the federal government to build consistency and evenness across hundreds of localities nationally.

As discussed in Chapter 7, local law enforcement authorities are required to perform homeland security duties when necessary, in addition to their traditional mission of enforcing the law. This new role requires proper training in preparation for carrying out domestic security tasks, such as protecting infrastructure, responding to terrorist events, and working local contacts as intelligence assets. Local emergency medical and fire authorities must likewise receive additional training in preparation for responding to warlike terrorist incidents. DHS and other federal agencies sponsor such training initiatives for local homeland security missions, either directly or through funded initiatives.

Additional training is sponsored by DHS and FEMA via the National Domestic Preparedness Consortium (NDPC). The NDPC program directly engages local emergency responders to provide practical training assistance. The mission of the NDPC is summarized as follows:

> The NDPC is a DHS/FEMA training partner providing high-quality training to emergency responders throughout the United States and its territories under DHS/FEMA's Homeland Security National Training Program Cooperative Agreement. Preparedness is a shared, national responsibility requiring our active participation to prepare America to address its threats.[18]

The consortium serves as a national forum for establishing collaborative and systemic protocols for responding to potentially worst-case scenarios. In this regard,

> The NDPC is a partnership of several nationally recognized organizations whose membership is based on the urgent need to address the counter-terrorism preparedness needs of the nation's emergency first responders within the context of all hazards including chemical, biological, radiological, and explosive Weapons of Mass Destruction (WMD) hazards.[19]

Case in Point: Health and Medical Preparedness Strategies

Proper medical preparedness planning requires that state and local authorities be adequately organized for terrorist incidents prior to when they occur, respond effectively after incidents occur, and ideally mitigate the eventual human cost of terrorist incidents. The federal homeland security bureaucracy has taken the lead in coordinating medical preparedness strategies by publishing plans and handbooks for medical personnel responsible for responding to homeland security incidents.

Homeland Security Presidential Directive 21

In October 2007, **Homeland Security Presidential Directive 21 (HSPD-21)** established a **National Strategy for Public Health and Preparedness**. Written as an all-hazards approach to homeland security, the purpose of HSPD-21 was to "transform our national approach to protecting the health of the American people against all disasters."[20] The document reviewed key principles from several policy documents, in particular the *National Strategy to Combat Weapons of Mass Destruction* (published in December 2002), *Biodefense for the 21st Century* (published in April 2004), and the *National Strategy for Homeland Security* (published in October 2007). Five key principles were synthesized, which are summarized as follows:

- Preparedness for all potential catastrophic health events

- Vertical and horizontal coordination across all levels of government, all jurisdictions, and all disciplines

- A regional approach to health preparedness

- Engagement of the private sector, academia, and other nongovernmental entities in preparedness and response efforts

- The important roles of individuals, families, and communities

HSPD-21 also identified four critical components of public health and medical preparedness: bio-surveillance, countermeasure distribution, mass-casualty care, and community resilience.[21] These components are cited as the central policy priorities for effective public health and medical preparedness and are discussed extensively in HSPD-21.

The Medical Surge Capacity and Capability Handbook

In 2007, the U.S. Department of Health and Human Services (HHS) published a management handbook promoting a system for integrating medical and health resources when responding to large-scale emergencies. Titled *Medical Surge Capacity and Capability Handbook: A Management System for Integrating Medical and Health Resources During Large-Scale Emergencies* **(MSCC)**,[22] it defines its purpose and the relatively new concept of **medical surge** as follows: "Medical and public health systems in the United States must prepare for major emergencies or disasters involving human casualties. Such events will severely challenge the ability of healthcare systems to adequately care for large numbers of patients (surge capacity) and/or victims with unusual or highly specialized medical needs (surge capability)."[23]

A management methodology for responding to large-scale medical incidents is recommended in the MSCC. The proposed medical incident management system requires tiered integration of federal, state, and local medical resources. To facilitate this goal, the MSCC identifies six tiers of responsibility, which are intended to function as part of an integrated response system rather than independently. These six tiers are as follows:

- *Management of individual healthcare assets* (Tier 1): A well-defined Incident Command System (ICS) to collect and process information, to develop incident plans, and to manage decisions is essential to maximizing MSCC.

Homeland Security Presidential Directive 21 (HSPD-21): A 2007 directive by President George W. Bush that established a National Strategy for Public Health and Preparedness. Written as an all-hazards approach to homeland security, the purpose of HSPD-21 was to "transform our national approach to protecting the health of the American people against all disasters."

National Strategy for Public Health and Preparedness: The National Strategy for Public Health and Preparedness was established in October 2007 by Homeland Security Presidential Directive 21 as part of an all-hazards approach to homeland security.

Medical Surge Capacity and Capability Handbook: A Management System for Integrating Medical and Health Resources During Large-Scale Emergencies **(MSCC):** A management handbook published by the U.S. Department of Health and Human Services that promoted a system for integrating medical and health resources when responding to large-scale emergencies.

medical surge: A management methodology for responding to large-scale medical incidents.

▶ **Photo 12.2**

Weaponized biological
hazards are a plausible
threat to homeland
security. The United
States has experienced
several bioterror attacks,
the most serious involving
salmonella and anthrax.

- *Management of a healthcare coalition* (Tier 2): Coordination among local healthcare assets is critical to providing adequate and consistent care across an affected jurisdiction.

- *Jurisdiction incident management* (Tier 3): A jurisdiction's ICS integrates healthcare assets with other response disciplines to provide the structure and support needed to maximize MSCC.

- *Management of state response* (Tier 4): State government participates in medical incident response across a range of capacities, depending on the specific event.

- *Interstate regional management coordination* (Tier 5): Effective mechanisms must be implemented to promote incident management coordination among affected states.

- *Federal support to state, tribal, and jurisdiction management* (Tier 6): Effective management processes at the state (Tier 4) and jurisdiction (Tier 3) levels facilitate the request, receipt, and integration of federal public health and medical resources to maximize MSCC.[24]

HHS promulgates the MSCC to medical emergency responders nationwide in an effort to standardize state, regional, and local response systems.

PREVENTION AND MITIGATION PLANNING

Prevention and mitigation planning refers to devising procedures and policies that reduce the severity of threats to homeland security. Many policies and procedures have been proposed, all of which are intended to mitigate danger arising from potential threats. They can be quite complex to implement and are often conceptually diverse. Nevertheless, prevention and mitigation planning is essential for overall homeland security preparedness planning.

Components of the Prevention and Mitigation Process

Prevention and mitigation policy options must focus on reducing the probability that a terrorist event will take place as well as lessening potential consequences when one occurs. To achieve this, the prevention and mitigation process includes the following three primary components: (1) identifying potential risks, (2) assessing the probability that a terrorist incident will occur, and (3) assessing the consequences to society should a terrorist event occur. Effective prevention and mitigation planning should ideally rely on data and conclusions derived from these components.

The first component, identifying potential risks, is a fundamental consideration for counterterrorism planners. Identifying terrorist risks requires assessing the intentions of extremist organizations, including projecting which locations constitute likely targets. Fortunately, extremist organizations often disseminate pronouncements about their intentions via the Internet or other media. Although there is little specific information about when, where, and how they intend to strike, terrorist intentions rarely require nuanced interpretation. The intelligence community is especially well organized and equipped to identify potential risks originating from violent extremists.

The second component, assessing the probability that an event will occur, requires measured applications of analytical skills and resources in order to project the likelihood of a terrorist incident. This is naturally an inexact and uncertain process, but modern analytical resources utilize new technologies and existing databases to detect patterns of behavior and interpret communications chatter among extremist operatives and organizations. Patterns of behavior and communications chatter among operationally active extremist movements do afford analysts with plausible scenarios with which they can reasonably assess the probability that the extremists involved intend to continue engaging in terrorist violence.

The third component, assessing the consequences to society should an event occur, is an evaluation of the human and financial costs of plausible scenarios. Projecting possible consequences of terrorist incidents is difficult because terrorists actively attempt to strike unexpected targets using unpredictable tactics—the essence of asymmetrical warfare. Nevertheless, considerable attention is given to carefully assessing the consequences of a terrorist incident. Although difficult, this process is feasible because there are unfortunately ample precedents for calculating the human and financial costs of terrorist incidents. For example, a 2009 unpublished study conducted by researchers at the University of Southern California projected the economic impact of a radiological bomb attack in downtown Los Angeles.[25]

Federal Mitigation Assistance

The mitigation of risks posed by violent extremists requires utilization of data derived from the foregoing three components. Although this is a complex process requiring collaborative information acquisition and analysis, counterterrorist analysis and implementation have successfully thwarted a significant number of terrorist conspiracies. In effect, the probability of incidents has been mitigated on numerous occasions.

Regarding the mitigation of consequences to society, extensive research has been devoted to mitigation of the effects of terrorist attacks on infrastructure. For example, DHS published the *Reference Manual to Mitigate Potential Terrorist Attacks Against Buildings*, which was previously discussed in Chapter 11.[26]

FEMA manages five **Hazard Mitigation Assistance (HMA) grant programs**. HMA grant programs "provide funding for eligible mitigation activities that reduce disaster losses and protect life and property from future disaster damages."[27] These programs promote FEMA's mission to provide all-hazards mitigation assistance. The following grants are available for mitigation assistance from FEMA:

Hazard Mitigation Assistance (HMA) grant programs: Grant programs that "provide funding for eligible mitigation activities that reduce disaster losses and protect life and property from future disaster damages."

- *Hazard Mitigation Grant Program* (HMGP). HMGP assists in implementing long-term hazard mitigation measures following presidential disaster declarations. Funding is available to implement projects in accordance with State, Tribal, and local priorities.

- *Pre-Disaster Mitigation* (PDM). PDM provides funds on an annual basis for hazard mitigation planning and the implementation of mitigation projects prior to a disaster. The goal of the PDM program is to reduce overall risk to the population and structures, while at the same time also reducing reliance on federal funding from actual disaster declarations.

- *Flood Mitigation Assistance* (FMA). FMA provides funds on an annual basis so that measures can be taken to reduce or eliminate risk of flood damage to buildings insured under the National Flood Insurance Program (NFIP).

- *Repetitive Flood Claims* (RFC). RFC provides funds on an annual basis to reduce the risk of flood damage to individual properties insured under the NFIP that have had one or more claim payments for flood damages. RFC provides up to 100% federal funding for projects in communities that meet the reduced capacity requirements.

- *Severe Repetitive Loss* (SRL). SRL provides funds on an annual basis to reduce the risk of flood damage to residential structures insured under the NFIP that are qualified as severe repetitive loss structures. SRL provides up to 90% federal funding for eligible projects.[28]

CHEMICAL AND BIOLOGICAL HAZARD PLANNING

Chemical and biological agents are capable of being transformed into weaponized devices that, if skillfully deployed, represent a credible risk of mass casualties and widespread disruption. Weaponized chemical and biological agents have been stockpiled and occasionally deployed by nations during wartime. There are also examples of extremist groups that have attempted to weaponize chemical and biological agents. Many terrorist threat scenarios posit that terrorists can feasibly obtain or manufacture chemical and biological weapons and that these weapons will be used as weapons of mass destruction (WMDs) against "soft" civilian targets. Tables 12.2 and 12.3 report general indicators of the possible use of chemical and biological agents.

Table 12.2   General Indicators of Possible Chemical Agent Use
Stated Threat to Release a Chemical Agent
Unusual Occurrence of Dead or Dying Animals
• For example, lack of insects, dead birds
Unexplained Casualties
• Multiple victims
• Surge of similar 911 calls
• Serious illnesses
• Nausea, disorientation, difficulty breathing, or convulsions
• Definite casualty patterns
Unusual Liquid, Spray, Vapor, or Powder
• Droplets, oily film
• Unexplained odor
• Low-lying clouds/fog unrelated to weather
Suspicious Devices, Packages, or Letters
• Unusual metal debris
• Abandoned spray devices
• Unexplained munitions

Source: U.S. Federal Emergency Management Agency.

Table 12.3 General Indicators of Possible Biological Agent Use
Stated Threat to Release a Biological Agent
Unusual Occurrence of Dead or Dying Animals
Unusual Casualties
• Unusual illness for region/area
• Definite pattern inconsistent with natural disease
Unusual Liquid, Spray, Vapor, or Powder
• Spraying; suspicious devices, packages, or letters

Source: U.S. Federal Emergency Management Agency.

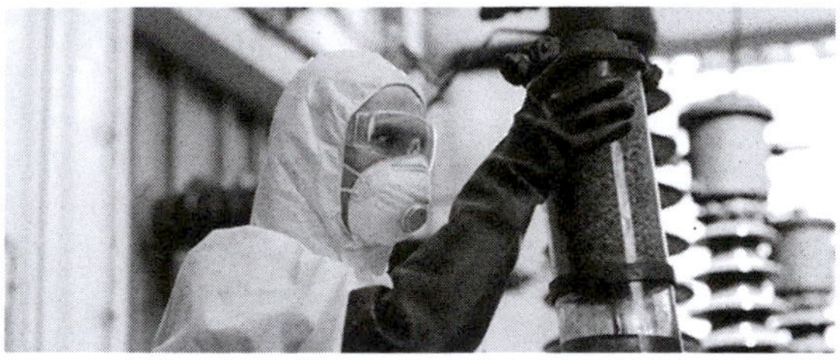

Real-time consequences from a mass-casualty chemical or biological attack have never been experienced in the United States, although, as discussed later in this chapter, the homeland was subjected to an anthrax incident. Should chemical or biological weapons be activated by terrorists, state and local authorities will necessarily be the first to respond to the incident. For this reason, many state and local agencies actively prepare for the eventuality when they may be tasked with mitigating the consequences of a chemical or biological attack. DHS regularly disburses grants and other funding to assist state and local authorities in responding to chemical or biological incidents. A large proportion of DHS support is specifically earmarked for the mitigation of WMD threats, and such earmarks often compose the majority of DHS disbursements. Other federal agencies also assist local authorities in preparing for chemical and biological incidents.

Marcin Balcerzak/Shutterstock

▶ **Photo 12.3** Chemical weapons have been used many times historically by adversaries. Planning effective response procedures in response to the deployment of weaponized chemical hazards by terrorists presents a significant challenge to homeland security systems.

Understanding the Threat: Chemical and Biological Hazards

As introduced previously in Chapter 2, plausible terrorist threat scenarios include the deliberate use of weaponized biological and chemical agents against perceived enemies. When analyzing scenarios involving the weaponization of these agents, it is important to remember to consider them within the overall context of weaponized radiological agents, nuclear weapons, and explosives. These CBRNE (chemical, biological, radiological, nuclear, and explosives) weapons potentially present serious risks of deployment against civilian and government targets.

Chemical Agents

Chemical agents are "chemical substances, whether gaseous, liquid, or solid, which are used for hostile purposes to cause disease or death in humans, animals, or plants, and which depend on direct toxicity for their primary effect."[29] Some chemical agents, such as pesticides, are commercially available. Other chemical agents can be manufactured by extremists using available instruction guides. Because of many plausible threat scenarios,[30] experts believe that chemical weapons in the possession of terrorists pose a more likely possibility than biological, radiological, or nuclear weapons.[31]

chemical agents: "Chemical substances, whether gaseous, liquid, or solid, which are used for hostile purposes to cause disease or death in humans, animals, or plants, and which depend on direct toxicity for their primary effect."

Examples of possible weaponized chemical agents in the arsenals of terrorists include the following:

- *Phosgene gas* causes the lungs to fill with water, choking the victim.

- *Chlorine gas* destroys the cells that line the respiratory tract.

- *Mustard gas* is actually a mist rather than a gas. It is a blistering agent that blisters skin, eyes, and the nose and can severely damage the lungs if inhaled.

- *Nerve gases*, such as sarin, tabun, and VX, block (or "short-circuit") nerve messages in the body. A single drop of a nerve agent, whether inhaled or absorbed through the skin, can shut down the body's neurotransmitters.

Biological Agents

biological agents:
"Living organisms . . . or infective material derived from them, which are intended to cause disease or death in man, animals, and plants, and which depend on their ability to multiply in the person, animal, or plant attacked."

Biological agents are "living organisms . . . or infective material derived from them, which are intended to cause disease or death in man, animals, and plants, and which depend on their ability to multiply in the person, animal, or plant attacked."[32] Viruses, fungi, and bacteria are all labeled *biological* weapons, although viruses are not living organisms and, hence, technically cannot be considered biological weapons. Once biological components are obtained, the problem of *weaponizing* them can be difficult.[33] Toxins such as botulism are easier to obtain or manufacture than other potential weapons-grade biological components, especially virulent diseases such as bubonic plague and Ebola (hemorrhagic fever). The threat from such attacks comes mostly from possible poisoning of food or water rather than a catastrophic epidemic. Poisoning attacks would produce limited but potentially severe casualties.

Experts generally agree that the biological agents (whether bacteria or not) most likely to be sought by terrorists would be the following:

- *Anthrax.* Anthrax is a disease that afflicts livestock and humans. It can exist as spores or be suspended in aerosols. Humans contract anthrax either through cuts in the skin (cutaneous anthrax), through the respiratory system (inhalation anthrax), or by eating contaminated meat. Obtaining lethal quantities of anthrax is difficult but not impossible. Anthrax-infected letters were sent through the mail in the eastern United States immediately after the September 11, 2001, attacks. Those who died from anthrax exposure suffered from inhalation anthrax.

- *Smallpox.* Eradicated in nature, smallpox is a virus that is very difficult to obtain because samples exist solely in laboratories, apparently only in the United States and Russia. Its symptoms appear after about 12 days of incubation and include flu-like symptoms and a skin condition that eventually leads to pus-filled lesions. It is a highly contagious disease and can be deadly if it progresses to a hemorrhagic (bleeding) stage known as the *black pox.*

- *Botulinum toxin (botulism).* Also known as botulism, botulinum toxin is a rather common form of food poisoning. It is a bacterium, rather than a virus or fungus, and can be deadly if inhaled or ingested, even in small quantities.

- *Bubonic plague.* A bacterium-induced disease known as the "Black Death" in medieval Europe, bubonic plague is spread by bacteria-infected fleas that infect hosts when bitten. The disease is highly infectious and often fatal.

- *Ebola (hemorrhagic fever).* A virulent tropical disease that results in a high fatality rate among infected individuals, often between 50 and 90 percent. It is a relatively rare disease, but when it occurs, outbreaks can be quite intensive in Africa and elsewhere. Infected areas are quarantined until the disease runs its course.

A Complex Dilemma: Mitigating Chemical and Biological Incidents

Responding to chemical and biological incidents is an inherently complex dilemma because there are many variables, multiple jurisdictions may be involved, and difficulties will likely arise when an attack occurs that does not have clearly delineated solutions. The complexity of the mitigation process is indicated by the following challenges:

- Confirmation that an attack has occurred and of the type of agent used may not occur for some time depending on the characteristics of the agent.

- Large-scale incidents may result in many more casualties than can be managed by emergency medical systems. This occurred during the sarin nerve gas attack in Tokyo on March 20, 1995 (discussed further in Chapter Perspective 12.1).

- The decision to deploy first responders must weigh whether or not contaminants at the incident site will endanger emergency personnel. This process may cause delays in emergency intervention.

- Decontamination of the incident site may be delayed or rendered infeasible if the attack is massive in scale.

- Medical facilities and emergency medical supplies may become severely stressed by mass-casualty incidents.

- If the geographic scope of the incident is widespread, multiple jurisdictions will be affected and possibly unable to respond to all emergencies.

Aum Shinrikyō: A cult based in Japan and led by Shoko Asahara that was responsible for releasing nerve gas into the Tokyo subway system, injuring 5,000 people.

The Asahi Shimbun/Getty Images

These considerations were evident in the aftermath of the 1995 sarin nerve gas attack on the Tokyo subway system by members of the **Aum Shinrikyō** cult. Chapter Perspective 12.1 is a case study discussing the severity of the Tokyo terrorist attack and the significant challenges encountered by Japanese emergency response agencies. The severity and challenges of this incident are instructive for similar scenarios that may occur in other urban areas.

The foregoing challenges and the Aum Shinrikyō case illustrate the complexity of responding to chemical and biological incidents. It is essential that thorough guidance and instruction be provided on how to build collaborative and coordinated response systems to address these incidents. In consideration of this necessity, federal research and publications are available that offer guidelines on response procedures.

▶ **Photo 12.4** The aftermath of the March 1995 sarin nerve gas attack on the Tokyo subway system. The case of the Aum Shinrikyō cult represents the feasibility of deploying such weapons by dedicated extremists.

CHAPTER PERSPECTIVE 12.1

Case Study: The Tokyo Subway Nerve Gas Attack and Aftermath

The Tokyo sarin nerve gas attack is an instructive case study of the manufacture and use of chemical agents by terrorists and the challenges posed to urban emergency response systems by such an incident.

Aum Shinrikyō is a Japan-based cult founded in 1987 by Shoko Asahara. Their goal under Asahara's leadership was to seize control of Japan and then the world. The core belief of the cult is that Armageddon—the final battle before the end of the world—is imminent. One component of this belief is that the United States will wage World War III against Japan.

At its peak membership, Aum Shinrikyō had perhaps 9,000 members in Japan and 40,000 members around the world, thousands of them in Russia. Asahara claimed to be the reincarnation of Jesus Christ and the Buddha and urged his followers to arm themselves if they were to survive Armageddon. This apocalyptic creed led to the stockpiling of chemical and biological weapons, including nerve gas, anthrax, and Q fever. One report indicated that Aum Shinrikyō members had traveled to Africa to acquire the deadly Ebola virus. Several mysterious biochemical incidents occurred in Japan, including one in June 1994 in the city of Matsumoto, where seven people died and 264 were injured from a release of gas into an apartment building.

On March 20, 1995, members of Aum Shinrikyō positioned several packages containing sarin nerve gas on five trains in the Tokyo subway system. The trains were scheduled to travel through Tokyo's Kasumigaseki train station. The containers were simultaneously punctured with umbrellas, thus releasing the gas into the subway system. Twelve people were killed, and an estimated 5,000 to 6,000 were injured.

Tokyo's emergency medical system was unable to respond adequately to the attack so that only about 500 victims were evacuated, with the remaining victims making their own way to local hospitals. In essence, the emergency response system of a major urban center was unable to adequately respond to the incident. Mass casualties overwhelmed the system.

The police were also surprised by the attack, and "it took . . . several weeks to narrow their search to the Aum sect, locate its leaders, and seize some of their arsenal, despite the fact that Aum was not a secret organization but one that paraded through the streets of Tokyo, albeit in masks that depicted the face of their guru and leader, Shoko Asahara."[a]

The police seized tons of chemicals stockpiled by the cult. Asahara was arrested and charged with 17 counts of murder and attempted murder, kidnapping, and drug trafficking. A new leader, Fumihiro Joyu, assumed control of Aum Shinrikyō in 2000 and renamed the group Aleph (the first letter in the Hebrew and Arabic alphabets). He has publicly renounced violence, and the cult's membership has enjoyed new growth in membership.

Aum Shinrikyō is an example of the potential terrorist threat from apocalyptic cults and sects that are completely insular and segregated from mainstream society. Some cults are content to simply prepare for the End of Days, but others—like Aum Shinrikyō—are not averse to giving the apocalypse a violent "push."

A Japanese court sentenced cult leader Shoko Asahara to death by hanging on February 27, 2004. He died in detention in July 2018.

Discussion Questions

1. What are the lessons of the challenges experienced by the Tokyo emergency response system during the attack?

2. How can emergency response systems prepare themselves to mitigate the impact of weapons such as Sarin nerve gas?

3. Do you think extremists adhering to certain ideological beliefs are more motivated to use such weapons than other extremists?

Note

a. Walter Laqueur, *The New Terrorism: Fanaticism and the Arms of Mass Destruction* (New York: Oxford University Press, 1999), 129.

Case in Point: The Centers for Disease Control and Prevention Guidelines for Biological and Chemical Preparedness and Response

The Centers for Disease Control and Prevention (CDC) publishes and disseminates preparedness and response guidelines for health professionals responding to mass-casualty events, including chemical and biological terrorist events. These guidelines address multiple terrorist scenarios involving a variety of weapons and provide specific information about the nature of prospective terrorist use of weaponized chemical and biological agents. CDC publications are readily available to state and local authorities and the general public.

In 2000, CDC published a strategic plan titled *Biological and Chemical Terrorism: Strategic Plan for Preparedness and Response*.[34] It is a comprehensive assessment of multiple considerations required for effective preparedness and response agendas. The purpose of the strategic plan is summarized as follows:

> The U.S. national civilian vulnerability to the deliberate use of biological and chemical agents has been highlighted by recognition of substantial biological weapons development programs and arsenals in foreign countries, attempts to acquire or possess biological agents by militants, and high-profile terrorist attacks. Evaluation of this vulnerability has focused on the role public health will have [in] detecting and managing the probable covert biological terrorist incident with the realization that the U.S. local, state, and federal infrastructure is already strained as a result of other important public health problems. . . . CDC has developed a strategic plan to address the deliberate dissemination of biological or chemical agents. The plan contains recommendations to reduce U.S. vulnerability to biological and chemical terrorism—preparedness planning, detection and surveillance, laboratory analysis, emergency response, and communication systems. Training and research are integral components for achieving these recommendations. Success of the plan hinges on strengthening the relationships between medical and public health professionals and on building new partnerships with emergency management, the military, and law enforcement professionals.[35]

CDC also publishes guidelines discussing how health officials should respond in the immediate aftermath of a terrorist incident involving chemical and biological agents. Posted online and titled "First Hours: Emergency Responses," these guidelines specifically recommend resources "to aid health officials as they communicate with the public in the first hours of an emergency."[36] Sections of the guidelines include discussions of bioterrorism and chemical agents. Also included is a template for emergency messages that CDC recommends be disseminated in the first few minutes after a terrorist incident.

RADIOLOGICAL AND NUCLEAR HAZARD PLANNING

Radiological and nuclear hazards refer to radioactive agents that are released or otherwise dispersed from storage facilities. Such releases or dispersals can occur accidentally or deliberately. Within the context of domestic security and counterterrorism, radiological releases or dispersals are deliberately and illegally caused, with the intent to expose people and infrastructure to the radioactive material. Because of credible warnings that

Table 12.4	General Indicators of Possible Nuclear Weapon/Radiological Agent Use
Stated Threat to Deploy a Nuclear or Radiological Device	
Presence of Nuclear or Radiological Equipment	
• Spent fuel canisters or nuclear transport vehicles	
Nuclear Placards/Warning Materials Along With Otherwise Unexplained Casualties	

Source: U.S. Federal Emergency Management Agency.

radiological agents: Materials that emit radiation that can harm living organisms when inhaled or otherwise ingested. Non-weapons-grade radiological agents could theoretically be used to construct a toxic *dirty bomb.*

dirty bomb: Non-weapons-grade radiological agents could theoretically be used to construct a toxic *dirty bomb* that would use conventional explosives to release a cloud of radioactive contaminants. Radioactive elements that could be used in a dirty bomb include plutonium, uranium, cobalt-60, strontium, and cesium-137.

terrorism-related scenarios are plausible, federal agencies have prepared security protocols to safeguard sources of radioactive agents. Table 12.4 reports general indicators of the possible use of nuclear weapons or radiological agents.

Threat scenarios anticipate the risk of terrorist activity against facilities that make use of radioactive agents. These include nuclear power plants, private industrial sites, medical facilities, and universities. Violent extremists have overtly indicated that nuclear power facilities, in particular, are desirable targets because of the potential for power grid disruption, environmental contamination, and mass casualties. Terrorist incidents could involve sabotage, the use of explosives, an attack by armed insurgents, or the use of aircraft. Nuclear power facilities have, in fact, been attacked by terrorists in South Africa, France, Spain, and Russia.

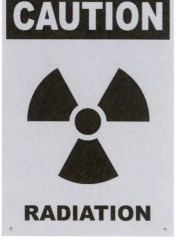

▶ **Photo 12.5** Radioactive material is usually well secured, but it is obtainable on the illicit market. Such material can be weaponized and transformed into a "dirty bomb" or other weapons.

Understanding the Threat: Radiological and Nuclear Hazards

Radiological Agents

Radiation emitted by certain materials are known to pose potential harm to living organisms. Such materials are known as **radiological agents**. To become threatening to life or health, these radioactive substances must be "ingested, inhaled, or absorbed through the skin" in sufficient quantities.[37] A toxic **dirty bomb** could theoretically be constructed from non-weapons-grade radiological agents. Plutonium, uranium, cobalt-60, strontium, and cesium-137 are examples of radioactive elements that could be used in a dirty bomb.[38] Conceptually, radiological weapons are not unlike chemical or biological weapons in the sense that the effectiveness of each is based on contaminating or infecting living organisms. Absent large quantities of radioactive materials, this type of weapon would likely cause minimal casualties outside of the blast radius of the bomb, but its psychological effect could be quite disruptive. Radiological materials are available, making the threat from a radiological weapon entirely plausible—much more than that from nuclear weapons.

Thus, the conversion process is relatively uncomplicated; simply inserting radiological agents into conventional explosive devices could transform them into radiological dirty bombs. Even if minimal casualties and subsequent radiation sickness resulted from a dirty bomb detonation, terrorists would succeed in causing mass disruption if rumors spread that a toxic radiological cloud had been released.

Nuclear Weapons

Nuclear weapons are high-explosive military weapons using weapons-grade plutonium and uranium. Explosions from nuclear bombs could potentially result in the following hazards: devastation of the area within their blast zone and irradiation of an area outside the blast zone. Nuclear explosions are also capable of discharging dangerous radioactive debris into the atmosphere, which would eventually descend to earth as toxic **fallout**. Nuclear devices are sophisticated weapons that are difficult to manufacture, even for highly motivated governments. Modern nuclear arsenals include large strategic weapons powerful enough to lay waste to large areas as well as smaller, relatively compact tactical nuclear weapons that were originally developed to support ground troops. Although it is conceivable that terrorists could construct a nuclear device, this would be a very difficult technical and logistical endeavor.[39] Therefore, most threat scenarios envision the acquisition of tactical nuclear weapons, such as artillery shells, by terrorists. These could be acquired from rogue states or transnational organized-crime groups. The Soviets apparently developed several so-called suitcase bombs—nuclear weapons that are quite compact.

In sum, plausible scenarios exist involving terrorists obtaining possession of nuclear weapons, from either rogue states or transnational organized crime groups. An additional risk is that violent extremists could find a way to penetrate domestic facilities where nuclear weapons are stored, designed, or manufactured. For this reason, domestic nuclear security preparedness and planning necessarily includes the hardening of facilities where nuclear weapons are produced and stored as well as laboratories contracted to design nuclear weapons. These facilities also represent attractive and valuable targets for terrorist organizations.

Coordinating Nuclear Security: The Role of the U.S. Nuclear Regulatory Commission

Radiological and nuclear preparedness and prevention include target hardening and the implementation of security protocols recommended by federal agencies. In this regard, authority has been delegated to the **U.S. Nuclear Regulatory Commission (NRC)** to regulate security and safety protocols at nuclear plants. The NRC's authority is explained as follows: "Sections 3(c) and (e) of the Atomic Energy Act of 1954, as revised, and Section 204(b) (1) of the Energy Reorganization Act of 1974 give NRC the responsibility for ensuring that the peaceful uses of nuclear energy 'make the maximum contribution to the common defense and security and the national welfare.'"[40]

A primary mission of the NRC is to regulate nuclear security and safeguards. Specific areas of NRC responsibility include the following:

- Domestic safeguards

- Information security

- Radioactive material security

- Conducting public meetings on nuclear security and safeguards[41]

The agency accomplishes its mission by enforcing required mandates and conducting oversight to "regulate accounting systems for special nuclear and source materials"[42] and "regulate security programs and contingency plans"[43] for licensees. For example, the NRC mandates and enforces the proper disposal, storage, and transporting of radioactive waste; such waste would be a very attractive and valuable target for violent extremists. The agency

nuclear weapons: High-explosive military weapons using weapons-grade plutonium and uranium. Nuclear explosions devastate the area within their blast zone, irradiate an area outside the blast zone, and are capable of sending dangerous radioactive debris into the atmosphere that descends to earth as toxic fallout.

fallout: Explosions from nuclear bombs devastate the area within their blast zone, irradiate an area outside the blast zone, and are capable of sending dangerous radioactive debris into the atmosphere that descends to earth as toxic fallout.

U.S. Nuclear Regulatory Commission (NRC): The U.S. Nuclear Regulatory Commission is authorized to regulate safety and security protocols at nuclear plants.

also regulates the use of radioactive materials in private industry, medical facilities, and universities. Very importantly, the NRC regulates the use of nuclear reactors used for research and the generation of power.

The NRC coordinates emergency preparedness programs that are designed to "enable emergency personnel to rapidly identify, evaluate, and react to a wide spectrum of emergencies, including those arising from terrorism."[44] These programs include an incident response program and coordination of the National Response Framework, which is intended to build collaboration among federal, state, and local emergency response authorities. NRC's **Office of Nuclear Security and Incident Response** has primary responsibility for emergency planning and preparedness. To carry out this mandate, the office "develops overall agency policy and provides management direction for evaluation and assessment of technical issues involving security at nuclear facilities, and is the agency safeguards and security interface with the Department of Homeland Security (DHS), the Intelligence and Law Enforcement Communities, Department of Energy (DOE), and other agencies."[45] The Office of Nuclear Security and Incident Response also "develops emergency preparedness policies, regulations, programs, and guidelines for both currently licensed nuclear reactors and potential new nuclear reactors; [p]rovides technical expertise regarding emergency preparedness issues and interpretations[;] conducts and directs the NRC program for response to incidents[;] and is the agency emergency preparedness and incident response interface with the DHS, Federal Emergency Management Agency (FEMA) and other Federal agencies."[46]

Office of Nuclear Security and Incident Response: The U.S. Nuclear Regulatory Commission office that has primary responsibility for emergency planning and preparedness.

GLOBAL PERSPECTIVE: THE EUROPEAN APPROACH TO HOMELAND SECURITY

European countries have a long history of responding to ideological, religious, and nationalist extremist violence. During the process of designing internal security policies to counter these challenges, West European democracies adopted administrative approaches that differ from the American homeland security system.

GLOBAL PERSPECTIVE

THE EUROPEAN APPROACH TO HOMELAND SECURITY PREPAREDNESS

Western European countries have not established centralized ministries or agencies similar to the U.S. Department of Homeland Security. Most homeland security and counterterrorism functions are designed to be administered across multiple ministries and agencies. This approach is derived from Europe's historical experience with terrorism in the modern era.

Modern terrorist activity in Europe was characterized by ideological, nationalist, and international spillover political violence. The intensity of this violence was often rather high and sustained over several decades. In essence, the European approach to the homeland security concept has evolved in a unique manner during decades of security challenges from organizations such as the Red Brigade, Red Army Faction, ETA, Action Direct, and nationalist spillovers from around the globe.

Most European nations allocated increased resources to counterterrorist law enforcement and intelligence efforts in response to the September 11 attack in the United States and the transportation system attacks in Spain and the UK. However, the primary focus has been on counterterrorist law enforcement and multiagency cooperative approaches. This is in contrast to the drastic reorganization and nearly total centralization of federal homeland security bureaucracies in the United States.

Thus, although the American approach has been to create a sweeping homeland security apparatus, the common European approach has been to operate from within existing bureaucracies.

CHAPTER SUMMARY

This chapter has developed your understanding and appreciation of homeland security preparedness and planning. Preparedness and planning require an interconnected and tiered system of response mechanisms at the federal, state, and local levels of the homeland security enterprise. In particular, federal coordination and leadership are critical for consistency among state and local homeland security response authorities.

Federal agencies have designed and disseminated detailed strategies for preparedness and planning. The Centers for Disease Control and Prevention, the Nuclear Regulatory Commission, the White House, the Federal Emergency Management Agency, and the

Department of Homeland Security have all published guidelines and strategies for integrating all levels of homeland security systems. For example, the CDC's health and medical preparedness strategies are designed to create a national system of preparedness. Federal assistance is also provided for prevention and mitigation planning. Such assistance is an essential element when preparing state and local government responses to chemical, biological, radiological, and nuclear terrorist incidents. It is the potential for terrorist deployment of weapons of mass destruction that drives much of the federal disbursement of financial and other assistance to state and local authorities.

DISCUSSION BOX

This chapter's Discussion Box is intended to stimulate critical debate about preparedness and planning for the use of weapons of mass destruction by terrorists.

Domestic Preparedness and Weapons of Mass Destruction

The use of weapons of mass destruction by terrorists represents a threat scenario of extreme concern for policymakers and the general public. Weaponized chemical, biological, radiological, and nuclear devices have the potential to cause widespread damage and mass casualties. Even in the event that the use of such weapons does not result in great loss of life or infrastructure damage, the mere use of these weapons would cause widespread fear and disruption. The challenge for policymakers is the extent to which such an incident can be prepared for and mitigated.

Discussion Questions

1. How effective would urban emergency management systems be if a mass-casualty attack occurs?

2. Can emergency medical systems cope with mass-casualty attacks?

3. Which domestic security measures are acceptable in the aftermath of a mass-casualty attack?

4. Which domestic security measures are unacceptable in the aftermath of a mass-casualty attack?

5. Should federal authority be expanded to directly manage state and local emergency response systems?

KEY TERMS AND CONCEPTS

The following topics were discussed in this chapter and can be found in the glossary:

Aum Shinrikyō   269

biological agents   268

chemical agents   267

dirty bomb   272

fallout   273

Hazard Mitigation Assistance (HMA) grant programs   265

Homeland Security Presidential Directive 5 (HSPD-5)   258

Homeland Security Presidential Directive 21 (HSPD-21)   263

Information Sharing and Analysis Centers (ISACs)   262

medical surge   263

Medical Surge Capacity and Capability Handbook: A Management System for Integrating Medical and Health Resources During Large-Scale Emergencies (MSCC)   263

National Incident Management System (NIMS)   258

National Infrastructure Protection Plan (NIPP)   261

National Strategy for Public Health and Preparedness   263

nuclear weapons   273

Office of Nuclear Security and Incident Response   274

radiological agents   272

U.S. Nuclear Regulatory Commission (NRC)   273

ON YOUR OWN

Get the tools you need to sharpen your study skills. SAGE edge offers a robust online environment featuring an impressive array of free tools and resources.

Access practice quizzes, eFlashcards, video, and multimedia at **edge.sagepub.com/martinhs3e**

RECOMMENDED WEBSITES

The following websites and documents provide information about homeland security preparedness and planning, including resources and documentary guidelines:

Centers for Disease Control and Prevention, *Biological and Chemical Terrorism: Strategic Plan for Preparedness and Response*: www.cdc.gov/mmwr/preview/mmwrhtml/rr4904a1.htm

Federal Emergency Management Agency: www.fema.gov/hazard-mitigation-assistance

U.S. Department of Health and Human Services, *Medical Surge Capacity and Capability Handbook*:

www.phe.gov/Preparedness/planning/mscc/handbook/Pages/default.aspx

U.S. Department of Homeland Security, *Reference Manual to Mitigate Potential Terrorist Attacks Against Buildings*: www.dhs.gov/xlibrary/assets/st/st-bips-06.pdf

U.S. Nuclear Regulatory Commission: www.nrc.gov/security.html

White House, *Homeland Security Presidential Directive 21*: www.fas.org/irp/offdocs/nspd/hspd-21.htm

WEB EXERCISE

Using this chapter's recommended websites and documents, conduct an online investigation of homeland preparedness and planning resources and policy challenges.

1. Are there certain federal or local agencies that are better configured than others to engage in homeland security preparedness and planning?

2. Read the executive summaries of the documents referenced in this chapter. Do they reflect objective and realistic approaches to coordinating preparedness and planning nationally?

3. In your opinion, how effective are proposals for homeland security preparedness and planning?

To conduct an online search on homeland security preparedness and planning, activate the search engine on your Web browser and enter the following keywords:

"Federal homeland security preparedness"

"Homeland security state agencies"

RECOMMENDED READINGS

The following publications provide discussions on prevention, preparedness, and planning:

Beyer, Cornelia and Michael Bauer, eds. 2009. *Effectively Countering Terrorism: The Challenges, Prevention, Preparedness and Response*. Eastbourne, East Sussex, UK: Sussex Academic Press.

Biersteker, Thomas J. and Sue E. Eckert, eds. 2008. *Countering the Financing of Terrorism*. London: Routledge.

Kamien, David G. 2005. *The McGraw-Hill Homeland Security Handbook*. New York: McGraw-Hill.

Radvanovsky, Robert S. and Allan McDougall. 2013. *Critical Infrastructure: Homeland Security and Emergency Preparedness*. 3rd ed. Boca Raton, FL: CRC Press.

CHAPTER THIRTEEN

CRITICAL OUTCOMES
Response and Recovery

Chapter Learning Objectives

This chapter will enable readers to do the following:

1. Define homeland security response and recovery

2. Analyze national homeland security response and recovery protocols

3. Evaluate the necessity of federal leadership in response and recovery

4. Apply federal leadership initiatives

Opening Viewpoint: The 1993 World Trade Center Bombing and the Case of Ramzi Yousef

On February 26, 1993, Ramzi Yousef detonated a bomb in a parking garage beneath Tower One of the World Trade Center in New York City. The bomb was a mobile truck bomb that Yousef and an associate had constructed in New Jersey from a converted Ford Econoline van. It was of a fairly simple design but extremely powerful. The detonation occurred as follows:

> The critical moment came at 12.17 and 37 seconds. One of the fuses burnt to its end and ignited the gunpowder in an Atlas Rockmaster blasting cap. In a split second the cap exploded with a pressure of around 15,000 lbs per square inch, igniting in turn the first nitro-glycerin container of the bomb, which erupted with a pressure of about 150,000 lbs per square inch—the equivalent of about 10,000 atmospheres. In turn, the nitro-glycerin ignited cardboard boxes containing a witches' brew of urea pellets and sulphuric acid.[a]

According to investigators and other officials, Yousef's objective was to topple Tower One onto Tower Two "like a pair of dominoes,"[b] release a cloud of toxic gas, and thus achieve a very high death toll.

Ramzi Yousef, apparently born in Kuwait and reared in Pakistan, was an activist educated in the United Kingdom. His education was interrupted during the Soviet war in Afghanistan, when he apparently "spent several months in Peshawar [Pakistan] in training camps funded by Osama bin Laden learning bomb-making skills."[c] After the war, Yousef returned to school in the United Kingdom and received a higher national diploma in computer-aided electrical engineering.

In the summer of 1991, Ramzi Yousef returned to the training camps in Peshawar for additional training in electronics and explosives. He arrived in New York City in September 1992 and, shortly thereafter, began planning to carry out a significant attack, having selected the World Trade Center as his target. Yousef established contacts with former associates already in the New York area and eventually became close to Muhammed Salameh, who assisted in the construction of the bomb. They purchased chemicals and other bomb-making components, stored them in a rented locker, and assembled the bomb in an apartment in Jersey City. They apparently tested considerably scaled-down versions of the bomb several times. After the attack, Yousef boarded a flight at JFK Airport and flew to Pakistan.

This case is a good example of the technical skill and criminal sophistication of some terrorists. Ramzi Yousef had connections with well-funded terrorists, was a sophisticated bomb maker, knew how to obtain the necessary components in a foreign country, was very adept at evasion, and obviously planned his actions in meticulous detail.

Several basic questions remain about the 1993 attack and the case of Ramzi Yousef. For example, could the 1993 attack have been thwarted? How serious is the threat of terrorism from extremists with a similar profile as Yousef? Very importantly, why have similar attacks not occurred with more frequency in the United States?

As a postscript, Ramzi Yousef remained very active among bin Laden's associates, and his travels within the movement took him far afield, including trips to Thailand and the Philippines. In an example of international law enforcement cooperation, he was eventually captured in Pakistan in February 1995 and sent to the United States to stand trial for the bombing. Yousef was tried, convicted, and sentenced to serve at least 240 years in prison.

a. Simon Reeve, *The New Jackals: Ramzi Yousef, Osama bin Laden and the Future of Terrorism* (Boston: Northeastern University Press, 1999), 10.

b. Comment by Judge Kevin Duffy, *United States of America v. Muhammed A. Salameh et al.*, S593 CR.180 (KTD).

c. Reeve, 120.

Response and recovery procedures for domestic security emergencies are a complex and multi-tiered endeavor. Federal, state, and local authorities must assess incidents, coordinate responses, and implement post-incident response processes. This can only be accomplished by means of an efficient and nimble national response system, coordinated by well-managed, overarching federal direction. Ideally, the national response enterprise will design and implement protocols that allow seamless multilevel implementation of response and recovery procedures. Without successful implementation of established procedures, the national response and recovery effort risks becoming dysfunctional and ineffective.

A fundamental concern of the national response effort is the feasibility of achieving the desired goal of a coordinated and tiered approach to incident response management.

The federal government has dedicated significant research and resources to creating a national framework for coordinating tiered responses to terrorist incidents. Known as the National Planning Frameworks, these documents institutionalize the long-term goal of creating integrated response systems at all levels of authority. State and local tiers of authority have adopted response mechanisms that ideally provide immediate services to stricken locations when an incident occurs. Integrated into these response mechanisms are joint operations centers where information and relief initiatives are coordinated. Thus, multi-tiered collaboration and coordination have become a central policy goal of the national homeland security response and recovery system.

The discussion in this chapter will review the following:

- National response and recovery: Fostering administrative coordination

- Federal response and recovery coordination

- State and local response and recovery planning

- Final consideration: Despite best efforts, is terrorism effective?

▶ **Photo 13.1** The aftermath of the bombing of the Alfred P. Murrah Federal Building on April 19, 1995. The attack was the most serious domestic terrorist incident at the time, and continues to be the most destructive domestic incident committed by American extremists.

NATIONAL RESPONSE AND RECOVERY: FOSTERING ADMINISTRATIVE COORDINATION

Homeland security doctrine necessarily accepts the premise that terrorist incidents may occur anywhere in the nation, at any time. For this reason, fostering overarching policy frameworks for efficient response and recovery is a critical component of national preparedness. This is especially important when such frameworks are implemented at the state and local levels of authority, for it is at these levels that first responses to terrorist incidents occur. Real-time activation of protocols contained in these policy frameworks can literally save lives and secure critical infrastructure when terrorist events take place.

As discussed in Chapter 12, the National Incident Management System (NIMS) was established to institutionalize collaborative protocols among federal, state, and local authorities. In the event of an incident, tiered coordination and collaboration must be actuated among these levels of government to successfully mitigate the effect of the incident on domestic security. For example, in the event of a major incident, the restoration of *C4I* (Command, Control, Communications, Computers, and Intelligence) is an enormously complex process, necessitating careful coordination and collaboration across governmental tiers.

Because of the national scope of federal administrative and financial resources, the federal tier of government is the authority most capable of coordinating efficient response and recovery management among different levels of authority. Federal guidance and intervention are intended to maximize efficiency, and without targeted federal assistance to state and local authorities, there would be a high risk of inefficient coordination in responding to serious

U.S. Federal Emergency Management Agency

terrorist incidents. For this reason, the federal tier of government is recognized as the leading authority for framing national response and recovery protocols.

The Problem of Reactive Planning

Federal officials and agencies do engage in a constant process of planning and implementing new procedures and innovations to protect aviation centers, but these upgrades are often challenged by newly available technologies, resources, and tactics adopted by violent extremists. Many successful terrorist attacks as well as thwarted terrorist preparations reveal that counterterrorist procedures and innovations are largely reactive in nature. In effect, they are often responsive to terrorist scenarios after those scenarios have been implemented by the terrorists.

For example, prior to the September 11, 2001, attacks, the federal government and agencies such as the Federal Aviation Administration were mostly reactive, rather than proactive, when planning for aviation security. Tragedies such as the 1988 bombing of Pan Am Flight 103 over Lockerbie, Scotland, and the 1996 crash of Trans World Airlines Flight 800 in New York resulted in legislative and procedural changes but only after the fact and only after public political outcry. This reactive policymaking mode ended in the aftermath of the September 11 attacks.

Federal National Planning Frameworks and the National Preparedness Goal

In synchrony with NIMS, five **National Planning Frameworks** were published by the Federal Emergency Management Agency (FEMA); each framework was dedicated to promoting efficient preparedness within the context of one of five **mission areas**. The purpose of each mission area and the National Planning Frameworks is to achieve the **National Preparedness Goal**. First approved in September 2011, and updated in September 2015, the National Preparedness Goal seeks to achieve "a secure and resilient nation with the capabilities required across the whole community to prevent, protect against, mitigate, respond to, and recover from the threats and hazards that pose the greatest risk."[1] The five mission areas are introduced and explained by the National Preparedness Goal. These mission areas have been identified as necessary elements for successful management of homeland security incidents; prevention, protection, mitigation, response, and recovery compose the five mission areas. They represent fundamental concepts for the creation of an overarching national planning system.

Each mission area is further explicated by a specific National Planning Framework. FEMA published the following documents as constituting the National Planning Frameworks:

- The National Prevention Framework

- The National Protection Framework

- The National Mitigation Framework

- The National Response Framework (second edition)

- The National Disaster Recovery Framework

The advantage of promulgating the National Planning Frameworks is that they characterize an ongoing and institutionalized effort to maintain integrated preparedness at all tiers of authority.

National Planning Frameworks: In synchrony with NIMS, the Federal Emergency Management Agency published five National Planning Frameworks, each dedicated to promoting efficient preparedness within the context of one of five mission areas. The purpose of the National Planning Frameworks and the identified mission areas is to achieve the National Preparedness Goal. The National Planning Frameworks comprise the National Prevention Framework, the National Protection Framework, the National Mitigation Framework, the National Response Framework (second edition), and the National Disaster Recovery Framework.

mission areas: The Federal Emergency Management Agency published five National Planning Frameworks, each dedicated to promoting efficient preparedness within the context of one of five mission areas. The five mission areas are prevention, protection, mitigation, response, and recovery.

National Preparedness Goal: The National Preparedness Goal, approved in September 2011, seeks to achieve "a secure and resilient nation with the capabilities required across the whole community to prevent, protect against, mitigate, respond to, and recover from the threats and hazards that pose the greatest risk." The National Preparedness Goal introduces and explains the five mission areas, which have been identified as necessary elements for successful management of homeland security incidents.

Specified **core capabilities** are contained in every document in the National Planning Frameworks. These core capabilities are defined as the elements, or *action items*, needed to achieve the National Preparedness Goal. Within each framework, the core capabilities are the central implementation mechanisms for achieving the National Preparedness Goal. In this regard, the core capabilities are described as

> the distinct critical elements necessary for our success. They are highly interdependent and will require us to use existing preparedness networks and activities, improve training and exercise programs, promote innovation, and ensure that the administrative, finance, and logistics systems are in place to support these capabilities. The capability targets—the performance threshold(s) for each core capability—will guide our allocation of resources in support of our national preparedness.[2]

Thus, the Core Capabilities are designed to effectuate the National Planning Frameworks, which, in turn, will theoretically achieve the National Preparedness Goal. Taken together, these concepts and protocols reflect an all-hazards conceptualization of homeland security, and they also conceptually embody *whole-community collaboration* within the homeland security enterprise. Under this policy framework, all tiers of authority are envisioned as an integrated system of collaborative partners who are able to respond quickly and aggressively when an incident occurs. Figure 13.1 reports the expected implementation processes for achieving the National Preparedness Goal.

The Incident Command System

Managing and coordinating response and recovery efforts is an inherently difficult process, especially when the tiered configuration of federal, state, and local response authorities must be quickly mobilized and deployed. As discussed previously, NIMS provides a foundational framework for managing disasters at all tiers of the homeland security enterprise. NIMS incorporated and advocated the enhancement of a preexisting concept, the Incident Command System (ICS).

ICS is a relatively well-developed emergency coordination concept that was originally adopted in 1970 after a wildfire emergency in California. Its central principle is the development of a viable incident command system to efficiently and effectively respond to incidents.

Figure 13.1 National Planning Frameworks and National Preparedness Goal

Core Capabilities → Implementation Mechanisms → Mission Areas
- Prevention
- Protection
- Mitigation
- Response
- Recovery

→ Overarching System → National Preparedness Goal

Potential problems include the following: multiple reporting lines within agencies, redundant or unclear lines of authority, unverified information reported during incidents, independent (and sometimes contradictory) responses, and poor communications. Thus, ICS's continuation as a viable concept is predicated on the ongoing challenges of enhancing coordination among emergency response agencies.

Federal agencies tasked with emergency response duties have incorporated ICS principles as integral components of their overall missions. Some agencies mandate the adoption of ICS. For example, the Occupational Safety and Health Administration (OSHA) explains its adoption of ICS as follows:

> ICS is a standardized on-scene incident management concept designed specifically to allow responders to adopt an integrated organizational structure equal to the complexity and demands of any single incident or multiple incidents without being hindered by jurisdictional boundaries. . . . An ICS enables integrated communication and planning by establishing a manageable span of control. An ICS divides an emergency response into five manageable functions essential for emergency response operations: Command, Operations, Planning, Logistics, and Finance and Administration.[3]

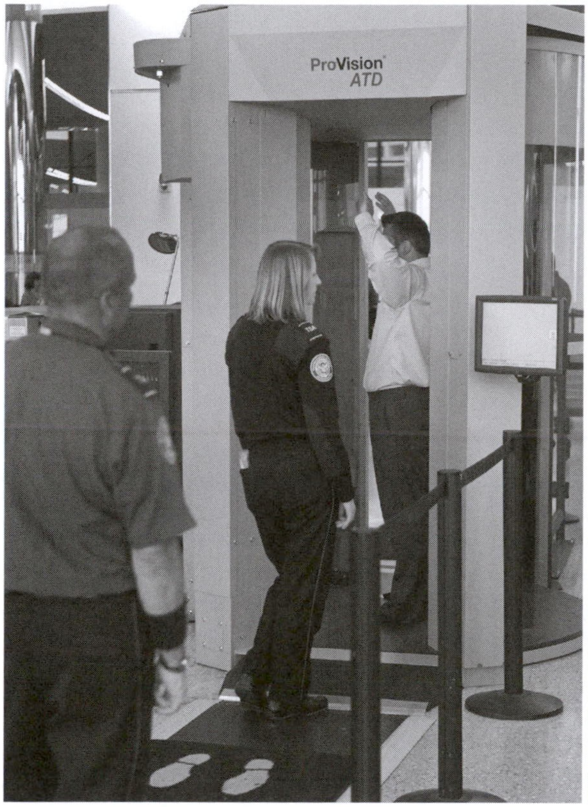

David L. Ryan/The Boston Globe via Getty Images

▶ **Photo 13.2** A body scanner in operation at an airport secured by the Transportation Security Administration. Increasing sophistication by extremists has necessitated the deployment of new security technologies.

The functional areas reported in the OSHA example are central coordinating centers under ICS. Similarly, FEMA's ICS training module discusses the following features:[4]

Standardization

- *Common Terminology*: Using common terminology helps to define organizational functions, incident facilities, resource descriptions, and position titles.

Command

- *Establishment and Transfer of Command*: The command function must be clearly established from the beginning of an incident. When command is transferred, the process must include a briefing that captures all essential information for continuing safe and effective operations.

- *Chain of Command and Unity of Command*: Chain of command refers to the orderly line of authority within the ranks of the incident management organization. Unity of command means that every individual has a designated supervisor to whom he or she reports at the scene of the incident. These principles clarify reporting relationships and eliminate the confusion caused by multiple, conflicting directives. Incident managers at all levels must be able to control the actions of all personnel under their supervision.

Planning/Organizational Structure

- *Management by Objectives*: Includes establishing overarching objectives; developing and issuing assignments, plans, procedures, and protocols; establishing specific, measurable objectives for various incident management functional activities; and directing efforts to attain the established objectives.

- *Modular Organization*: The Incident Command organizational structure develops in a top-down, modular fashion that is based on the size and complexity of the incident, as well as the specifics of the hazard environment created by the incident.

- *Incident Action Planning*: Incident Action Plans (IAPs) provide a coherent means of communicating the overall incident objectives in the contexts of both operational and support activities.

- *Manageable Span of Control*: Span of control is key to effective and efficient incident management. Within ICS, the span of control of any individual with incident management supervisory responsibility should range from three to seven subordinates.

Facilities and Resources

- *Incident Locations and Facilities*: Various types of operational locations and support facilities are established in the vicinity of an incident to accomplish a variety of purposes. Typical predesignated facilities include Incident Command Posts, Bases, Camps, Staging Areas, Mass Casualty Triage Areas, and others as required.

- *Comprehensive Resource Management*: Resource management includes processes for categorizing, ordering, dispatching, tracking, and recovering resources. It also includes processes for reimbursement for resources, as appropriate. Resources are defined as personnel, teams, equipment, supplies, and facilities available or potentially available for assignment or allocation in support of incident management and emergency response activities.

Communications/Information Management

- *Integrated Communications*: Incident communications are facilitated through the development and use of a common communications plan and interoperable communications processes and architectures.

- *Information and Intelligence Management*: The incident management organization must establish a process for gathering, sharing, and managing incident-related information and intelligence.

Professionalism

- *Accountability*: Effective accountability at all jurisdictional levels and within individual functional areas during incident operations is essential. To that end, the following principles must be adhered to:
  - *Check-In*: All responders, regardless of agency affiliation, must report in to receive an assignment in accordance with the procedures established by the Incident Commander.
  - *Incident Action Plan*: Response operations must be directed and coordinated as outlined in the IAP.
  - *Unity of Command*: Each individual involved in incident operations will be assigned to only one supervisor.

- ○ *Span of Control*: Supervisors must be able to adequately supervise and control their subordinates, as well as communicate with and manage all resources under their supervision.
- ○ *Resource Tracking*: Supervisors must record and report resource status changes as they occur.
- ○ *Dispatch/Deployment*: Personnel and equipment should respond only when requested or when dispatched by an appropriate authority.

As indicated in the foregoing OSHA and FEMA examples, ICS functions are intended to be broadly applicable to response coordination for all incidents. Figure 13.2 depicts a representative coordinating system under ICS.

Case in Point: Presidential Declarations

When an incident involving terrorist activity occurs, a critical mechanism for activating coordinated and intensive intervention originates from **presidential declarations**. Presidential declarations bring into effect an array of options for federal assistance to state and local authorities; they are significant exercises of the authority of the executive branch of the federal government.

There are two types of presidential declarations: **presidential major disaster declarations** and **emergency declarations**. Both are important tools for releasing necessary resources and specialized personnel for response and recovery. These declarations are differentiated as follows:

> The **Robert T. Stafford Disaster Relief and Emergency Assistance Act** (referred to as the Stafford Act - 42 U.S.C. 5721 et seq.) authorizes the president to issue "major disaster" or "emergency" declarations before or after catastrophes occur. Emergency declarations trigger aid that protects property, public health, and safety and lessens or averts the threat of an incident becoming a catastrophic event. A major disaster declaration, issued after catastrophes occur, constitutes broader authority for federal agencies to provide supplemental assistance to help state and local governments, families and individuals, and certain nonprofit organizations recover from the incident.[5]

The bottom line is that presidential declarations authorize national intervention when serious natural or man-made disasters and emergencies occur.

Presidential Declarations: Standard Procedures

Presidential declarations are usually activated only after local and state levels of authority request assistance when responding to disasters and emergencies. Standard procedures lead to presidential declarations after the following series of events occur within the tiered framework of authority:

- When an incident occurs, local authorities are typically the first responders to the event and are tasked with quickly resolving the crisis if possible.

- If local authorities are unable to contain the consequences of the event, the governor and state government are called by local officials to assist local authorities in responding to the incident. State resources are considerable and include, for example, the ability to deploy National Guard units and to investigate events statewide.

presidential declarations: Significant exercises of the authority of the executive branch of the federal government that bring into effect an array of options for federal assistance to state and local authorities.

presidential major disaster declarations: Presidential declarations, issued after catastrophes occur, that give broad authority to federal agencies "to provide supplemental assistance to help state and local governments, families and individuals, and certain nonprofit organizations recover from the incident."

emergency declarations: Presidential declarations that trigger aid that "protects property, public health, and safety and lessens or averts the threat of an incident becoming a catastrophic event."

Robert T. Stafford Disaster Relief and Emergency Assistance Act: Legislation that "authorizes the President to issue 'major disaster' or 'emergency' declarations before or after catastrophes occur."

Figure 13.2   FEMA: Recommended Incident Command System (ICS)

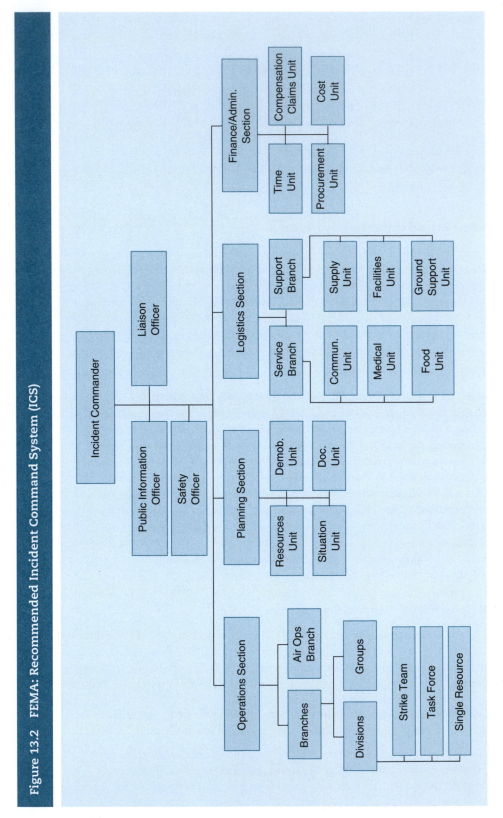

Source: U.S. Federal Emergency Management Agency.

- If state authorities are unable to contain the consequences of the event, the governor may issue a formal request to the president for a declaration. The request is delivered through the relevant regional office of FEMA. The president then decides whether to issue a declaration, and if so, whether the declaration will be a major disaster declaration or an emergency declaration.

This process is intended to ensure maximum collaboration and coordination among responsible authorities and to deliver assistance to the affected locale as soon as possible.

Presidential Declarations: Exceptional Procedures

As mentioned previously, one document included in the National Planning Frameworks is the National Response Framework (discussed in detail in the next section). An essential feature of the National Response Framework is that it acknowledges emergency scenarios where the president may unilaterally announce a disaster declaration without awaiting implementation of the standard procedure or receiving a prior request from a governor. The National Response Framework also states that federal agencies may provide relief intervention on their own initiative without a formal presidential declaration, to the extent permitted within the purview of their specific missions.

These provisions reflect practical recognition that extraordinary circumstances may exist that require immediate proactive intervention. In essence, there is a realistic understanding that awaiting the outcome of standard procedures could exacerbate the consequences of a disaster or emergency.

FEDERAL RESPONSE AND RECOVERY COORDINATION

The federal tier of the homeland security enterprise has been configured to activate designated response protocols when incidents occur. These protocols are included in the five documents of the National Planning Frameworks and their core capabilities, all of which are necessary to achieve the National Planning Goal. A concerted effort has been made in the National Planning Frameworks to incorporate the state and local tiers of authority as collaborative partners. This effort has met with some success. For example, in the post–September 11 era, state authorities adapted to the new security environment by establishing state-level equivalents of the U.S. Department of Homeland Security. There is also general acceptance by state and local authorities of federal grant assistance for funding homeland security initiatives. In this way, the goal of multilevel collaboration has become a prevalent concept within the homeland security enterprise.

Background: The Progression of Federal Response Planning

The present tiered system is the culmination of a successive progression in federal policy development, in which federal response frameworks were periodically written to promote coordination of a multilevel incident response system. These response frameworks reflected the nature of the contemporary threat environment existing at the time they were written. For example, prior to the September 11 attacks, the 1993 Federal Response Plan presented the procedural framework for federal response coordination should an incident occur. When activated in response to actual terrorist incidents, elements of the 1993 Federal Response Plan provided demonstrably effective guidance—for instance, in response to the April 19, 1995, bombing of the Alfred P. Murrah Federal Building in Oklahoma City and the September 11, 2001, homeland attacks.

National Response Framework: A procedural roadmap for federal coordination of a national effort to institute an efficient disaster and emergency response system. The document "is built on scalable, flexible, and adaptable concepts identified in the National Incident Management System to align key roles and responsibilities across the Nation."

Federal Response Plan: A procedural framework drafted in 1993 for federal response coordination should an incident occur. When activated, elements of the plan provided demonstrably effective guidance when actual terrorist incidents occurred.

After the September 11 attacks, a new National Response Plan was written to provide a framework for federal coordination. The National Response Plan was superseded in 2008 by the first edition of the National Response Framework, which was updated further in 2013.

Case in Point: Implementing the National Response Framework

The third edition of the National Response Framework was disseminated in June 2016. Building upon concepts presented in prior editions, it continued to outline a procedural roadmap for federal coordination of a national effort to institute an efficient disaster and emergency response system. The document "is built on scalable, flexible, and adaptable concepts identified in the National Incident Management System to align key roles and responsibilities across the Nation."[6] It also states that "[t]he priorities of the Response mission area are to save lives, protect property and the environment, stabilize the incident, and provide for basic human needs."[7] The National Response Framework is a national recognition of the need to design and maintain a systemic mechanism that promotes preparedness at every level of authority.

The National Response Framework: The Core Capabilities

As discussed previously, each document included in the National Planning Frameworks contains several core capabilities. The 15 core capabilities featured in the National Response Framework are designed as practical elements in achieving the National Preparedness Goal. **Critical tasks** are included in each of the 15 core capabilities; identified critical tasks are intended to be undertaken by response authorities tasked with implementing the framework. The core capabilities for the National Response Framework are as follows:

critical tasks: The National Response Framework features 15 core capabilities that are designed as practical elements in achieving the National Preparedness Goal. Each of the 15 core capabilities includes identified *critical tasks* to be undertaken by response authorities tasked with implementing the framework.

- Planning
- Public information and warning
- Operational coordination
- Critical transportation
- Environmental response/health and safety
- Fatality management services
- Fire management and suppression
- Infrastructure systems
- Logistics and supply chain management
- Mass care services
- Mass search and rescue operations
- On-scene security, protection, and law enforcement
- Operational communications
- Public health, healthcare, and medical services
- Situational assessment[8]

These core capabilities were developed after extensive consultation among expert policy planners for each of the five mission areas developed in the National Planning Frameworks. They provide common terms and systematized collaboration on how to achieve the National Preparedness Goal.

National Response Framework: The Emergency Support Functions

To ensure delivery of the 15 core capabilities, the National Response Framework identifies **emergency support functions** (ESFs). Like the core capabilities, the emergency support functions represent systematized collaboration among policy planners. ESFs also represent federal coordinating structures used to achieve the core capabilities. The underlying purposes of the emergency support functions are summarized as follows:

> The Federal Government and many state governments organize their response resources and capabilities under the ESF construct. ESFs have proven to be an effective way to bundle and manage resources to deliver core capabilities. The Federal ESFs are the primary, but not exclusive, Federal coordinating structures for building, sustaining, and delivering the response core capabilities. Most Federal ESFs support a number of the response core capabilities. In addition, there are responsibilities and actions associated with Federal ESFs that extend beyond the core capabilities and support other response activities, as well as department and agency responsibilities.[9]

Emergency support functions are designed to assemble resources from diverse sources and "package" them for dissemination to critical disaster and emergency relief centers.

National Response Framework: Unified Coordination and the Joint Field Office

Unified coordination is a concept applied by the National Response Framework to planning initiatives for response and recovery in the aftermath of domestic security incidents. *Unified coordination* is defined as "the term used to describe the primary state/tribal/Federal incident management activities conducted at the incident level."[10] An underlying purpose of unified coordination is to provide support to agencies and personnel who are directly involved with operations at the incident site. Thus, unified coordination "does not manage on-scene operations. Instead, it focuses on providing support to on-scene response efforts and conducting broader support operations that may extend beyond the incident site."[11]

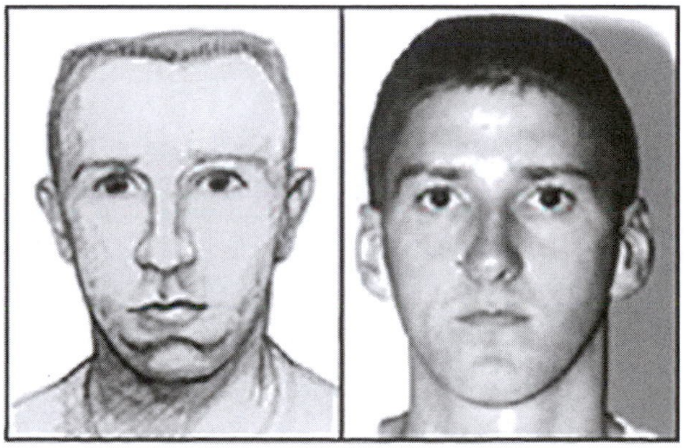

The implementation of state and federal incident management activities is directed by, and coordinated from, **Joint Field Offices** (JFOs), which are established locally to respond to an incident and to coordinate recovery activities. Each Joint Field Office is established as "a temporary Federal facility that provides a central location for coordination of response efforts by the private sector, NGOs, and all levels of government."[12] Its temporary status means that it will necessarily be scaled back and eventually dissolved as the response and recovery efforts return the incident site to its pre-incident conditions. Nevertheless, depending on the scale of response and recovery, Joint Field Offices can conceivably have a relatively long duration.

emergency support functions: To ensure delivery of the 15 core capabilities, the National Response Framework identifies emergency support functions. The emergency support functions represent federal coordinating structures used to achieve the core capabilities, and like the core capabilities, they represent systematized collaboration among policy planners.

unified coordination: Conceptually defined as "the term used to describe the primary state/Federal incident management activities conducted at the incident level." Its underlying purpose is to provide support to personnel and agencies directly involved with operations at the incident site.

U.S. Federal Bureau of Investigation

▶ **Photo 13.3** A comparison of the sketched witness description of Oklahoma City bomber Timothy McVeigh to his mug shot. It is an excellent example of the value of cooperation between witnesses and law enforcement officials.

Because Joint Field Offices are charged with delivering unified coordination, the National Response Framework recommends the creation of a **Unified Coordination Group** (UCG) to manage each Joint Field Office. The central purpose of the Unified Coordination Group is to assist Joint Field Offices charged with delivery of unified coordination of relief efforts. A Unified Coordination Group

> is composed of senior leaders representing state, tribal, and Federal interests and, in certain circumstances, local jurisdictions and the private sector. UCG members must have significant jurisdictional responsibility and authority. The composition of the UCG varies from incident to incident depending on the scope and nature of the disaster. The UCG leads the unified coordination staff. Personnel from state, tribal, and Federal departments and agencies, other jurisdictional entities, the private sector, and NGOs may be assigned to the unified coordination staff at various incident facilities (e.g., JFO, staging areas, and other field offices). The UCG determines staffing of the unified coordination staff based on incident requirements.[13]

Figure 13.3 illustrates the recommended composition of a Unified Coordination Group staff. If successfully implemented, the Joint Field Office and Unified Coordination Group will provide on-site collaboration of all tiers of authority for the management of potential incidents.

STATE AND LOCAL RESPONSE AND RECOVERY PLANNING

The state and local tiers of homeland security authority represent the front line of responsibility when an incident occurs. Agencies and personnel employed by state and local authorities are tasked with the occasionally overwhelming responsibility of providing police, fire, and medical assistance to the affected location. Depending on the scale of the emergency, first responders may be stretched to their operational limits, which is an underlying justification for integrating all levels of authority in a collaborative response and recovery system. Thus, the creation of viable incident command structures at the state and local levels of authority—predicated on the principles of ICS—is an indispensable element of effective response and recovery. Figure 13.4 illustrates the recommended incident command structure for state and local authorities.

U.S. Federal Emergency Management Agency

▶ **Photo 13.4** The front line of homeland security. Members of local police agencies serve as first responders to secure the immediate area near terrorist events.

Response and Recovery Planning by State Authorities

The state-level tier of authority provides statewide emergency management support for local authorities responding to terrorist incidents. State emergency management offices offer preparedness and mitigation support on an ongoing basis to local jurisdictions and generally do not become directly involved in emergency response until a request is received from local authorities. Such requests normally occur when local authorities are unable to mitigate the immediate consequences of an incident. Integration of state and local response operations is common in emergency scenarios, with state agencies mobilizing resources from statewide institutions that would normally not be available to local authorities.

Figure 13.3 FEMA: Recommended Unified Coordination Staff (UCS)

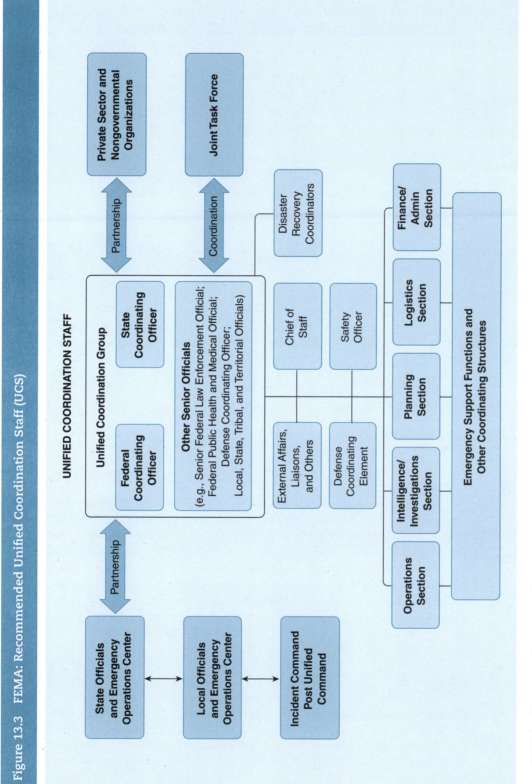

Source: U.S. Federal Emergency Management Agency.

Figure 13.4   FEMA: Recommended State and Local Incident Command Structure

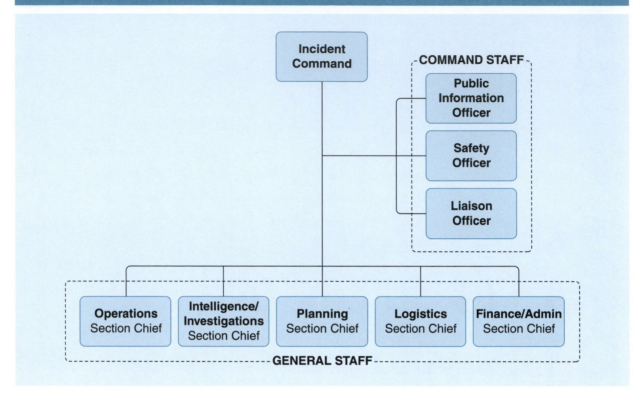

Source: U.S. Federal Emergency Management Agency.

State governments assign homeland security coordination duties to state emergency management offices, which operate as the state-level equivalents of the U.S. Department of Homeland Security. As discussed in Chapter 7, there is no uniformity among states on reporting lines for emergency management offices, largely because of political and budgetary considerations. For example, some offices report to the governor's office, others to community affairs offices, and others to adjutant generals of state National Guards. Reporting lines are apt to be redesigned when new state governments are elected and reorganized. Such reorganization can affect response efficiency in the short term while reporting lines of authority are reconfigured. Figure 13.5 illustrates the state and local response structure recommended when an incident occurs.

Regardless of the preferred reporting lines, budgets for state emergency management offices are determined by the responsible state authority (i.e., state legislatures) and funded through a combination of state revenue and federal assistance. FEMA and other federal agencies provide a large proportion of funding for state emergency management offices, disbursing more than $300 million each year. Federal funding sources are used to provide a broad range of necessities, including equipment, training, and salaries for personnel. State funding sources are variable, with some states disbursing a significant amount of financial resources

Figure 13.5   FEMA: Recommended State and Local Response Structure

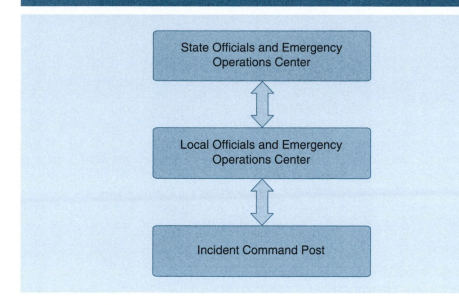

State Officials and Emergency Operations Center

Local Officials and Emergency Operations Center

Incident Command Post

Source: U.S. Federal Emergency Management Agency.

emergency operations plans: Plans that explain the mission and duties of local response personnel and agencies.

and others committing much less financial support. The reason for this inconsistency among states is that some states have a relatively low incidence of emergency events. States that unfortunately experience a relatively high incidence of emergency events will necessarily allocate a larger proportion of their financial resources to disaster relief.

Response and Recovery Planning by Local Authorities

The local tier of authority provides immediate police, fire, and medical emergency intervention. Because of their proximity to incident sites, personnel from local agencies are commonly the first to deploy to emergency events. These agencies train continuously to respond to routine emergencies, and indeed, they do respond daily throughout the nation. Routine emergencies can be quite serious and deadly, but local responders are generally adept at managing these events in a methodical and organized manner. Because of this system- atized responsibility profile, local authorities are also prepared to serve as first responders to extraordinary emergencies, such as terrorist incidents. Local **emergency operations plans** commonly explain the duties and mission of local response agencies and personnel. Emergency managers drawn from local personnel and agencies are tasked with designing,

U.S. Federal Emergency Management Agency

▶ **Photo 13.5** Members of the Los Angeles County Fire Department preparing their equipment. Firefighters are tasked to stabilize fires and other destruction caused during terrorist incidents.

updating, and implementing local emergency operations plans, ideally in collaboration with state, federal, and other local authorities.

Local responders constitute a very large segment of the homeland security enterprise. For example, nationwide there are

- more than 1 million firefighters,
- more than 435,000 sworn law enforcement officers (county sheriffs and local police), and
- more than 155,000 emergency medical personnel.

All of these personnel have been trained and are operationally prepared to serve as first responders when routine and extraordinary emergencies occur. Federal financial support is an essential factor in local ability to maintain a high degree of operational readiness. Extensive federal funding is regularly disbursed to local authorities, with funds earmarked specifically for planning, equipment, training, and exercises.

FINAL CONSIDERATION: DESPITE BEST EFFORTS, IS TERRORISM EFFECTIVE?

The best available response and recovery systems and procedures are not an absolute deterrent to the execution of terrorist attacks. Nor do they guarantee that the aftermath will not have long-term effects on society and the overall political environment. In essence, even with the adoption of modern response and recovery systems, the question remains: Does terrorism work?

When we consider the effectiveness of terrorism, we must examine attacks from the perspective of the extremist perpetrators. Within this context, the basic question to be answered is, do the methods used by terrorists against their selected targets promote their goals and objectives? Terrorism is arguably effective—however defined—in some manner to someone.[14] The key (for terrorists) is to establish a link between terrorist methods used in incidents and desirable outcomes. Of course, success and effectiveness can be very subjective considerations. In this regard, there is a tendency for terrorists to use unconventional factors as measures of their effectiveness. For example, terrorists have been known to declare victory using the following criteria:

- Acquiring global media and political attention
- Having an impact on a target audience or championed constituency
- Forcing concessions from an enemy interest
- Disrupting the normal routines of a society
- Provoking the state to overreact

This is not an exhaustive evaluation of measures of effectiveness, but it demonstrates commonalities found among modern terrorists globally. Table 13.1 summarizes measures of effectiveness, international cases, and outcomes.

Table 13.1 Measures of Effectiveness

Response and recovery systems and procedures do not necessarily define whether terrorist incidents are effective. From the perspective of terrorists, perpetrators measure the effectiveness of their violent behavior by linking the incident to identifiable outcomes. These measures of effectiveness are unconventional in the sense that they are frequently media oriented and audience oriented.

The following table summarizes measures of effectiveness by illustrating the linkage between terrorist incidents and outcomes.

Measure of Effectiveness	Activity Profile	
	Incident	Outcome
Media and political attention	ISIS executions posted on Internet websites	Global media and political attention achieved
Impact on an audience	January and November 2015 Paris attacks by ISIS	Escalation of French counterterror efforts
Concessions from an enemy interest	2004 Madrid train bombings	Spanish troops withdrawn from Iraq and ouster of conservative government
Disruption of societal routines	Suicidal hijackings of four airliners on September 11, 2001	Fewer Americans traveled via airlines; industry suffered revenue losses
Provoke the state	2014 and 2015 Hamas rocket attacks and infiltration into Israel	Israeli military incursions into Gaza and arrests of suspects

CHAPTER SUMMARY

Response and recovery is a restorative and rebuilding process that occurs in the aftermath of a domestic security incident. The baseline for successful recovery involves restoring affected infrastructure and affected people to at least the conditions that existed prior to the incident. The recovery process begins in the immediate aftermath of domestic security incidents and can be a complex and prolonged process, possibly lasting for months or years. All tiers of authority engage in the recovery process, which may require costly allocations of resources and technical support. Because the federal government is more capable than states of providing wide-ranging recovery funds and technical backing, federal authorities regularly coordinate recovery efforts in collaboration with state and local authorities.

National Planning Frameworks are designed to institutionalize collaboration among the federal, state, and local tiers of authority. Core capabilities and mission areas are incorporated into the frameworks as national concepts for promoting efficient response systems. The National Response Framework addresses short-term and long-term recovery and recommends that the recovery effort be coordinated and implemented in a manner similar to the response effort. In compliance with the guiding principles of unified coordination, recovery is directed from a federal–state Joint Field Office managed by a Unified Coordination Group.

DISCUSSION BOX

This chapter's Discussion Box is intended to stimulate critical debate about the operational integration of recommended command and response structures.

The Viability of Recommended Command and Response Structures

In the *National Response Framework*, the Federal Emergency Management Agency encourages the creation of state and local response and incident command structures. FEMA also recommends the adoption of a joint federal, state, and local unified coordination staff configuration. These recommendations represent a concerted effort to establish nationwide response and recovery protocols and structures.

Absent the adoption of nationally accepted protocols and command structures, the administrative complexities of coordinating command and response measures could become disjointed following terrorist events. For these reasons, recommended procedures under the *National Response Framework* are an important resource for state and local authorities to create coordinated protocols.

Discussion Questions

1. How should state and local governments coordinate their command and response structures when an attack occurs?

2. How can federal agencies best be used to assist state and local authorities to prepare for and respond to terrorist threats?

3. What is an acceptable degree of federal supervision of and intervention in state and local configuration of their command and response structures?

4. Should federal resources be used to mandate compliance with *Framework* recommendations?

5. To what extent should agencies at the state and local levels be independent of federal oversight?

KEY TERMS AND CONCEPTS

The following topics were discussed in this chapter and can be found in the glossary:

ON YOUR OWN

Get the tools you need to sharpen your study skills. SAGE edge offers a robust online environment featuring an impressive array of free tools and resources.

Access practice quizzes, eFlashcards, video, and multimedia at **edge.sagepub.com/martinhs3e**

RECOMMENDED WEBSITES

The following websites and documents provide information about national response and recovery institutions and planning protocols:

Department of Justice Joint Terrorism Task Force: www.justice.gov/jttf

FEMA, National Planning Frameworks: www
.fema.gov/national-planning-frameworks

FEMA, National Preparedness Goal: www.fema
.gov/national-preparedness-goal

U.S. Department of Homeland Security, *Joint
Field Office Activation and Operations*: www.fema
.gov/pdf/emergency/nims/jfo_sop.pdf

WEB EXERCISE

Using this chapter's recommended websites, conduct an
online critical assessment of national response and recovery protocols.

1. Are there certain national or local response
   and recovery protocols that appear to be more
   practical or viable than others?

2. Critically assess the National Planning
   Frameworks and National Preparedness Goal. Do
   they reflect objective and reasonable approaches
   for enhancing response and recovery systems?

3. In your opinion, how effective are national
   protocols?

To conduct an online search on state terrorism, activate
the search engine on your Web browser and enter the
following keywords:

"Homeland security response and recovery"

"National disaster response and recovery"

RECOMMENDED READINGS

The following publications provide discussions on
response and recovery:

McCarthy, Francis X. 2011. *FEMA's Disaster Declaration Process: A
Primer*. Washington, DC: Congressional Research Service.

Schneider, Saundra K. 2011. *Dealing With Disaster: Public
Management in Crisis Situations*. Armonk, NY: M. E. Sharpe.

Smith, Gavin. 2012. *Planning for Post-Disaster Recovery: A Review
of the United States Disaster Assistance Framework*. Washington,
DC: Island Press.

U.S. Department of Homeland Security, Federal Emergency
Management Agency. 2013. *National Response Framework*.
Washington, DC: U.S. Department of Homeland Security.

Wells, April, Charlyn Walker, Timothy Walker, and David
Abarca. 2006. *Disaster Recovery: Principles and Practices*. New York:
Prentice Hall.

A Predator drone aircraft
departs on a mission. Robotic
aircraft are deployed globally
and serve as reconnaissance
and firing platforms.

PART FIVE

HOMELAND SECURITY

An Evolving Concept

Chapter 14 **The Future of Homeland Security**

THE FUTURE OF HOMELAND SECURITY

Chapter Learning Objectives

This chapter will enable readers to do the following:

1. Analyze near-term projections for the future of homeland security

2. Evaluate the feasibility of adapting theoretical and practical counterterrorist options to dynamic security environments

3. Identify continuing risks to domestic security from violent extremists

4. Evaluate the need for ongoing international collaboration

Opening Viewpoint: Terrorist "Waves" and Near-Term Risks to Homeland Security

Professor David C. Rapoport designed a theory that holds that modern terrorism has progressed through three "waves" that lasted for roughly 40 years each and that we now live in a fourth wave. His four waves are as follows:

- The anarchist wave: 1880s to the end of World War I
- The anticolonial wave: End of World War I until the late 1960s
- The New Left wave: Late 1960s to the near present
- The religious wave: About 1980 to the present

If Professor Rapoport's theory is correct, the current terrorist environment will be characterized by the New Terrorism for the foreseeable future. Having made this observation, it can also be argued that the sources of extremist behavior in the modern era will generally remain unchanged in the near future. The modern era of terrorism will likely continue to occur for the following reasons:

- People who have been relegated to the social and political margins—or who believe that they have been so relegated—often form factions that resort to violence.
- Movements and nations sometimes adopt religious or ethnonational supremacist doctrines that they use to justify aggressive political behavior.
- Many states continue to value the "utility" of domestic and foreign terrorism.

These factors are not, of course, the only sources of terrorism and extremist sentiment, but they are certainly among the most enduring sources. However, these enduring sources have precipitated new trends in terrorist behavior that began to spread during the 1990s and early 2000s. These new trends include the following:

- Increasing use of communications and information technologies by extremists

- Adaptations of cell-based organizational and operational strategies by global revolutionary movements

- The use of relatively low-tech tactics such as suicide bombers and improvised explosive devices

- Efforts to construct or obtain relatively high-tech weapons of mass destruction or, alternatively, to convert existing technologies into high-yield weapons

Projecting the future of homeland security is an imprecise and potentially speculative exercise. However, it is possible to project likely developments and general trends. For example, two developments can be predicted with some certainty: First, the threat from terrorism will continue into the foreseeable future, and second, the homeland security enterprise must be prepared and configured to constantly adapt to new threat scenarios and terrorist environments. These developments are at the core of ensuring the viability of a strong homeland security enterprise.

Previous chapters have provided a great deal of information about homeland security, federal and local collaboration, laws and legal issues related to homeland security, and the characteristics of the modern terrorist environment. Previous chapters also challenged readers to evaluate domestic security, preparedness, response, and recovery. Many examples of homeland security regulations and their application by government authorities were presented to illustrate collaborative concepts and trends. These discussions fostered critical analysis of perspectives on how to promote a well-integrated homeland security system.

The homeland security enterprise must successfully adjust to and intervene in an ever-evolving terrorist environment. The era of the New Terrorism is a dynamic historical period of conflict distinguished by a resilient and committed foe. Successful homeland security in the near term depends on accepting the reality that violent extremists will consistently attempt to thwart defensive measures by planning and carrying out unexpected attacks or other types of political violence. For this reason, the homeland security enterprise must operationalize the mission of creating an adaptive and collaborative system.

The discussion in this concluding chapter synthesizes many of these considerations, and it projects emerging trends that are likely to characterize homeland security and terrorist risks in the near future. It discusses foreseeable threats from domestic and international terrorists and presents options on how to respond to these threats. The underlying assumption of this presentation is that threats similar to those that caused the creation of the modern homeland security environment will continue to influence its permutation into the near future.

The discussion in this chapter will review the following:

- Near-term projections
- Threat environments in the twenty-first century

NEAR-TERM PROJECTIONS

Homeland security experts in the modern era will be challenged to successfully integrate several domestic security objectives into homeland security policy initiatives. These objectives can realistically—and in a practical sense—simply mitigate, rather than eliminate, unpredictable risks from the New Terrorism and domestic extremists. Nevertheless, long-term domestic security initiatives must incorporate the following objectives:

- Ensure capable federal leadership in developing common standards for homeland security systems at state and local levels

- Enact adaptive laws, regulations, and guidelines to form a legally and procedurally sound homeland security enterprise

- Disrupt and prevent violent extremists from operationalizing terrorist conspiracies

- Decrease the probability that terrorist attacks will result in extensive infrastructure destruction, social disruption, or human casualties

Integrating these objectives into complete policymaking agendas requires linking adaptive policy measures to dynamic homeland security environments. For example, the global community is constantly challenged by new and creative terrorist threats, and for this reason, it is critically necessary that flexibility be incorporated into future policymaking agendas.

Adaptive Measures: Theoretical and Practical Considerations

It is clear that no single model or system for promoting homeland security will apply across different timelines or terrorist environments. Creative policy design and implementation is an effective tool for anticipating future trends. Because of this reality, the process for projecting successful homeland security models must include a longitudinal (forward-thinking) framework based on both theory and practical necessity.

The *theoretical* models used in the near future will continue to reflect the same concerns about domestic security seen in the recent past. These include the fluid operational profile of the New Terrorism, legal and civil liberties challenges, and coordination of complex tiered initiatives and collaborative systems at all levels of the homeland security enterprise. Theoretical considerations must take into account the dynamic nature of the New Terrorism and, thereby, adapt practical policies to unanticipated terrorist threats. The fundamentally asymmetrical nature of the New Terrorism requires constant theoretical adaptation by homeland security policymakers.

The *practicality* of applying theoretical models to verified security risks depends on whether these models realistically assess the terrorist environment. Accurate assessment will require theoretical models to be continually updated and adapted to new threats. If these updated models and adaptations are successful, the probability increases that a secure domestic environment will be created as an enduring norm by mitigating foreseeable terrorist risks. Practical adaptations are measured as successful if they apply theoretical models that prevent extremists from destabilizing the domestic environment.

Controlling Terrorism: Understanding Fundamental Challenges

To firmly appreciate near-term projections for homeland security, it is necessary to accept the notion that promoting domestic security is an exercise in preempting and countering terrorism. There exist several fundamental challenges to effective control of terrorist threats, in particular the challenges of countering domestic and international extremism, building collaborative consensus, and opposing the New Terrorism.

The Challenge of Countering Extremism

Extremist ideologies and beliefs are the fertile soil for politically violent behavior. Although a common motivation for political violence in the era of the New Terrorism is religious extremism, there remain extremists who are motivated by ethnocentrism, nationalism, ideological intolerance, and racism. For example, ideology and race frequently underlie domestic extremist tendencies in the United States.

Modern history has shown that multi-tiered and collaborative measures can successfully mitigate risks emanating from these tendencies and that coercive measures alone are often only marginally successful. It is difficult to forcibly reverse these tendencies, and thus, the skillful adoption of other responses is required, along with an underlying appreciation that sheer repression is a risky long-term solution. Nevertheless, the use of force as a policy option does retain continued utility in the international domain, as discussed in Chapter Perspective 14.1 and summarized in Table 14.1.

CHAPTER PERSPECTIVE 14.1

The Continued Utility of Force

Violent coercion will continue to be a viable counterterrorist option. The dismantling of terrorist cells, especially in disputed regions where they enjoy popular support, cannot be accomplished solely by the use of law enforcement, intelligence, or nonmilitary assets. Situations sometimes require a warlike response by military assets ranging in scale from small special operations units to large deployments of significant air, naval, and ground forces. The stark use of force, when successfully used against terrorists, has a demonstrated record that is relevant for coercive counterterrorist policies in the near future.

This record includes the following successes:

- *Elimination of terrorist threats.* This occurred, for example, in the successful hostage rescue operations by West German and Israeli Special Forces in Mogadishu and Entebbe, respectively, in 1977.

- *General deterrence by creating a generalized climate in which the risks of political violence outweigh the benefits.* An example of this is Saddam Hussein's use of the Iraqi military to suppress armed opposition from Iraqi Kurds in the north and the so-called Marsh Arabs in the south—this sent an unmistakable message to other would-be opponents.

- *Specific deterrence against a specific adversary that communicates the high risks of further acts of political violence.* One example is the American

air raids against Libya during Operation El Dorado Canyon in 1986.

- *Demonstrations of national will.* We see this, for example, in the deployment of hundreds of thousands of Indian troops to Kashmir in 2002 after a series of terrorist attacks and provocations by Kashmiri extremists, some of whom acted as Pakistani proxies.

History has shown, of course, that military and paramilitary operations are not always successful. Some of these operations have ended in outright disaster. Others have been marginally successful. It is, therefore, likely that future uses of force will likewise fail on occasion. Nevertheless, the past utility of this option and its symbolic value are certain to encourage its continued use throughout the modern counterterrorist era. Absent a viable threat of force, states are highly unlikely to dissuade committed revolutionaries or aggressive states from committing acts of political violence.

Discussion Questions

1. What is the likelihood that there will always exist a low-intensity war against violent extremists?

2. Who should ultimately decide when military force should be deployed against terrorists?

3. In what circumstances should preemptive force be used?

Table 14.1 The Use of Force

The purpose of violent responses is to attack and degrade the operational capabilities of terrorists. This can be done by directly confronting terrorists or destabilizing their organizations. The following table summarizes basic elements of the use of force.

Counterterrorist Option	Activity Profile		
	Rationale	Practical Objectives	Typical Resources Used
Suppression campaigns	Symbolic strength Punitive measures Preemption	Destruction of the terrorists Disruption of the terrorists	Military assets Paramilitary assets
Coercive covert operations	Symbolic strength Destabilization Preemption	Disruption of the terrorists Deterrent effect on potential terrorists	Military assets Paramilitary assets
Special operations forces	Coercive covert operations Destabilization Preemption	Disruption of the terrorists Deterrent effect on potential terrorists	Military and police assets

Because violent extremism has historically originated primarily from domestic conflict, future prevention and mitigation efforts must incorporate inclusive societal and cultural considerations as counterbalances to extremist ideas. Policy-driven programs of inclusion are more likely to succeed in democratic systems and less likely to have appreciable success in repressive societies. This is because new societal and cultural norms can be peacefully assimilated by democratic systems, thus mitigating the impact of demographic changes and political shifts.

The Challenge of Building Collaborative Consensus

If policymakers accept the utility of adhering to a collaborative homeland security enterprise, federal, state, and local authorities will be better positioned to engage in consensus-building processes. The resulting collaboration will strengthen homeland security prevention and mitigation systems. As discussed, this is a complex and tiered policy goal with many administrative and political challenges. Federal agencies are uniquely positioned to provide support for a national approach to issues within their jurisdiction—for example, the Nuclear Regulatory Commission's application of national standards at the state and local levels of authority. At the same time, state and local authorities are uniquely positioned to articulate how federal outlays should be prioritized and disbursed to local institutions. This collaborative goal is incorporated in the conceptualization of the Federal Emergency Management Agency's whole-community approach, discussed previously in Chapter 12.

If skillfully administered, future adaptations of collaborative initiatives will serve to reduce the likelihood of disorganization and disarray in the event of a terrorist incident. In the recent past, efforts intending to build collaborative consensus were undertaken with the presumption that some degree of common purpose would be an underlying motivation for successful outcomes. The continuation of this presumption will enhance the probability of near-term success.

The Challenge of Countering the New Terrorism

One projection for near-term homeland security policy stands out: Counterterrorist models must be flexible enough to respond to new terrorist environments and must avoid stubborn reliance on methods that "fight the last war." This reality is particularly pertinent to the new era of terrorism and the war against terrorism. This war, unlike previous wars, was declared against behavior as much as against terrorist groups and revolutionary cadres.

The "fronts" in the new war will continue to be amorphous and include the following elements:

- *Homeland security protection.* Homeland security procedures have become ubiquitous in the modern terrorist environment, and they are required to harden targets, deter attacks, and thwart conspiracies. Domestic security requires the extensive use of nonmilitary security personnel, such as customs officials, law enforcement agencies, and immigration authorities.

- *Covert shadow wars.* The use of force as a policy option may include **shadow wars**, which are fought outside public scrutiny using unconventional methods. Shadow wars require deployment of military, paramilitary, and coercive covert assets to far regions of the world.

  shadow wars: Covert campaigns to suppress terrorism.

- *Counterterrorist financial operations.* Covert financial operations are directed against bank accounts, private foundations, businesses, and other potential sources of revenue for terrorist networks. Although intelligence agencies are able to hack into financial databases, a broad-based coalition of government and private financial institutions is necessary for this task.

- *Global surveillance of communications technologies.* Electronic surveillance requires the tracking of telephones, cell phones, browser traffic, and e-mail. Agencies specializing in electronic surveillance, such as the U.S. National Security Agency (NSA), are the institutions most capable of carrying out this mission. However, electronic surveillance operations can be quite controversial and often lead to intense political criticism. For example, the United States was strongly criticized by its European allies when electronic surveillance by the NSA was reported in October 2013. French and German leaders were particularly incensed by NSA monitoring of communications of national political leaders, including monitoring of the cell phone of German chancellor Angela Merkel.

- *Identifying and disrupting transnational terrorist cells and support networks.* Preempting and disrupting terrorist conspiracies requires international cooperation to track extremist operatives and "connect the dots" on a global scale. Primary responsibility for this task lies with intelligence communities and law enforcement agencies.

The new fronts in the new war clearly highlight the need to continuously upgrade homeland security policies and procedures, particularly physical, organizational, and operational counterterrorist measures. Flexibility, creativity, and tiered collaboration are essential for effective domestic security. Failure in any regard will hinder adaptation to the terrorist environment of the twenty-first century. The inability to control and redress long-standing homeland security bureaucratic and international rivalries, to take one example, could be disastrous.

Case in Point: International Collaboration on Law Enforcement and Intelligence

The world's criminal justice systems and intelligence communities have always been important counterterrorist instruments. After the September 11 attacks, these institutions were tasked with increased responsibilities in the global war against terrorism, largely because of their demonstrated ability to preempt, incapacitate, and punish terrorists. These institutions, perhaps more so than military institutions, are also able to apply steady and long-term pressure on terrorist networks. Law enforcement and intelligence agencies are, in many cases, more adept than military assets at keeping terrorists "on the run" over time. This is not to say that these institutions are a panacea for future terrorism, but international cooperation between law enforcement and intelligence agencies does provide the means to track terrorist operatives, identify networks, and interdict other assets on a global scale.

International law enforcement cooperation, in particular, provides worldwide access to extensive criminal justice systems that have well-established terminal institutions (such as prisons) for use against terrorists. Counterterrorist terminal institutions—under the jurisdiction of criminal justice and military justice systems—provide final resolution to individual terrorists' careers after they have been captured, prosecuted, convicted, and imprisoned. Applying a concept familiar to administration-of-justice practitioners, these institutions can effectively incapacitate terrorists by ending their ability to engage in political violence or propaganda. When faced with the prospect of lifelong incarceration, terrorists are likely to become susceptible to manipulation wherein, for example, favors can be exchanged for intelligence information. In a cooperative environment, these intelligence data may be shared among allied governments.

Several important objectives are attainable through enhanced international cooperation among law enforcement, intelligence, and homeland security institutions. These objectives, which are certain to be central considerations for homeland security policymakers and counterterrorist analysts well into the near future, include the following:

- Destabilization of terrorist networks
- Disruption of terrorist conspiracies
- Collection of intelligence from captured terrorist operatives
- Incapacitation of imprisoned operatives

THREAT ENVIRONMENTS IN THE TWENTY-FIRST CENTURY

The death of Osama bin Laden in May 2011 was a victory in the war on terrorism, but terrorism will continue to be a problem as long as violent dissidents find safe haven in supportive environments, such as in Pakistan's semiautonomous regions. There is also a distinct probability that stateless revolutionaries and other independent internationalist movements will engage in terrorist violence on behalf of a variety of vague causes, such as al-Qaeda- and ISIS-inspired religious terrorism. Terrorism will also be used as a violent option in countries where ethnonational, religious, racial, and political conflicts remain unresolved.

Figure 14.1 summarizes recent patterns of domestic hate (bias) crimes.

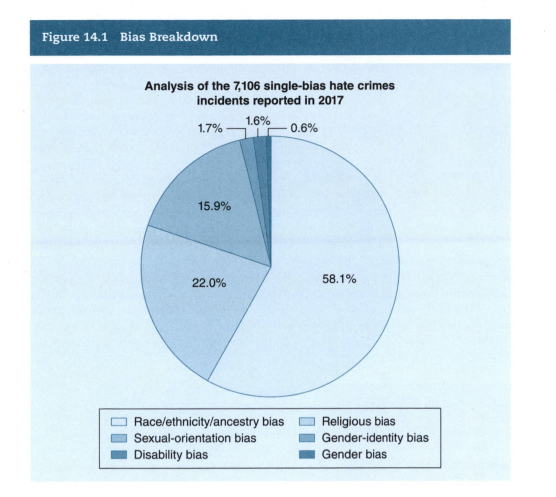

Figure 14.1 Bias Breakdown

Analysis of the 7,106 single-bias hate crimes incidents reported in 2017

- 1.7%
- 1.6%
- 0.6%
- 15.9%
- 22.0%
- 58.1%

☐ Race/ethnicity/ancestry bias   ☐ Religious bias
☐ Sexual-orientation bias   ☐ Gender-identity bias
☐ Disability bias   ☐ Gender bias

Source: U.S. Department of Justice.

Defending the Homeland: Ongoing Patterns of Political Violence

The threat of terrorism in the United States emanates from domestic and international sources. In the modern era, the United States and the West have become targets of the practitioners of the New Terrorism. Terrorists still associate them with international exploitation, but they add other dimensions to the new terrorist environment, such as fundamentalist religion, anti-Semitism, and a willingness to use new technologies and weapons of mass destruction. Symbolic targets have been attacked worldwide, and these include embassies, military installations, religious sites, tourists, and business visitors to foreign countries. In the 1990s and 2000s, attacks against Western interests were symbolic, high profile, and very lethal. In the modern era, attacks became highly destructive and caused a significant number of casualties.

The Threat From Ideological Terrorism

Domestic sources of ideological terrorism include threats from right-wing racial supremacist groups and antigovernment movements originating from the Patriot movement. Potential threats from left-wing sources come primarily from single-issue groups, such as radical environmentalists.

▶ **Photo 14.1** Extremist activism in the United States. Members of the neo-Nazi National Socialist Movement march in Washington, D.C., from the Washington Monument to the U.S. Capitol Building.

The Future of the Violent Left in the United States

The future of the terrorist left in the United States is not encouraging for would-be violent extremists. The modern political and social environment does not exhibit the same mass fervor as existed during the era that spawned the New Left, civil rights, and Black Power movements. Nor is there strong nationalism within American ethnonational communities. Perhaps most importantly, radical Marxism and its contemporary applications are no longer relevant to the American activist community.

Single-issue extremism is certain to be a feature of the radical left. Radical environmentalists have attracted a small but loyal constituency. New movements have also shown themselves to be adept at attracting new followers. For example, a nascent anarchist movement has taken root in the United States and other Western democracies. This movement is loosely rooted in an antiglobalist ideology that opposes alleged exploitation by prosperous nations of poorer nations in the new global economy. It remains to be seen whether violent tendencies will develop within this trend. Sporadic incidents from single-issue terrorists are likely to occur from time to time.

▶ **Photo 14.2** A ritualistic Ku Klux Klan "cross lighting" ceremony in the United States. The KKK is a long-standing racist movement that lives according to a code of racial supremacy. Its ceremonies invoke mystical symbols such as hooded gowns and the burning cross, as well as the adoption of bizarre titles such as Imperial Wizard and Exalted Cyclops.

The Future of the Violent Right in the United States

Trends found in the recent past suggest that extremists on the fringe and far right will continue to promulgate conspiracy theories and attract a few true believers to their causes. Reactionary activists attracted a number of people to their causes during the 1990s and 2000s, and continued to do so thereafter. Violence emanating from rightist tendencies has not abated. For example, racial and anti-Semitic violence continued to be inflicted by lone-wolf extremists. Similarly, extremist anti-abortion activism occurred sporadically, and its viability was sustained by a core of dedicated true believers.

The future of the politically violent right also comes from a relatively long-standing antigovernment scenario. By advocating leaderless resistance, the violent Patriot and neo-Nazi right learned the lessons of 1980s cases of terrorist groups such as The Order. Thus, conspiracies that were uncovered by law enforcement authorities beginning in the mid-1990s exhibited a covert and cell-based organizational philosophy. Possible threats also exist from religious extremists who could reinvigorate the violent moralist movement. An extremist pool still resides within the Patriot movement and racial supremacist communities. There has also been a continued proliferation of antigovernment and racial conspiracy theories. Publications such as *The Turner Diaries*[1] and *The Myth of the Six Million*[2] continue to spread racial and anti-Semitic extremism. The promulgation of these theories keeps reactionary tendencies alive on the right, illustrating the conspiracy mythology that continues to be characteristic of racial supremacist extremism.

From these trends, it is reasonable to project that threat scenarios from the extremist right include the probability of occasional lone-wolf violence. Other plausible scenarios include vehicular attacks, bombings, and mass shooting incidents.

The Threat From International Terrorism

International sources of terrorism originate primarily from cell-based religious extremist networks that replicate the al-Qaeda model. This is the result of an evolution of terrorist tactics and motivations. Patterns of international terrorism during the 1990s and 2000s were decreasingly ideological and increasingly "cultural." Ethnonationalist terrorism continued to occur on a sometimes grand scale, and religious terrorism spread among radical Islamic groups. In addition, stateless international terrorism began to emerge as the predominant model in the global arena. These trends are likely to continue, as vestiges of the East–West ideological competition give way to patterns of seemingly interminable communal conflicts and religious extremism. This scenario, termed the **clash of civilizations**, has been extensively debated since it was theorized by Professor Samuel Huntington.[3]

clash of civilizations:
Theoretical trends that occur as vestiges of the East–West ideological competition give way to patterns of religious extremism and seemingly interminable communal conflicts. This clash of civilizations scenario has been extensively debated since it was theorized by Professor Samuel Huntington.

CHAPTER PERSPECTIVE 14.2

The Changing Environment of International Terrorism

International terrorism is, in many ways, a reflection of global politics, so that the international terrorist environment is dominated at different times by different terrorist typologies. Thus, one may conclude that the profile of international terrorism has progressed through several phases in the postwar era, which can be roughly summarized as follows:[a]

- From the 1960s through the early 1980s, *left-wing* terrorists figured prominently in international incidents. For example, West European groups frequently attacked international symbols "in solidarity" with defined oppressed groups. Only a few leftist groups remain, and most of them (such as Colombia's FARC) do not often practice international terrorism.

- From the beginning of the modern era of international terrorism in the late 1960s, *Palestinian nationalists* were perhaps the leading practitioners of international terrorism. Participating in their struggle were Western European and Middle Eastern extremists who struck targets in solidarity with the Palestinian cause. By the late 1990s, with the creation of the governing authority on the West Bank and Gaza, Palestinian-initiated terrorism focused primarily on targets inside Israel and the occupied territories. Their radical Western comrades ceased their violent support by the late 1980s, but many Middle Eastern extremists continued to cite the Palestinian cause as a reason for their violent activism.

- Throughout the postwar era, *ethnonational* terrorism occupied an important presence in the international arena. Its incidence has ebbed and flowed in scale and frequency, but ethnonationalist violence has never completely disappeared. By the late 1990s, these groups operated primarily inside their home countries but continued to occasionally attack international symbols to bring attention to their domestic agendas.

- By the end of the 20th century, the most prominent practitioners of international terrorism were *religious* extremists. Islamist movements such as al-Qaeda and ISIS became the most prolific international religious terrorists, with similar movements adopting the al-Qaeda and ISIS "brands" for themselves.

Note

a. Paul R. Pillar, *Terrorism and U.S. Foreign Policy* (Washington, DC: Brookings Institution, 2001), 44–45.

Chapter Perspective 14.2 summarizes the changing environments of international terrorism, focusing on the predominant dissident profiles during the latter part of the twentieth century.

International Terrorism in the United States

During the years immediately preceding and following the turn of the twenty-first century, it has become very clear that the near future of international terrorism in the United States will be considerably threatening. Trends suggest that the United States will be a preferred target for international terrorists both domestically and abroad. Although this is not a new phenomenon, the modern terrorist environment has made the American homeland vulnerable to attack for the first time in its history. The asymmetrical nature of new threats and the destructive magnitude of newly obtainable weapons are unlike the threats inherent in previous terrorist environments. The existence of prepositioned **sleeper agents** has proven to be a real possibility. For example, an alleged Hezbollah cell was broken up in July 2000 when 18 people were arrested in Atlanta, Georgia.

Violence emanating from international sources will continue to come from Middle Eastern spillovers during the beginning of the twenty-first century. It is unlikely that the previous activity profile in the United States—threats from groups such as Omega 7, the Provos, and the Jewish Defense League—will reemerge on the same scale. The most significant spillover threat of the early twenty-first century comes from religious extremists. It is also conceivable that spillover activity from nationalists with an anti-American agenda will occur in the American homeland; these would likely be cases of the contagion effect, with newcomers imitating previous homeland incidents committed by other terrorists. Regardless of the source of international terrorism, the activity profile is almost certain to be that of the New Terrorism.

sleeper agents: A tactic used by international terrorist movements in which operatives theoretically establish residence in another country to await a time when they will be activated by orders to carry out a terrorist attack.

U.S. Department of Justice

▶ **Photo 14.3** The aftermath of September 11, 2001. Rescue workers amidst the smoke and debris of the World Trade Center in New York City.

Counterpoints to International Terrorism

Counterpoints exist to the continuation of dissident political violence in the near future, including the following three examples.

First, the 2011 Arab Spring uprisings demonstrated that support for and the momentum of domestic terrorist movements can be undercut by genuine social reform. Examples of successful Arab Spring reforms suggest that these reforms can reduce and even end political violence, as long as they are pursued with determination despite setbacks, sabotage, and authoritarian tendencies.

A second counterpoint is the utility of international counterterrorist cooperation. Several terrorist conspiracies were thwarted in the Middle East and Europe—and a number of terrorists were captured or kept on the run—as a direct result of cooperation among intelligence, law enforcement, and security agencies.

A third counterpoint is that there is no single binding or common ideological foundation for political violence. There is no modern equivalent to revolutionary Marxism during the heyday of ideological terrorism. Religious extremism is self-isolating, and there is no longer an international "solidarity movement" in the West for ethnonationalist (or religious) violence. It remains to be seen what the long-term impact will be on religious terrorist violence, but it is unlikely that there will be an ideological "glue" to bind together extremist sentiment in the near future.

CHAPTER SUMMARY

This chapter summarized salient issues and challenges for homeland security policies and procedures that are likely to be encountered in the near future. Significant challenges are posed from adaptive and resilient extremists who will continue to plan new tactics to circumvent homeland security protections. Unlike previous terrorist environments, the era of the New Terrorism requires an ongoing process of designing theoretical homeland security models that have practical counterterrorist applications. This process is complex and dynamic.

Collaboration at all levels of the homeland security enterprise is a necessary precondition for promoting policy consistency over time. In this regard, federal leadership is needed to promote collaboration. Coordinated national guidance by federal authorities is a

critical process for establishing effective homeland security systems at state and local levels of authority. At the same time, an international perspective must be maintained when defining and assessing the roles of law enforcement and intelligence authorities.

Future terrorist risks to the United States will emanate from domestic ideological extremists and international sources. Domestic extremists in the United States represent lingering right-wing and leftist extremist belief systems. International terrorist threats are characterized by the New Terrorism of primarily stateless religious extremist movements. All of these risks will challenge the homeland security enterprise to persistently adapt to new and emergent threat scenarios.

DISCUSSION BOX

This chapter's Discussion Box is intended to stimulate critical debate about the international dimension of homeland security.

Countering Future Terror: The Case for International Cooperation

Cooperation between nations has always been essential to counterterrorist operations, and it is also now necessary for homeland security systems. International treaties, laws, and informal agreements were enacted during the postwar era to create a semblance of formality and consistency to global counterterrorist efforts. However, cooperation at the operational level was not always consistent or mutually beneficial. In the era of homeland security, the New Terrorism, and international counterterrorist warfare, international cooperation at the operational level has become a central priority for policymakers. A good example of this priority is found in the new frontline missions of intelligence and criminal justice agencies.

Discussion Questions

1. How should international cooperation be implemented for law enforcement purposes?

2. How can intelligence at the international level be disseminated to state and local authorities?

3. In what ways should future homeland security initiatives be broadly adapted to the international terrorist environment?

4. Should future homeland security policies task federal law enforcement agencies to operate internationally?

5. To what extent should allied nations be briefed about domestic homeland security initiatives?

KEY TERMS AND CONCEPTS

The following topics were discussed in this chapter and can be found in the glossary:

clash of civilizations   309   shadow wars   305   sleeper agents   310

ON YOUR OWN

Get the tools you need to sharpen your study skills. SAGE edge offers a robust online environment featuring an impressive array of free tools and resources.

Access practice quizzes, eFlashcards, video, and multimedia at **edge.sagepub.com/martinhs3e**

WEB EXERCISE

Conduct an online investigation of the future of homeland security and terrorism.

1. Are there certain trends in the modern homeland security environment that portend successful homeland security vigilance against terrorist incidents?

2. Are there certain trends in the modern terrorist environment that portend success by terrorists against homeland security procedures?

3. In your opinion, how effective are homeland security procedures in anticipating terrorist behavior?

To conduct an online search on the future of homeland security, activate the search engine on your Web browser and enter the following keywords:

"Future of homeland security"

"Future of terrorism"

RECOMMENDED READINGS

The following publications project homeland security challenges facing the United States and world community in the present and in the near future:

Heiberg, Marianne, Brendan O'Leary, and John Tirman, eds. 2007. *Terror, Insurgency, and the State: Ending Protracted Conflicts*. Philadelphia: University of Pennsylvania Press.

Holmes, Jennifer S. 2008. *Terrorism and Democratic Stability Revisited*. Manchester, UK: Manchester University Press.

APPENDIX A

Office of the Director of National Intelligence. 2019.
National Intelligence Strategy of the United States of America.
Washington, DC: The White House.

The National Intelligence Strategy
of the United States of America

IC Vision

A Nation made more secure by a fully integrated, agile, resilient, and innovative Intelligence Community that exemplifies America's values.

IC Mission

Provide timely, insightful, objective, and relevant intelligence and support to inform national security decisions and to protect our Nation and its interests.

>>

This National Intelligence Strategy (NIS) provides the Intelligence Community (IC) with strategic direction from the Director of National Intelligence (DNI) for the next four years. It supports the national security priorities outlined in the *National Security Strategy* as well as other national strategies. In executing the NIS, all IC activities must be responsive to national security priorities and must comply with the Constitution, applicable laws and statutes, and Congressional oversight requirements.

All our activities will be conducted consistent with our guiding principles: We advance our national security, economic strength, and technological superiority by delivering distinctive, timely insights with clarity, objectivity, and independence; we achieve unparalleled access to protected information and exquisite understanding of our adversaries' intentions and capabilities; we maintain global awareness for strategic warning; and we leverage what others do well, adding unique value for the Nation.

From the Director
of National Intelligence

As the Director of National Intelligence, I am fortunate to lead an Intelligence Community (IC) composed of the best and brightest professionals who have committed their careers and their lives to protecting our national security. The IC is a 24/7/365 organization, scanning the globe and delivering the most distinctive, timely insights with clarity, objectivity, and independence to advance our national security, economic strength, and technological superiority.

This, the fourth iteration of the National Intelligence Strategy (NIS), is our guide for the next four years to better serve the needs of our customers, to help them make informed decisions on national security issues, and to ultimately keep our Nation safe. The NIS is designed to advance our mission and align our objectives with national strategies, and it provides an opportunity to communicate national priority objectives to our workforce, partners, oversight, customers, and also to our fellow citizens.

We face significant changes in the domestic and global environment; we must be ready to meet 21st century challenges and to recognize emerging threats and opportunities. To navigate today's turbulent and complex strategic environment, we must do things differently. This means we must:

- Increase integration and coordination of our intelligence activities to achieve best effect and value in executing our mission,
- Bolster innovation to constantly improve our work,
- Better leverage strong, unique, and valuable partnerships to support and enable national security outcomes, and
- Increase transparency while protecting national security information to enhance accountability and public trust.

This National Intelligence Strategy increases emphasis in these areas. It better integrates counterintelligence and security, better focuses the IC on addressing cyber threats, and sets clear direction on privacy, civil liberties and transparency.

We have crucial work before us. Our customers depend on us to help them to make wise national security decisions, and Americans count on us to help protect the Nation, all while protecting their privacy and civil liberties. We must provide the best intelligence possible to support these objectives; doing so is a collective responsibility of all of our dedicated IC professionals and, together with our partners, we can realize our vision.

Our ongoing goal is to continue to be the very best intelligence community in the world. Thank you for your service and for bringing your talent and commitment to the work of keeping our Nation safe each and every day. Thank you for your dedication to our mission and to the security of our fellow citizens as we take this journey together.

Daniel R. Coats
Director of National Intelligence

Strategic Environment

The strategic environment is changing rapidly, and the United States faces an increasingly complex and uncertain world in which threats are becoming ever more diverse and interconnected. While the IC remains focused on confronting a number of conventional challenges to U.S. national security posed by our adversaries, advances in technology are driving evolutionary and revolutionary change across multiple fronts. The IC will have to become more agile, innovative, and resilient to deal effectively with these threats and the ever more volatile world that shapes them. The increasingly complex, interconnected, and transnational nature of these threats also underscores the importance of continuing and advancing IC outreach and cooperation with international partners and allies.

Traditional adversaries will continue attempts to gain and assert influence, taking advantage of changing conditions in the international environment—including the weakening of the post-WWII international order and dominance of Western democratic ideals, increasingly isolationist tendencies in the West, and shifts in the global economy. These adversaries pose challenges within traditional, non-traditional, hybrid, and asymmetric military, economic, and political spheres. Russian efforts to increase its influence and authority are likely to continue and may conflict with U.S. goals and priorities in multiple regions. Chinese military modernization and continued pursuit of economic and territorial predominance in the Pacific region and beyond remain a concern, though opportunities exist to work with Beijing on issues of mutual concern, such as North Korean aggression and continued pursuit of nuclear and ballistic missile technology. Despite its 2015 commitment to a peaceful nuclear program, Iran's pursuit of more advanced missile and military capabilities and continued support for terrorist groups, militants, and other U.S. opponents will continue to threaten U.S. interests. Multiple adversaries continue to pursue capabilities to inflict potentially catastrophic damage to U.S. interests through the acquisition and use of weapons of mass destruction (WMD), which includes biological, chemical, and nuclear weapons.

In addition to these familiar threats, our adversaries are increasingly leveraging rapid advances in technology to pose new and **evolving threats**— particularly in the realm of space, cyberspace, computing, and other emerging, disruptive technologies. Technological advances will enable a wider range of actors to acquire sophisticated capabilities that were previously available only to well-resourced states.

No longer a solely U.S. domain, the democratization of **space** poses significant challenges for the United States and the IC. Adversaries are increasing their presence in this domain with plans to reach or exceed parity in some areas. For example, Russia and China will continue to pursue a full range of anti-satellite weapons as a means to reduce U.S. military effectiveness and overall security. Increasing commercialization of space now provides capabilities that were once limited to global powers to anyone that can afford to buy them. Many aspects of modern society—to include our ability to conduct military operations—rely on our access to and equipment in space.

Cyber threats are already challenging public confidence in our global institutions, governance, and norms, while imposing numerous economic costs domestically and globally. As the cyber capabilities of our adversaries grow, they will pose increasing threats to U.S. security, including critical infrastructure, public health and safety, economic prosperity, and stability.

Emerging technologies, such as artificial intelligence, automation, and high performance computing are advancing computational capabilities that can be economically beneficial, however these advances also enable new and improved military and intelligence capabilities for our adversaries. Advances in nano- and bio-technologies have the potential to cure diseases and modify human performance, but without common ethical standards and shared interests to govern these developments, they have the potential to pose significant threats to U.S. interests and security. In addition, the development and spread of such technologies remain uneven, increasing the potential to drastically widen the divide between so-called "haves" and "have-nots."

Advances in communications and the democratization of other technologies have also generated an ability to create and share vast and exponentially growing amounts of information farther and faster than ever before. This **abundance of data** provides significant opportunities for the IC, including new avenues for collection and the potential for greater insight, but it also challenges the IC's ability to collect, process, evaluate, and analyze such enormous volumes of data quickly enough to provide relevant and useful insight to its customers.

These advances in communications and the democratization of other technologies have empowered non-state actors and will continue to exponentially expand the potential to influence people and events, both domestically and globally.

The ability of individuals and groups to have a larger impact than ever before—politically, militarily, economically, and ideologically—is undermining traditional institutions. This empowerment of groups and individuals is increasing the influence of ethnic, religious, and other sources of identity, changing the nature of conflict, and challenging the ability of traditional governments to satisfy the increasing demands of their populations, increasing

the potential for greater instability. Some **violent extremist groups** will continue to take advantage of these sources and drivers of instability to hold territory, further insurgencies, plan external attacks, and inspire followers to launch attacks wherever they are around the world.

Increasing migration and urbanization of populations are also further straining the capacities of governments around the world and are likely to result in further fracturing of societies, potentially creating breeding grounds for radicalization. Pressure points include growing influxes of migrants, refugees, and internally displaced persons fleeing conflict zones; areas of intense economic or other resource scarcity; and areas threatened by climate changes, infectious disease outbreaks, or transnational criminal organizations.

All of these issues will continue to drive global change on an unprecedented scale and the IC must be able to warn of their strategic effects and adapt to meet the changing mission needs in this increasingly unstable environment. There will likely be demand for greater intelligence support to domestic security, driven in part by concerns over the threat of terrorism, the threat posed by transnational illicit drug and human trafficking networks, and the threat to U.S. critical infrastructure. Intelligence support to counter these threats must be conducted in accordance with IC authorities, with appropriate levels of transparency to the public, and with adequate protection for civil liberties and privacy.

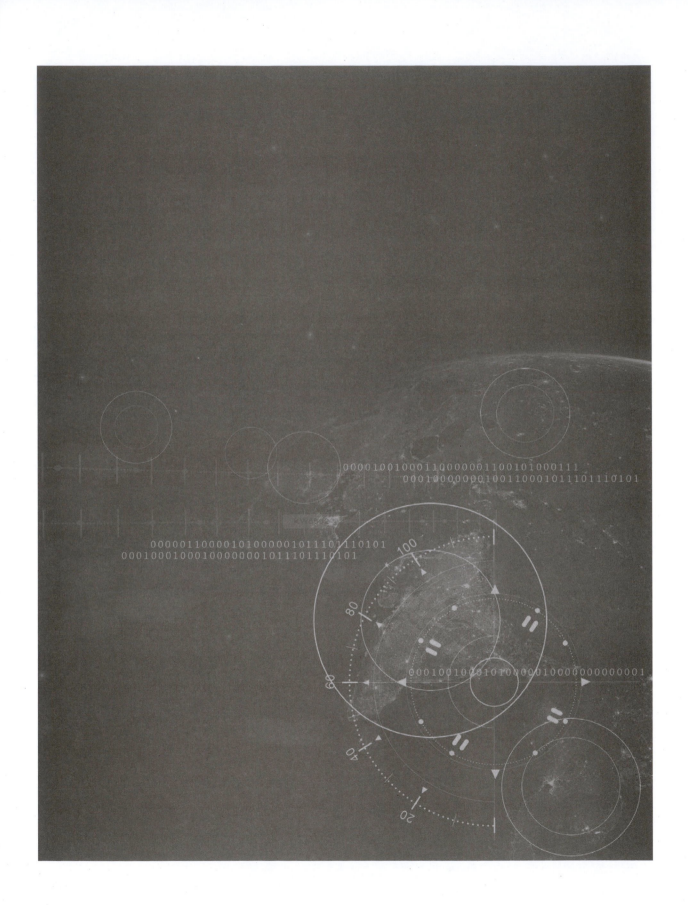

Mission
Objectives

The seven mission objectives broadly describe the activities and outcomes necessary for the IC to deliver timely, insightful, objective, and relevant intelligence and support to its customers. Mission objectives address a broad range of regional and functional topics facing the IC and their prioritization is communicated to the IC through the National Intelligence Priorities Framework.

The first three mission objectives address foundational missions of the IC which transcend individual threats, topics, or geographic regions. This is different from foundational military intelligence, which is intelligence on foreign military capabilities. As such, **foundational mission objectives** collectively represent the broadest and most fundamental of the IC's intelligence missions.

1 **Strategic Intelligence**
addresses issues of enduring national security interest.

2 **Anticipatory Intelligence**
addresses new and emerging trends, changing conditions, and underappreciated developments.

3 **Current Operations Intelligence**
supports planned and ongoing operations.

The next four mission objectives address specific, topical missions of the IC. The topical mission objectives are supported by the three foundational mission objectives and may contain elements of these. Other specific regional and functional issues, such as conflict areas and transnational criminal organizations, are implicitly covered by the mission objectives.

4 **Cyber Threat Intelligence**
addresses state and non-state actors engaged in malicious cyber activities.

5 **Counterterrorism**
addresses state and non-state actors engaged in terrorism and related activities.

6 **Counterproliferation**
addresses state and non-state actors engaged in the proliferation of weapons of mass destruction and their means of delivery.

7 **Counterintelligence and Security**
addresses threats from foreign intelligence entities and insiders.

National Intelligence and Intelligence Related to National Security means all intelligence, regardless of the source from which derived and including information gathered within or outside the United States, that pertains, as determined consistent with any guidance issued by the President, or that is determined for the purpose of access to information by the Director to pertain to more than one United States Government agency; and that involves threats to the United States, its people, property, or interests; the development, proliferation, or use of weapons of mass destruction; or any other matter bearing on United States national or homeland security.

(Executive Order 12333)

Strategic Intelligence

1

Identify and assess the capabilities, activities, and intentions of states and non-state entities to develop a deep understanding of the strategic environment, warn of future developments on issues of enduring interest, and support U.S. national security policy and strategy decisions.

Strategic intelligence is the process and product of developing the context, knowledge, and understanding of the strategic environment required to support U.S. national security policy and planning decisions. This work includes identifying and assessing the capabilities, activities, and intentions of states and non-state entities to identify risks to and opportunities for U.S. national security interests. Strategic intelligence involves assimilating a variety of information—including knowledge of political, diplomatic, economic, and security developments—to create a deep understanding of issues of enduring importance to the United States. Strategic intelligence also provides in-depth assessments of trends and developments to recognize and warn of changes related to these issues that will affect the future strategic environment.

The foundation for strategic intelligence requires developing and maintaining a deep understanding of the strategic environment, to include transnational issues such as terrorism and transnational organized crime, and the capabilities, activities, and intentions of states and non-state entities necessary to support U.S. national security policy and planning decisions. The IC must master strategic intelligence issues through research, knowledge development, collaboration with experts within the IC, and outreach to experts in academia and industry, as well as the use of advanced analytics and tradecraft to provide in-depth assessments and the strategic context for a wide variety of policy and strategy communities.

TO MEET THIS OBJECTIVE, THE IC WILL:
>>

- Develop and maintain capabilities to acquire and evaluate data to obtain a deep understanding of the global political, diplomatic, military, economic, security, and informational environment.

- Build and maintain expertise and knowledge of issues of enduring strategic importance to the United States, and assess trends and developments related to these issues to identify changes that would affect U.S. national security interests and to identify strategic risks and opportunities.

- Provide in-depth assessments, context, and expertise about the strategic environment, including the capabilities, activities, and intentions of key state and non-state entities, to inform U.S. national security policy and strategy development.

8

Anticipatory Intelligence

Identify and assess new, emerging trends, changing conditions, and underappreciated developments to challenge long-standing assumptions, encourage new perspectives, identify new opportunities, and provide warning of threats to U.S. interests.

Anticipatory intelligence involves collecting and analyzing information to identify new, emerging trends, changing conditions, and undervalued developments, which challenge long-standing assumptions and encourage new perspectives, as well as identify new opportunities and warn of threats to U.S. interests. Anticipatory intelligence usually leverages a cross-disciplinary approach, and often utilizes specialized tradecraft to identify emerging issues from "weak signals," cope with high degrees of uncertainty, and consider alternative futures.

Anticipatory intelligence looks to the future as foresight (identifying emerging issues), forecasting (developing potential scenarios), or warning. Anticipatory intelligence explores the potential for cascading events or activities to reinforce, amplify, or accelerate conflict. It may uncover previously unconnected groups or regions and include indicators or benchmarks to identify key developments as trends change over time. Anticipatory intelligence assesses risk, intelligence gaps, and uncertainties by evaluating the probability of occurrence and potential effects of a given development on U.S. national security.

The complexity, scale, and pace of changes developing around the world test the IC's ability to deliver insightful and actionable intelligence with optimal fidelity, scope, and speed required to mitigate threats and exploit opportunities. The IC will expand its use of quantitative analytic methods while reinforcing qualitative methods, especially those that encourage new perspectives and challenge long-standing assumptions. With evolving intelligence requirements, anticipatory intelligence is critical for efficient IC resource allocation. The IC will improve its ability to foresee, forecast, and alert regarding potential issues of concern and provide the best possible opportunities for action to our national security customers.

TO MEET THIS OBJECTIVE, THE IC WILL:
>>>>>>>>>>>>>>>>>>>>>>>>>>>>>>>>>>>>>

- Develop quantitative methods and data analysis techniques and tradecraft to improve the IC's ability to identify, analyze, and forecast changing conditions and emerging trends across multiple portfolios.

- Increase common understanding of the scope, definition, tradecraft, and methods of anticipatory intelligence across the community to develop workforce proficiency in these skills.

- Identify and work to remove the cultural, technological, human capital, and other barriers to incorporate anticipatory intelligence into the IC's routine analytic efforts.

- Produce and provide intelligence information and products that highlight emerging trends, changing conditions, and opportunities or threats in areas of limited customer focus to maximize decision advantage.

- Develop integrated capabilities to create alerts within the IC to provide timely and relevant warning to our customers, as well as to apprise them of opportunities.

Current Operations Intelligence

Provide timely intelligence support to enable planned and ongoing operations.

Current operations intelligence is the collection, analysis, operations, and planning support that IC elements conduct to enable successful planned and ongoing operations. Current operations intelligence includes the intelligence necessary to support the time-sensitive needs of military, diplomatic, homeland security, and policy customers in times of conflict or crisis, but also provides opportunities to shape future operations and desired operational outcomes.

The IC will adapt to evolving operational requirements, maintain the robust support customers expect, and further enhance capabilities. Faced with a wide spectrum of operations in support of military, diplomatic, and homeland security activities, to include addressing transnational organized crime, the IC will prioritize its efforts and mitigate risk, operate in denied areas, balance forward presence with robust reach-back, and provide operational resiliency to more fully integrate intelligence with operations.

TO MEET THIS OBJECTIVE, THE IC WILL:

- Provide timely intelligence support to enable planned and ongoing operations.

- Develop and maintain a robust, IC-wide intelligence architecture that delivers actionable, timely, and agile intelligence support to achieve and maintain operational decision advantage.

- Expand and enhance collaboration with domestic and global partners to maximize the effectiveness and reach of intelligence capabilities in support of operations.

- Conduct sensitive intelligence operations to support effective national security action.

10

Cyber Threat Intelligence

4

Detect and understand cyber threats from state and non-state actors engaged in malicious cyber activity to inform and enable national security decisionmaking, cybersecurity, and the full range of response activities.

> **Cyber threat intelligence** is the collection, processing, analysis, and dissemination of information from all sources of intelligence on foreign actors' cyber programs, intentions, capabilities, research and development, tactics, targets, operational activities and indicators, and their impact or potential effects on U.S. national security interests. Cyber threat intelligence also includes information on cyber threat actor information systems, infrastructure, and data; and network characterization, or insight into the components, structures, use, and vulnerabilities of foreign cyber program information systems.

Despite growing awareness of cyber threats and improving cyber defenses, nearly all information, communication networks, and systems will be at risk for years to come. Our adversaries are becoming more adept at using cyberspace capabilities to threaten our interests and advance their own strategic and economic objectives. Cyber threats will pose an increasing risk to public health, safety, and prosperity as information technologies are integrated into critical infrastructure, vital national networks, and consumer devices. The IC must continue to grow its intelligence capabilities to meet these evolving cyber threats as a part of a comprehensive cyber posture positioning the Nation for strategic and tactical response.

TO MEET THIS OBJECTIVE, THE IC WILL:
>>>>>>>>>>>>>>>>>>>>>>>>>>>>>>>>>>>>

- Increase our awareness and understanding of adversaries' use of cyber operations—including leadership plans, intentions, capabilities, and operations—to inform decisions and enable action.

- Expand tailored production and appropriate dissemination and release of actionable cyber threat intelligence to support the defense of vital information networks and critical infrastructure.

- Expand our ability to enable diplomatic, information, military, economic, financial, intelligence, and law enforcement plans and operations to deter and counter malicious cyber actors and activities.

Counterterrorism

5

Identify, understand, monitor, and disrupt state and non-state actors engaged in terrorism and related activities to defeat threats to the United States, our people, interests, and partners.

The dynamic nature of the terrorist threat facing the United States requires continued emphasis on intelligence collection and analysis. The IC is therefore an integral part of the national whole-of-government effort to protect our country from terrorist attacks. The IC works across agencies and with domestic and foreign partners to disrupt, dismantle, and defeat terrorists who threaten our homeland, our people, our interests, and our partners overseas.

The IC identifies and helps to eliminate terrorist safe havens and degrade the illicit financial networks that fund terrorist activities. The IC supports broader U.S. Government efforts to counter the spread of violent extremist ideology that drives terrorist actions and to leverage domestic and foreign partnerships and capabilities to strengthen our own capacity and resilience. The enduring and evolving nature of the threat, to include the threat of WMD terrorism, means that the IC must continue to pursue innovative approaches to collection and analysis to ensure counterterrorism (CT) efforts remain effective, efficient, and fully integrated.

TO MEET THIS OBJECTIVE, THE IC WILL:

>>>>>>>>>>>>>>>>>>>>>>>>>>>>>>>>>>>>>>

- Collect and analyze intelligence to enable the disruption of terrorist attacks and attack planning, as well as terrorism-related activities.

- Identify and warn of emerging and changing threats, trends, and violent extremist ideologies to develop opportunities to counter them.

- Broaden and deepen strategic knowledge of the global terrorism landscape to provide context to customers.

12

Counterproliferation

6

Detect, characterize, and disrupt activities of state and non-state actors engaged in the proliferation of weapons of mass destruction (WMD) and their means of delivery to defeat WMD threats to the United States, our people, interests, and partners.

Proliferation is the development and spread of WMD, related technologies, materials, or expertise, and their means of delivery, including both indigenous development and transfers.

Counterproliferation discourages interest in WMD, denies or disrupts acquisition, degrades programs and capabilities, deters use, and mitigates consequences.

The IC must continue to implement a whole-of-government approach to advancing the enduring U.S. counterproliferation policy goals of discouraging interest in WMD, denying or disrupting acquisition, degrading programs and capabilities, including financial networks that fund proliferation activities, deterring use, and mitigating consequences. This issue is increasingly important as regional security dynamics evolve and as states, terrorists, and proliferators take advantage of rapidly emerging technological advances.

Many adversaries continue to pursue capabilities to inflict catastrophic damage to U.S. interests through the acquisition and use of WMD. Their possession of these capabilities can have major impacts on U.S. national security, overseas interests, allies, and the global order. The intelligence challenges to countering the proliferation of WMD and advanced conventional weapons are increasing as actors become more sophisticated, WMD-related information becomes broadly available, proliferation mechanisms increase, and as political instability erodes the security of WMD stockpiles.

- Strengthen U.S. Government efforts to secure global WMD stockpiles, disrupt adversaries' programs, and prevent the transfer of WMD, related technologies, materials, or expertise.

- Bolster U.S. ability to anticipate and manage WMD crises, including potential disclosure, loss, theft, or use of WMD-related materials or weapons.

- Develop, maintain, and enhance intelligence capabilities to advance understanding of foreign WMD programs, related technologies, materials, or expertise to effectively inform interagency counterproliferation strategic planning and operations.

2019 NIS - MISSION OBJECTIVES

13

Counterintelligence and Security

7

Detect, understand, deter, disrupt, and defend against threats from foreign intelligence entities and insiders to protect U.S. national and economic security.

Foreign Intelligence Entity (FIE) is any known or suspected foreign state or non-state organization or person that conducts intelligence activities to acquire U.S. information, block or impair U.S. intelligence collection, influence U.S. policy, or disrupt U.S. systems and programs. The term includes foreign intelligence and security services and international terrorists.

Insider Threat is the threat that an insider—any person with authorized access to any U.S. Government resource, to include personnel, facilities, information, equipment, networks, or systems—will use his/her authorized access, wittingly or unwittingly, to do harm to the security of the United States. This threat can include damage to the United States through espionage, terrorism, unauthorized disclosure of national security information, or through the loss or degradation of departmental resources or capabilities.

The United States faces an increasingly complex and diverse set of counterintelligence (CI) and security challenges. Rapid technological advances are allowing a broad range of FIEs to field increasingly sophisticated capabilities and aggressively target the government, private sector partners, and academia. FIEs are proactive and use creative approaches—including the use of cyber tools, malicious insiders, espionage, and supply chain exploitation —to advance their interests and gain advantage over the United States. These activities intensify traditional FIE threats, place U.S. critical infrastructure at risk, erode U.S. competitive advantage, and weaken our global influence. To mitigate these threats, the IC must drive innovative CI and security solutions, further integrate CI and security disciplines into IC business practices, and effectively resource such efforts. While the authorities that govern CI and security and the programs they drive are distinct, their respective actions must be synchronized, coordinated, and integrated.

TO MEET THIS OBJECTIVE, THE IC WILL:

- Drive integrated IC activities to anticipate and advance our understanding of evolving FIE threats and security vulnerabilities.

- Develop and implement new capabilities to detect, deter, and disrupt FIE activities and insider threats.

- Advance CI and security efforts to protect our people, technologies, information, infrastructure, and facilities from FIEs and insider threats.

- Strengthen the exchange of FIE threat and security vulnerability information among key partners and stakeholders to promote coordinated approaches to mitigation.

14

Enterprise Objectives

The seven enterprise objectives provide the foundation for integrated, effective, and efficient management of mission capabilities and business functions.

>>>

The first two enterprise objectives focus on general mission and business practices of the IC.

1 **Integrated Mission Management**
addresses IC mission capabilities, activities, and resources to achieve unity of effort.

2 **Integrated Business Management**
addresses IC business functions and practices to enable mission success.

The next five enterprise objectives focus on integration of IC efforts in specific areas for the successful completion of the mission objectives.

3 **People**
seeks to forge and retain a diverse, inclusive, and expert workforce.

4 **Innovation**
addresses the improvement of mission and business processes through new technologies, innovative thought, and advancements in tradecraft.

5 **Information Sharing and Safeguarding**
improves collaboration and integration while protecting information.

6 **Partnerships**
seeks to enhance intelligence through partnerships.

7 **Privacy, Civil Liberties, and Transparency**
seeks to protect U.S. values and enhance public trust.

Integrated Mission Management

1

Prioritize, coordinate, align, and de-conflict IC mission capabilities, activities, and resources to achieve unity of effort and the best effect in executing the IC's mission objectives.

Effective mission execution requires flexible, responsive, and resilient efforts to appropriately share knowledge, information, and capabilities across organizational boundaries; mission-focused centers have proven effective in achieving these ends. IC leaders will integrate, collaborate, and exchange feedback across priority areas to meet customer needs.

The IC must strike a balance between unity of effort and specialization, using the best of each to meet mission objectives. Integrated mission management drives collaboration, creates efficiencies, and minimizes redundancies, allowing the IC to effectively use available resources.

TO MEET THIS OBJECTIVE, THE IC WILL:

>>>>>>>>>>>>>>>>>>>>>>>>>>>>>>>>>>>>>>>

- Provide leadership and community management to foster collaboration, streamline processes, and effectively manage resources to achieve IC mission objectives.

- Conduct integrated planning, analysis, collection, production, and dissemination to synchronize intelligence activities.

- Leverage IC capabilities and multi-disciplinary expertise to collaboratively anticipate intelligence problems, implement solutions, and enable innovation.

- Strengthen and integrate IC governance bodies to increase transparency, prioritize and optimize resources, balance tradeoffs, and manage risk.

- Drive integrated investment decisions and the delivery of multi-disciplinary, integrated capabilities to assure mission success.

18

Integrated Business Management

Provide and optimize IC business functions and practices to enable mission success.

IC business functions and practices enable the community to perform its missions, activities, and operations. These functions and practices include the coordinated development, alignment, de-confliction, execution, and monitoring of strategies, policies, plans, and procedures needed to manage and secure the IC and its people, information technology, and physical infrastructure.

Effectively managing business functions and practices across the IC contributes to communication and collaboration, supports the efficient use of resources, enables resilience, and strengthens integration. Common standards, shared services, and best practices within IC authorities can increase efficiency, effectiveness, and accountability; successfully manage and mitigate risk; and improve business processes through data-driven reviews and performance measurements. The IC will promote and identify best business practices and functions to optimize solutions and increase collaboration to create a culture of continuous learning, innovation, and partnerships across the community. The IC must also develop flexible, risk-managed acquisition processes that deliver innovative capabilities, data, and expertise at mission pace.

TO MEET THIS OBJECTIVE, THE IC WILL:

>>>>>>>>>>>>>>>>>>>>>>>>>>>>>>>>>>>>

- Advance a dynamic approach to continuous evaluation and common security practices and standards to strengthen the security of the IC infrastructure.

- Pursue common strategies and best practices for acquisition and procurement across the IC to enhance the cost-effectiveness, efficiency, and agility of acquiring and procuring IC products and services.

- Implement IC-wide financial standards, processes, tools, and services to achieve fiscal efficiency, accountability, and security.

- Enhance strategy-based performance evaluation across the IC that leverages both government and industry best practices to enable informed IC business decisions and guide efficient application of resources.

- Explore innovative means to advance IC facilities, logistics, environmental, and energy programs to enable joint-use functionality; increase efficiency, sustainability, and supportability; and achieve total asset management.

- Manage risk to intelligence capabilities through IC-wide continuity efforts to foster resilience under all conditions.

People

Forge and retain a diverse, inclusive, and expert workforce to address enduring and emerging requirements and enable mission success.

Diversity is a collection of individual attributes that together help IC elements pursue organizational objectives efficiently and effectively. These attributes include but are not limited to characteristics such as national origin, language, race, color, mental or physical disability, ethnicity, sex, age, religion, sexual orientation, gender identity or expression, socioeconomic status, veteran status, and family structure. **Inclusion** is a culture that connects each employee to the organization; encourages collaboration, flexibility, and fairness; and leverages diversity throughout the organization so that all individuals are able to participate and contribute to their full potential.

Linked together, diversity and inclusion drive innovation and enable the IC to attract and retain the highly-skilled workforce needed to meet mission requirements.

The IC is united in protecting and preserving national security, an objective that can only be met with the right, trusted, agile, and well-led workforce. IC personnel, including all civilians, military, and contractors, must adhere to the *Principles of Professional Ethics for the IC*. Effective approaches are needed to recruit, retain, develop, and motivate employees who possess skills that are fundamental to the intelligence mission, including critical thinking, foreign language, science, technology, engineering, and mathematics. The responsibility to lead and integrate the IC workforce extends beyond the IC's human capital, equal employment opportunity, and diversity and inclusion community to span the entire enterprise. Similarly, all IC employees are accountable for cultivating a performance-driven culture that encourages collaboration, flexibility, and fairness.

The IC must have effective tools and resources that integrate workforce planning, transformational leadership, continuous learning, information sharing, performance management, and accountability. Additionally, the IC will make long-term strategic investments in the workforce to promote agility and mobility throughout employees' careers, including joint duty rotations, and ensure that benefits, compensation, and work-life balance initiatives are fully considered and implemented wherever feasible.

TO MEET THIS OBJECTIVE, THE IC WILL:

- Create an inclusive environment empowering managers and employees at all levels to take responsibility and ownership for the diversity of the organization.

- Take measures to proactively prevent discrimination, harassment, and fear of reprisal, enabling the workforce to perform at its highest potential.

- Shape a diverse workforce with the skills and capabilities needed to address enduring and emerging requirements.

- Invest in mid-level managers and leaders to ensure they are appropriately trained, supported, and held accountable.

- Pursue common business functions and practices for human capital, diversity and inclusion, and equal employment opportunity (EEO) compliance programs to enable informed IC human resource investments and decisions.

20

Innovation

Find, create, and deploy scientific discoveries and new technologies, nurture innovative thought, advance tradecraft, and constantly improve mission and business processes to advance the IC in a rapidly changing landscape.

Innovation—through technological advancements and improved business practices—is critical to ensuring that the IC can provide the strategic and tactical decision advantage that policymakers and warfighters require. To continue meeting future challenges, the IC must drive new levels of innovation by proactively developing and rapidly incorporating breakthrough and incremental technologies, ideas, and constructs. The IC must also foster unconventional thinking and experimentation that address new, better ways of accomplishing the IC's mission, especially those approaches that emphasize acceleration, simplicity, and efficiency without sacrificing quality and outcomes. These approaches should increase insight, knowledge, and speed through artificial intelligence, automation, and augmentation, where applicable. To achieve this, IC leaders must be prepared to boldly accept calculated risks to attain high-value results, and accept the fact that initial failures may precede a successful outcome.

TO MEET THIS OBJECTIVE, THE IC WILL:

>>>>>>>>>>>>>>>>>>>>>>>>>>>>>>>>>>>>>>

- Conduct, leverage, protect, and operationalize groundbreaking research to create agile and revolutionary IC capabilities.

- Nurture an enterprise-wide atmosphere of innovation capable of rapidly and dynamically adapting to new challenges and opportunities.

- Explore novel operational applications of technology and other resources to advance tradecraft and achieve mission advantage.

- Continuously develop and adopt cutting-edge mission and business processes to improve intelligence capabilities and services.

Information Sharing and Safeguarding

5

Develop, enhance, integrate, and leverage IC capabilities and activities to improve collaboration and the lawful discovery, access, retrieval, and safeguarding of information.

The IC **Information Environment (IE)** includes the individuals, organizations, and information technology (IT) capabilities that collect, process, or share Sensitive Compartmented Information, or that, regardless of classification, are managed by the IC.

Mission success depends on the right people getting the right information at the right time to inform decisionmaking. To do this, the IC will take a cutting-edge approach to appropriately access information, regardless of where the information resides. Information that is better organized into appropriate data formats and tagged with metadata to increase its quality and usability will aid the transition to information-centered intelligence processes. An integrated IC IE will enable the IC to protect against external and insider threats, maintain the public trust, protect privacy and civil liberties, and carry out its mission. To do this, the IC must continue to adopt modern data management practices to make IC data discoverable, accessible, and usable through secure, modernized systems and standards.

TO MEET THIS OBJECTIVE, THE IC WILL:

>>>

- Identify, validate, prioritize, and address capability and policy requirements for the IC IE to enhance intelligence integration and operate as a secure, effective, and efficient IC enterprise.

- Increase the speed, portability, and trust of IC information system risk assessments to instill stakeholder confidence in the IC IE, and accelerate delivery of mission capability to users.

- Enhance foundational IC IT capabilities and infrastructure to increase mission effectiveness and reduce duplication.

- Develop and implement innovative means to manage, share, and protect intelligence information in accordance with law and policy.

- Leverage advanced analytics with modern data extraction, correlation, and enrichment capabilities to maximize the value of IC data.

22

Partnerships

6

Optimize partnerships to enhance intelligence and better inform decisionmaking.

Partners consist of organizations and entities working with us to advance national security priorities, including the U.S. military, our allies, foreign intelligence and security services, other federal departments and agencies, as well as state, local, and tribal officials and private sector entities, as appropriate.

The IC's partnerships are fundamental to our national security. Effectively leveraging their collective capabilities, data, expertise, and insights make our partners force multipliers. The IC will optimize existing partnerships and forge new relationships to enhance intelligence and inform decisions.

TO MEET THIS OBJECTIVE, THE IC WILL:

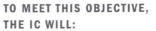

- Deepen mutual understanding and collaboration among partners to effectively inform decisions and enable action.

- Strengthen existing and develop new partnerships to increase access to information to meet mission needs, in accordance with applicable law.

- Institutionalize a strategic approach to partner engagement to facilitate collaboration and understanding.

Privacy, Civil Liberties, and Transparency

7

Safeguard privacy and civil liberties and practice appropriate transparency to enhance accountability and public trust in all we do.

The Principles of Intelligence Transparency for the Intelligence Community provide general norms for the IC to follow in making information publicly available that enhances public understanding of intelligence activities while continuing to protect information when disclosure would harm national security.

The IC must be accountable to the American people in carrying out its national security mission in a way that upholds the country's values. The core principles of protecting privacy and civil liberties in our work and of providing appropriate transparency about our work, both internally and to the public, must be integrated into the IC's programs and activities. Doing so is necessary to earn and retain public trust in the IC, which directly impacts IC authorities, capabilities, and resources. Mission success depends on the IC's commitment to these core principles.

TO MEET THIS OBJECTIVE, THE IC WILL:

>>>>>>>>>>>>>>>>>>>>>>>>>>>>>>>>>

- Incorporate privacy and civil liberties requirements into IC policy and programs to ensure that national values inform the intelligence mission.

- Engage proactively with oversight institutions and our partners to enhance public understanding and trust in the IC.

- Practice and promote appropriate transparency in the IC to make information publicly available without jeopardizing national security.

24

PRINCIPLES
OF
INTELLIGENCE TRANSPARENCY
FOR THE
INTELLIGENCE COMMUNITY

The Principles of Intelligence Transparency for the Intelligence Community (IC) are intended to facilitate IC decisions on making information publicly available in a manner that enhances public understanding of intelligence activities, while continuing to protect information when disclosure would harm national security. These Principles do not modify or supersede applicable laws, executive orders, and directives, including Executive Order 13526, Classified National Security Information. Instead, they articulate the general norms that elements of the IC should follow in implementing those authorities and requirements.

The Intelligence Community will:

1. Provide appropriate transparency to enhance public understanding about:
   a. the IC's mission and what the IC does to accomplish it (including its structure and effectiveness);
   b. the laws, directives, authorities, and policies that govern the IC's activities; and
   c. the compliance and oversight framework that ensures intelligence activities are conducted in accordance with applicable rules.

2. Be proactive and clear in making information publicly available through authorized channels, including taking affirmative steps to:
   a. provide timely transparency on matters of public interest;
   b. prepare information with sufficient clarity and context, so that it is readily understandable;
   c. make information accessible to the public through a range of communications channels, such as those enabled by new technology;
   d. engage with stakeholders to better explain information and to understand diverse perspectives; and
   e. in appropriate circumstances, describe why information cannot be made public.

3. In protecting information about intelligence sources, methods, and activities from unauthorized disclosure, ensure that IC professionals consistently and diligently execute their responsibilities to:
   a. classify only that information which, if disclosed without authorization, could be expected to cause identifiable or describable damage to the national security;
   b. never classify information to conceal violations of law, inefficiency, or administrative error, or to prevent embarrassment;
   c. distinguish, through portion marking and similar means, classified and unclassified information; and
   d. consider the public interest to the maximum extent feasible when making classification determinations, while continuing to protect information as necessary to maintain intelligence effectiveness, protect the safety of those who work for or with the IC, or otherwise protect national security.

4. Align IC roles, resources, processes, and policies to support robust implementation of these principles, consistent with applicable laws, executive orders, and directives.

25

Factors Affecting IC Performance:
Accomplishments, Risks, and Challenges

The IC is an increasingly integrated intelligence enterprise working toward the common vision of a Nation made more secure by a fully integrated, agile, resilient, and innovative IC that exemplifies America's values. To this end, the IC has made significant accomplishments towards NIS objectives, but much work remains.

Accomplishments

Through integration of effort, workforce initiatives, IC partnerships, transparency, and technological innovation, IC leaders and managers have promoted a more efficient, effective, and agile intelligence enterprise that enables the United States to safeguard our national interests in a challenging world.

Intelligence Integration

Increased intelligence integration has enabled the IC to better optimize mutually supporting collection and analysis activities, and more effectively manage its resources, resulting in intelligence support that has been critical to successful military, diplomatic, humanitarian, and other relevant operations. Examples of integrated IC mission successes include the monitoring of the Iran nuclear program, the investigation of downed Malaysian Airlines Flight MH-17, the monitoring of North Korean nuclear weapons development, and the response to the Ebola virus outbreak in West Africa.

IC Workforce

IC mission success is enabled by an inclusive work environment and a talented and diverse workforce with opportunities to cultivate career growth through a continued focus on developing and leveraging diversity. As a result, the IC has consistently been ranked at the top of the best places to work in the Federal Government.

Through the IC Joint Duty Program, thousands of civilian personnel have gained expertise, serving alongside their colleagues in partner agencies, broadening their professional development,

enhancing collaboration and information sharing, and promoting transparency and cooperation. Through this enriching program, our intelligence professionals have grown and gained additional experience and expertise and helped accomplish the IC's mission.

IC Partnerships

Significant progress has been made in building capacity, standardizing practices, and sharing information with partners in and outside the United States to help defend against and respond to foreign and foreign-inspired threats to U.S. interests, both at home and abroad. The IC advanced intelligence sharing with foreign partners, notably in countering terrorism and supporting military operations, and engaged in intelligence sharing activities with key domestic partners.

Transparency

The *Principles of Intelligence Transparency for the IC* facilitated decisions on making information publicly available to promote general understanding of intelligence activities, while continuing to protect information when disclosure would harm national security. As a result, the IC established a publicly available online repository for declassified documents, official statements, speeches, and testimony and has officially released thousands of pages of documents and posted them in this repository.

Technological Innovation

By deploying new scientific discoveries and technologies, nurturing innovative thought, and improving tradecraft and processes, the IC has achieved greater mission advantage on important issues. For example, IC leaders and managers have promoted a culture of collaboration and integration along with unification of intelligence activities to deliver shared IT services and capabilities across the IC. This integrated approach will enable the IC to appropriately access information and tools, regardless of where in the IC they reside. Information that is better organized and enriched by metadata will enable the transition to information-centered intelligence processes, while streamlining integrated data, applications, and services.

A pioneering program has developed new methods for generating accurate and timely probabilistic forecasts on a wide range of intelligence questions. These forecasts have been used for National Intelligence Estimates and other intelligence products. To improve tradecraft, machine learning research has led to new automated methods for forecasting political instability from social media, news, financial data, web search queries, and thousands of other data streams. These innovative applications are beginning to revolutionize the future of intelligence to better inform our decisionmakers.

Risks and Challenges

The NIS addresses various sources of risk and challenges that IC elements are called upon to mitigate. Risk is an uncertain event or condition that has a negative effect on the IC's ability to accomplish its mission. Risk factors may have internal (areas the IC can control) and/or external (areas beyond the IC's control) causes, requiring tailored mitigation strategies.

- Strategic risk refers to factors that affect the IC's ability to provide sufficient intelligence to inform decisions on national security issues. Factors such as potentially degraded operational environments and the number and complexity of national security threats, together with resource constraints, may challenge the IC's ability to fully monitor situations and warn policymakers of all developments. The IC needs to pursue new approaches to better identify and communicate areas to accept risk while maintaining an anticipatory and agile posture against emerging threats.

- Institutional risk refers to the factors that affect the IC's ability to execute effective mission and business management practices. For example, the likelihood of future unauthorized disclosures is a known risk that, if realized, may negatively affect intelligence collection, relations with domestic and foreign partners, and public trust. Better governance, auditing, and security procedures are needed to mitigate the risk and minimize the impact. The IC must attract and retain the right, trusted, agile workforce that possesses skills such as the critical analytic, scientific, technological, engineering, math, cyber, and foreign language skills required to support current and future mission challenges. This includes continuing to make progress recruiting and hiring a more diverse workforce more comparable to external benchmarks, such as those of the Federal workforce, the private sector workforce, and the U.S. population.

- Programmatic (fiscal) risk refers to the consequences of losing IC capabilities and resources due to unplanned or unforeseen factors that impact the effective use of available funds. Continued federal budget uncertainty strains the IC's ability to make deliberative and responsive resource decisions. The outcome may be overextended budgets or lack of cost-effective solutions to address intelligence issues. The IC needs to develop methods to efficiently shift resources to mitigate programmatic (fiscal) risk and avoid loss of vital programs, capabilities, and resource investments.

- Technological risk refers to the factors that affect the IC's scientific and innovative methods, practices, tools, and skills. Some factors are known, while others may be unforeseen. An inability to stay current with rapid changes in technology and industry standards may affect the IC's competitive advantage. Mitigation may require improving data strategies, software, and infrastructure to retain state-of-the-art capabilities.

Organization of the Intelligence Community

The Intelligence Community is an integrated enterprise comprised of 17 Executive Branch agencies and organizations (generally referred to as "IC elements") that conduct a variety of intelligence activities and work together to promote national security. The DNI is the leader of the IC and sets IC strategic priorities through the National Intelligence Strategy. Each IC member contributes through the execution of its organization's mission in accordance with statutory responsibilities.

The IC is comprised of the following 17 elements:

- Two independent agencies – the Office of the Director of National Intelligence (ODNI) and the Central Intelligence Agency (CIA);

- Eight Department of Defense elements – the Defense Intelligence Agency (DIA), the National Security Agency (NSA), the National Geospatial-Intelligence Agency (NGA), the National Reconnaissance Office (NRO), and the intelligence and counterintelligence elements of the military services—U.S. Air Force Intelligence, U.S. Navy Intelligence, U.S. Army Intelligence, and U.S. Marine Corps Intelligence, which also receive guidance and oversight from the Under Secretary of Defense for Intelligence (USDI).

- Seven elements of other departments and agencies – the Department of Energy's Office of Intelligence and Counterintelligence; the Department of Homeland Security's Office of Intelligence and Analysis and the intelligence and counterintelligence elements of the U.S. Coast Guard; the Department of Justice's Federal Bureau of Investigation and the Drug Enforcement Administration's Office of National Security Intelligence; the Department of State's Bureau of Intelligence and Research; and the Department of the Treasury's Office of Intelligence and Analysis.

In addition to collection, analysis, and production, IC elements serve in other roles. Functional managers oversee and coordinate a specific intelligence discipline or capability and advise the DNI on the performance of their functions within and across IC elements. National Intelligence Managers serve as the DNI's principal advisers on all aspects of intelligence collection, analysis, and counterintelligence against a specific area of concern. Program managers are IC element heads responsible for executing the mission and overseeing their elements' budget activities. IC enterprise managers align capabilities and business functions to enable the mission.

28

Implementing the
National Intelligence Strategy

The DNI, through the Office of the Director of National Intelligence, provides the IC with overarching oversight, direction, guidance, and coordination. IC elements execute their missions consistent with their statutory authorities. All members of the IC workforce are responsible for understanding how they contribute to the mission of the IC and executing their specific role to the best of their ability, while safeguarding privacy and civil liberties and practicing appropriate transparency.

DNI

Drive the Strategic Direction for the IC.
Through the NIS, the DNI sets strategic direction for the IC, bringing together the IC elements to address challenges that individual elements cannot solve on their own. The DNI provides direction for establishing and sustaining capabilities to enable mission success.

Lead Intelligence Integration. Under the direction of the DNI, the core mission of the ODNI is to lead and support IC integration; delivering insights, driving capabilities, and investing in the future. IC governance is the management of mission and enterprise activities through intelligence integration. Intelligence integration means coordinating and synchronizing collection, analysis, and counterintelligence so that they are fused, effectively operating as one team. The DNI establishes policies and standards to enable intelligence integration.

Enable IC Mission Execution. The DNI leads IC mission execution through decisionmaking bodies, IC strategies, IC budget and resource management, development of IC capabilities, information sharing and safeguarding, and partnering with domestic and foreign partners.

Direct the IC's Budget. The National Intelligence Program (NIP) is the IC's budget. The NIS serves as the DNI's mechanism to align NIP resources and report resource expenditures and performance to Congress. The DNI leads an IC-wide effort to develop an integrated NIP budget, maintaining strategic focus and cross-IC budget awareness, to assure that NIP investments best support national security goals and objectives. The DNI also participates in the development of annual budgets for the Department of Defense IC elements under the Military Intelligence Program (MIP). The DNI serves as a voice and advocate for the IC to Congress and other external entities.

IC Elements

Align Strategies, Plans, and Actions. The NIS informs the strategic plans of the IC elements. The mission and enterprise objectives in the NIS shall be incorporated and cascaded into the strategies and plans of the IC elements. Functional managers, National Intelligence Managers, Program managers, and IC enterprise managers will align, synchronize, and integrate their activities to the NIS.

Inform Resource Allocation. Program managers are heads of IC elements responsible for executing the missions of the IC. They oversee their element's budget activities, make investments in capabilities, and execute expenditures within the NIP and the MIP in the IC budget process. Each year they provide a strategic program briefing to the DNI and report to Congress on their respective programs.

Assess Outcomes. Activities, initiatives, and operations addressing NIS mission and enterprise objectives require constant and consistent evaluation. IC elements will document the specific impact of their activities, initiatives and operations, the extent to which this impact contributes to broader NIS objectives, and any factors that impede their ability to advance NIS objectives. Measuring progress toward meeting NIS objectives is crucial to improving the overall performance of the IC.

Conclusion

The NIS provides the IC with the DNI's strategic direction for the next four years, aligns IC priorities with other national strategies, and supports the IC's mission to provide timely, insightful, objective, and relevant intelligence and support to inform national security decisions and to protect our Nation and its interests. The IC must fully reflect the NIS in agency strategic plans, annual budget requests, and justifications for the NIP. The DNI will assess IC element proposals, projects, and programs toward the objectives of the NIS to realize the IC's vision of a Nation made more secure by a fully integrated, agile, resilient, and innovative Intelligence Community that exemplifies America's values.

" *We have to become much more agile, more innovative, more creative.* "

— Daniel R. Coats, *Director of National Intelligence*

30

PRINCIPLES of PROFESSIONAL ETHICS for the INTELLIGENCE COMMUNITY

As members of the intelligence profession, we conduct ourselves in accordance with certain basic principles. These principles are stated below, and reflect the standard of ethical conduct expected of all Intelligence Community personnel, regardless of individual role or agency affiliation. Many of these principles are also reflected in other documents that we look to for guidance, such as statements of core values, and the *Code of Conduct: Principles of Ethical Conduct for Government Officers and Employees*; it is nonetheless important for the Intelligence Community to set forth in a single statement the fundamental ethical principles that unite us and distinguish us as intelligence professionals.

MISSION We serve the American people, and understand that our mission requires selfless dedication to the security of our Nation.

TRUTH We seek the truth; speak truth to power; and obtain, analyze, and provide intelligence objectively.

LAWFULNESS We support and defend the Constitution, and comply with the laws of the United States, ensuring that we carry out our mission in a manner that respects privacy, civil liberties, and human rights obligations.

INTEGRITY We demonstrate integrity in our conduct, mindful that all our actions, whether public or not, should reflect positively on the Intelligence Community at large.

STEWARDSHIP We are responsible stewards of the public trust; we use intelligence authorities and resources prudently, protect intelligence sources and methods diligently, report wrongdoing through appropriate channels; and remain accountable to ourselves, our oversight institutions, and through those institutions, ultimately to the American people.

EXCELLENCE We seek to improve our performance and our craft continuously, share information responsibly, collaborate with our colleagues, and demonstrate innovation and agility when meeting new challenges.

DIVERSITY We embrace the diversity of our Nation, promote diversity and inclusion in our work force, and encourage diversity in our thinking.

31

The National Intelligence Strategy
of the United States of America

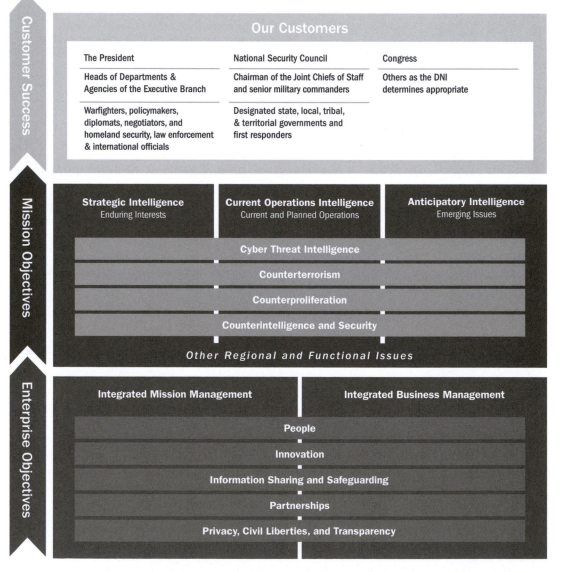

IC Vision *A Nation made more secure by a fully integrated, agile, resilient, and innovative Intelligence Community that exemplifies America's values.*

IC Mission *Provide timely, insightful, objective, and relevant intelligence and support to inform national security decisions and to protect our Nation and its interests.*

Customer Success

Our Customers

The President	National Security Council	Congress
Heads of Departments & Agencies of the Executive Branch	Chairman of the Joint Chiefs of Staff and senior military commanders	Others as the DNI determines appropriate
Warfighters, policymakers, diplomats, negotiators, and homeland security, law enforcement & international officials	Designated state, local, tribal, & territorial governments and first responders	

Mission Objectives

Strategic Intelligence Enduring Interests	**Current Operations Intelligence** Current and Planned Operations	**Anticipatory Intelligence** Emerging Issues

Cyber Threat Intelligence

Counterterrorism

Counterproliferation

Counterintelligence and Security

Other Regional and Functional Issues

Enterprise Objectives

Integrated Mission Management	Integrated Business Management

People

Innovation

Information Sharing and Safeguarding

Partnerships

Privacy, Civil Liberties, and Transparency

APPENDIX B

U.S. Department of Homeland Security. 2015. *National Preparedness Goal*, Second Edition. Washington, DC: Department of Homeland Security.

National Preparedness Goal

Second Edition
September 2015

Homeland
Security

Table of Contents

Introduction

Preparedness is the shared responsibility of our entire nation. The whole community contributes, beginning with individuals and communities, the private and nonprofit sectors, faith-based organizations, and all governments (local, regional/metropolitan, state, tribal[1], territorial, insular area[2], and Federal). This second edition of the National Preparedness Goal reflects the insights and lessons learned from four years of real world events and implementation of the National Preparedness System.[3]

We describe our security and resilience posture through the core capabilities (see Table 1) that are necessary to deal with the risks we face. We use an integrated, layered, and all-of-Nation approach as our foundation for building and sustaining core capabilities and preparing to deliver them effectively. The National Preparedness Goal is:

> A secure and resilient Nation with the capabilities required across the whole community to prevent, protect against, mitigate, respond to, and recover from the threats and hazards that pose the greatest risk.

Using the core capabilities, we achieve the National Preparedness Goal by:

- Preventing, avoiding, or stopping a threatened or an actual act of terrorism.

- Protecting our citizens, residents, visitors, assets, systems, and networks against the greatest threats and hazards in a manner that allows our interests, aspirations, and way of life to thrive.

- Mitigating the loss of life and property by lessening the impact of future disasters.

- Responding quickly to save lives, protect property and the environment, and meet basic human needs in the aftermath of an incident.

- Recovering through a focus on the timely restoration, strengthening, and revitalization of infrastructure, housing, and the economy, as well as the health, social, cultural, historic, and environmental fabric of communities affected by an incident.

The core capabilities contained in the Goal are the distinct critical elements necessary for our success. They are highly interdependent and require us to use existing preparedness networks and activities, coordinate and unify efforts, improve training and exercise programs, promote innovation, leverage and enhance our science and technology capacity, and ensure that administrative, finance, and logistics systems are in place to support these capabilities. The core capabilities serve as both preparedness tools and a means of structured implementation. All manner of incidents across the whole community have proven the usefulness of the core capabilities and the coordinating structures that sustain and deliver them. These range from

---

[1] The Federal Government recognizes that the tribal right of self-government flows from the inherent sovereignty of American Indian and Alaska Native Tribes as nations and that federally recognized tribes have a unique and direct relationship with the Federal Government.

[2] Per the Stafford Act, insular areas include Guam, the Commonwealth of the Northern Mariana Islands, American Samoa, and the U.S. Virgin Islands. Other statutes or departments and agencies may define the term insular area differently.

[3] The National Preparedness System outlines an organized process for the whole community to move forward with its preparedness activities and achieve the National Preparedness Goal.

localized and regional incidents, larger events involving both Stafford Act[4] disaster and emergency declarations, and operations conducted under other authorities (e.g., response to an emerging infectious disease outbreak).

Individual and community preparedness is fundamental to our National success. Providing individuals and communities with information and resources will facilitate actions to adapt to and withstand an emergency or disaster. As we have seen in tragic incidents both at home and abroad, anyone can contribute to safeguarding the Nation from harm. Our national resilience can be improved, for example, by raising awareness of the techniques that can save lives through such basic actions as stopping life-threatening bleeding. By providing the necessary knowledge and skills, we seek to enable the whole community to contribute to and benefit from national preparedness. Whole community contributors include children[5]; older adults; individuals with disabilities and others with access and functional needs; those from religious, racial, and ethnically diverse backgrounds; people with limited English proficiency; and owners of animals including household pets and service animals. Their needs and contributions must be integrated into our efforts. Each community contributes to the Goal by individually preparing for the risks that are most relevant and urgent for them individually. By empowering individuals and communities with knowledge and skills they can contribute to achieving the National Preparedness Goal.

We continue to make progress in building and sustaining our national preparedness. The Goal builds on these achievements, but our aspirations must be even higher to match the greatest risks facing our Nation. As we prepare for these challenges, our core capabilities will evolve to meet those challenges.

---

[4] The Robert T. Stafford Disaster Relief and Emergency Assistance Act (Stafford Act) authorizes the President to provide financial and other assistance to local, state, tribal, territorial, and insular area governments, as well as Federal agencies, to support Response and Recovery efforts in the wake of emergency or major disaster declarations.

[5] Children require a unique set of considerations across the core capabilities contained within this document. Their needs must be taken into consideration as part of any integrated planning effort.

2

Core Capabilities

Overview

Core capabilities are essential for the execution of each of the five mission areas: Prevention, Protection, Mitigation, Response, and Recovery (see Table 1). The core capabilities are not exclusive to any single government or organization, but rather require the combined efforts of the whole community.

Table 1: Core Capabilities by Mission Area[6]

Prevention	Protection	Mitigation	Response	Recovery
Planning				
Public Information and Warning				
Operational Coordination				
Intelligence and Information Sharing		Community Resilience	Infrastructure Systems	
Interdiction and Disruption		Long-term Vulnerability Reduction	Critical Transportation	Economic Recovery
Screening, Search, and Detection		Risk and Disaster Resilience Assessment	Environmental Response/Health and Safety	Health and Social Services
Forensics and Attribution	Access Control and Identity Verification	Threats and Hazards Identification	Fatality Management Services	Housing
	Cybersecurity		Fire Management and Suppression	Natural and Cultural Resources
	Physical Protective Measures		Logistics and Supply Chain Management	
	Risk Management for Protection Programs and Activities		Mass Care Services	
	Supply Chain Integrity and Security		Mass Search and Rescue Operations	
			On-scene Security, Protection, and Law Enforcement	
			Operational Communications	
			Public Health, Healthcare, and Emergency Medical Services	
			Situational Assessment	

---

[6] Planning, Public Information and Warning, and Operational Coordination are common to all mission areas.

3

These five mission areas serve as an aid in organizing our national preparedness activities and enabling integration and coordination across core capabilities. The mission areas are interrelated and require collaboration in order to be effective. The National Planning Frameworks and Federal Interagency Operational Plans expand on these relationships, to include how the mission areas and core capabilities are used to achieve the goal of a secure and resilient nation.

Three core capabilities: Planning, Public Information and Warning, and Operational Coordination span all five mission areas. They serve to unify the mission areas and, in many ways, are necessary for the success of the remaining core capabilities. Additionally, a number of core capabilities directly involve more than one mission area and are listed in each mission area as appropriate.

The core capabilities, like the risks we face, are not static. They will be vetted and refined, taking into consideration the evolving risk and changing resource requirements. Further, there is an expectation that each of the core capabilities will leverage advances in science and technology and be improved through post-event evaluation and assessment.

Risk and the Core Capabilities

Understanding the greatest risks to the Nation's security and resilience is a critical step in identifying the core capabilities. All levels of government and the whole community should assess and present risk in a similar manner to provide a common understanding of the threats and hazards confronting our Nation. The information gathered during a risk assessment also enables a prioritization of preparedness efforts and an ability to identify our capability requirements across the whole community.

The Strategic National Risk Assessment indicates that a wide range of threats and hazards continue to pose a significant risk to the Nation, affirming the need for an all-hazards, capability-based approach to preparedness planning. Key findings include:

- Natural hazards, including hurricanes, earthquakes, tornadoes, droughts, wildfires, winter storms, and floods, present a significant and varied risk across the country. Climate change has the potential to cause the consequence of weather-related hazards to become more severe.

- A virulent strain of pandemic influenza could kill hundreds of thousands of Americans, affect millions more, and result in economic loss. Additional human and animal infectious diseases, including those undiscovered, may present significant risks.

- Technological and accidental hazards, such as transportation system failures, dam failures, chemical spills or releases, have the potential to cause extensive fatalities and severe economic impacts. In addition, these hazards may increase due to aging infrastructure.

- Terrorist organizations or affiliates may seek to acquire, build, and use weapons of mass destruction (WMD). Conventional terrorist attacks, including those by "lone actors" employing physical threats such as explosives and armed attacks, present a continued risk to the Nation.

- Cyber-attacks can have catastrophic consequences, which in turn, can lead to other hazards, such as power grid failures or financial system failures. These cascading hazards increase the potential impact of cyber incidents. Cybersecurity threats exploit the increased complexity

4

and connectivity of critical infrastructure systems, placing the Nation's security, economy, and public safety and health at risk.

- Some incidents, such as explosives attacks or earthquakes, generally cause more localized impacts, while other incidents, such as human pandemics, may cause impacts that are dispersed throughout the Nation, thus creating different types of impacts for preparedness planners to consider.

In addition to these findings, climate change has the potential to adversely impact a number of threats and hazards. Rising sea levels, increasingly powerful storms, and heavier downpours are already contributing to an increased risk of flooding. Droughts and wildfires are becoming more frequent and severe in some areas of the country.

Cybersecurity poses its own unique challenges. In addition to the risk that cyber-threats pose to the nation, cybersecurity represents a core capability integral to preparedness efforts across the whole community. In order to meet the threat, preparedness planners must not only consider the unique core capability outlined in the Protection mission area, but must also consider integrating cyber preparedness throughout core capabilities in every mission area.

These findings supported the update of the core capabilities. Additionally, the Response and Recovery mission areas go one step further by focusing on a set of core capabilities based on the impact of a series of cascading incidents. Such incidents would likely stress the abilities of our Nation. A developed set of planning factors, intended to mimic this cascading incident and identify the necessary core capabilities, draws upon three hazards identified by the Strategic National Risk Assessment (i.e., large-scale earthquake, major hurricane, WMD attack).

The risks faced by a community can directly impact those responsible for delivering core capabilities. The whole community must maintain the ability to conduct mission essential functions during an actual hazard or incident to ensure delivery of core capabilities for all mission areas. The scope and magnitude of a catastrophic incident may result in a resource scarce environment. Because such incidents may affect an organization's ability to provide assets, assistance, and services, continuity planning and operations are an inherent component of each core capability and the coordinating structures that provide them. Continuity operations increase resilience and the likelihood that organizations can perform essential functions and deliver core capabilities that support the mission areas.

Mission Area: Prevention

Prevention includes those capabilities necessary to avoid, prevent, or stop a threatened or actual act of terrorism. Unlike other mission areas, which are all-hazards by design, Prevention core capabilities are focused specifically on imminent terrorist threats, including on-going attacks or stopping imminent follow-on attacks.

In addition, preventing an imminent terrorist threat will trigger a robust counterterrorism response wherein all instruments of national power may be used to resolve threats and save lives. Prevention also includes activities such as intelligence, law enforcement, and homeland defense as examples of activities conducted to address and resolve the threat.

The terrorist threat is dynamic and complex, and combating it is not the sole responsibility of a single entity or community. Ensuring the security of the homeland requires terrorism prevention through extensive collaboration with government and nongovernmental entities, international partners, and the private sector. We will foster a rapid, coordinated, all-of-Nation, effective terrorism prevention effort that reflects the full range of capabilities critical to avoid, prevent, or stop a threatened or actual act of terrorism in the homeland.

The Prevention mission area relies on ongoing support activities from across all mission areas that prepare the whole community to execute the core capabilities for preventing an imminent terrorist threat. These activities include information sharing efforts that directly support local communities in preventing terrorism and other activities that are precursors or indicators of terrorist activity and violent extremism.

Table 2 defines and details the Prevention core capabilities and the preliminary targets associated with each.

Table 2: Prevention Mission Area Core Capabilities and Preliminary Targets

Prevention Mission Area Core Capabilities and Preliminary Targets	
Planning	Conduct a systematic process engaging the whole community as appropriate in the development of executable strategic, operational, and/or tactical-level approaches to meet defined objectives.
1. Identify critical objectives during the planning process, provide a complete and integrated picture of the sequence and scope of the tasks to achieve the objectives, and ensure the objectives are implementable within the time frame contemplated within the plan using available resources for prevention-related plans. 2. Develop and execute appropriate courses of action in coordination with local, state, tribal, territorial, Federal, and private sector entities in order to prevent an imminent terrorist attack within the United States.	
Public Information and Warning	Deliver coordinated, prompt, reliable, and actionable information to the whole community through the use of clear, consistent, accessible, and culturally and linguistically appropriate methods to effectively relay information regarding any threat or hazard, as well as the actions being taken and the assistance being made available, as appropriate.
1. Share prompt and actionable messages, to include National Terrorism Advisory System alerts, with the public and other stakeholders, as appropriate, to aid in the prevention of imminent or follow-on terrorist attacks, consistent with the timelines specified by existing processes and protocols. 2. Provide public awareness information to inform the general public on how to identify and provide terrorism-related information to the appropriate law enforcement authorities, thereby enabling the public to act as a force multiplier in the prevention of imminent or follow-on acts of terrorism.	
Operational Coordination	Establish and maintain a unified and coordinated operational structure and process that appropriately integrates all critical stakeholders and supports the execution of core capabilities.
1. Execute operations with functional and integrated communications among appropriate entities to prevent initial or follow-on terrorist attacks within the United States in accordance with established protocols.	

6

Prevention Mission Area Core Capabilities and Preliminary Targets	
Forensics and Attribution	Conduct forensic analysis and attribute terrorist acts (including the means and methods of terrorism) to their source, to include forensic analysis as well as attribution for an attack and for the preparation for an attack in an effort to prevent initial or follow-on acts and/or swiftly develop counter-options.

1. Prioritize physical evidence collection and analysis to assist in preventing initial or follow-on terrorist acts.
2. Prioritize chemical, biological, radiological, nuclear, and explosive (CBRNE) material (bulk and trace) collection and analysis to assist in preventing initial or follow-on terrorist acts.
3. Prioritize biometric collection and analysis to assist in preventing initial or follow-on terrorist acts.
4. Prioritize digital media, network exploitation, and cyber technical analysis to assist in preventing initial or follow-on terrorist acts.

Intelligence and Information Sharing	Provide timely, accurate, and actionable information resulting from the planning, direction, collection, exploitation, processing, analysis, production, dissemination, evaluation, and feedback of available information concerning physical and cyber threats to the United States, its people, property, or interests; the development, proliferation, or use of WMDs; or any other matter bearing on U.S. national or homeland security by local, state, tribal, territorial, Federal, and other stakeholders. Information sharing is the ability to exchange intelligence, information, data, or knowledge among government or private sector entities, as appropriate.

1. Anticipate and identify emerging and/or imminent threats through the intelligence cycle.
2. Share relevant, timely, and actionable information and analysis with local, state, tribal, territorial, Federal, private sector, and international partners and develop and disseminate appropriate classified/unclassified products.
3. Ensure local, state, tribal, territorial, Federal, and private sector partners possess or have access to a mechanism to submit terrorism-related information and/or suspicious activity reports to law enforcement.

Interdiction and Disruption	Delay, divert, intercept, halt, apprehend, or secure threats and/or hazards.

1. Maximize our ability to interdict specific conveyances, cargo, and persons associated with an imminent terrorist threat or act in the land, air, and maritime domains to prevent entry into the United States or to prevent an incident from occurring in the Nation.
2. Conduct operations to render safe and dispose of CBRNE hazards in multiple locations and in all environments, consistent with established protocols.
3. Prevent terrorism financial/material support from reaching its target, consistent with established protocols.
4. Prevent terrorist acquisition of and the transfer of CBRNE materials, precursors, and related technology, consistent with established protocols.
5. Conduct tactical counterterrorism operations in multiple locations and in all environments, consistent with established protocols.

7

Prevention Mission Area Core Capabilities and Preliminary Targets	
Screening, Search, and Detection	Identify, discover, or locate threats and/or hazards through active and passive surveillance and search procedures. This may include the use of systematic examinations and assessments, biosurveillance, sensor technologies, or physical investigation and intelligence.

1. Maximize the screening of targeted cargo, conveyances, mail, baggage, and people associated with an imminent terrorist threat or act using technical, non-technical, intrusive, or non-intrusive means.
2. Initiate operations immediately to locate persons and networks associated with an imminent terrorist threat or act.
3. Conduct CBRNE search/detection operations in multiple locations and in all environments, consistent with established protocols.

Mission Area: Protection

Protection includes the capabilities to safeguard the homeland against acts of terrorism and man-made or natural disasters. It focuses on actions to protect our people, our vital interests, and our way of life.

Protection core capabilities are the product of diverse activities. These activities include defense against WMD threats; defense of agriculture and food; critical infrastructure protection[7]; protection of key leadership and events; border security; maritime security; transportation security; immigration security; and cybersecurity.

Table 3 defines and details the Protection core capabilities and the preliminary targets associated with each.

Table 3: Protection Mission Area Core Capabilities and Preliminary Targets

Protection Mission Area Core Capabilities and Preliminary Targets	
Planning	Conduct a systematic process engaging the whole community, as appropriate, in the development of executable strategic, operational, and/or tactical-level approaches to meet defined objectives.

1. Develop protection plans that identify critical objectives based on planning requirements, provide a complete and integrated picture of the sequence and scope of the tasks to achieve the planning objectives, and implement planning requirements within the time frame contemplated within the plan using available resources for protection-related plans.
2. Implement, exercise, and maintain plans to ensure continuity of operations.

Public Information and Warning	Deliver coordinated, prompt, reliable, and actionable information to the whole community through the use of clear, consistent, accessible, and culturally and linguistically appropriate methods to effectively relay information regarding any threat or hazard and, as appropriate, the actions being taken and the assistance being made available.

1. Use effective and accessible indication and warning systems to communicate significant hazards to involved operators, security officials, and the public (including alerts, detection capabilities, and other necessary and appropriate assets).

---

[7] See Critical Infrastructure in Appendix A for a full explanation.

8

Protection Mission Area Core Capabilities and Preliminary Targets	
Operational Coordination	Establish and maintain a unified and coordinated operational structure and process that appropriately integrates all critical stakeholders and supports the execution of core capabilities.

1. Establish and maintain partnership structures among Protection elements to support networking, planning, and coordination.

Access Control and Identity Verification	Apply and support necessary physical, technological, and cyber measures to control admittance to critical locations and systems.

1. Implement and maintain protocols to verify identity and authorize, grant, or deny physical and cyber access to specific locations, information, and networks.

Cybersecurity	Protect (and, if needed, restore) electronic communications systems, information, and services from damage, unauthorized use, and exploitation.

1. Implement risk-informed guidelines, regulations, and standards to ensure the security, reliability, integrity, and availability of critical information, records, and communications systems and services through collaborative cybersecurity initiatives and efforts.
2. Implement and maintain procedures to detect malicious activity and to conduct technical and investigative-based countermeasures, mitigations, and operations against malicious actors to counter existing and emerging cyber-based threats, consistent with established protocols.

Intelligence and Information Sharing	Provide timely, accurate, and actionable information resulting from the planning, direction, collection, exploitation, processing, analysis, production, dissemination, evaluation, and feedback of available information concerning threats to the United States, its people, property, or interests; the development, proliferation, or use of WMDs; or any other matter bearing on U.S. national or homeland security by local, state, tribal, territorial, Federal, and other stakeholders. Information sharing is the ability to exchange intelligence, information, data, or knowledge among government or private sector entities, as appropriate.

1. Anticipate and identify emerging and/or imminent threats through the intelligence cycle.
2. Share relevant, timely, and actionable information and analysis with local, state, tribal, territorial, Federal, private sector, and international partners and develop and disseminate appropriate classified/unclassified products.
3. Provide local, state, tribal, territorial, Federal, and private sector partners with or access to a mechanism to submit terrorism-related information and/or suspicious activity reports to law enforcement.

Interdiction and Disruption	Delay, divert, intercept, halt, apprehend, or secure threats and/or hazards.

1. Deter, detect, interdict, and protect against domestic and transnational criminal and terrorist activities that threaten the security of the homeland across key operational activities and critical infrastructure sectors.
2. Intercept the malicious movement and acquisition/transfer of CBRNE materials and related technologies.

9

Protection Mission Area Core Capabilities and Preliminary Targets	
Physical Protective Measures	Implement and maintain risk-informed countermeasures, and policies protecting people, borders, structures, materials, products, and systems associated with key operational activities and critical infrastructure sectors.

1. Identify, assess, and mitigate vulnerabilities to incidents through the deployment of physical protective measures.
2. Deploy protective measures commensurate with the risk of an incident and balanced with the complementary aims of enabling commerce and maintaining the civil rights of citizens.

Risk Management for Protection Programs and Activities	Identify, assess, and prioritize risks to inform Protection activities, countermeasures, and investments.

1. Ensure critical infrastructure sectors and Protection elements have and maintain risk assessment processes to identify and prioritize assets, systems, networks, and functions.
2. Ensure operational activities and critical infrastructure sectors have and maintain appropriate threat, vulnerability, and consequence tools to identify and assess threats, vulnerabilities, and consequences.

Screening, Search, and Detection	Identify, discover, or locate threats and/or hazards through active and passive surveillance and search procedures. This may include the use of systematic examinations and assessments, biosurveillance, sensor technologies, or physical investigation and intelligence.

1. Screen cargo, conveyances, mail, baggage, and people using information-based and physical screening technology and processes.
2. Detect WMD, traditional, and emerging threats and hazards of concern using:
    a. A laboratory diagnostic capability and the capacity for food, agricultural (plant/animal), environmental, medical products, and clinical samples
    b. Bio-surveillance systems
    c. CBRNE detection systems
    d. Trained healthcare, emergency medical, veterinary, and environmental laboratory professionals.

Supply Chain Integrity and Security	Strengthen the security and resilience of the supply chain.

1. Secure and make resilient key nodes, methods of transport between nodes, and materials in transit.

Mission Area: Mitigation

Mitigation includes those capabilities necessary to reduce loss of life and property by lessening the impact of disasters. It is focused on the premise that individuals, the private and nonprofit sectors, communities, critical infrastructure, and the Nation as a whole are made more resilient when the consequences and impacts, the duration, and the financial and human costs to respond to and recover from adverse incidents are all reduced.

Given the trend of increasing impacts from extreme events and catastrophic incidents, hazard mitigation stands as a critical linchpin to reduce or eliminate the long-term risks to life, property, and well-being. Spanning across community planning, housing, information systems, critical infrastructure, public health, healthcare, and future land use, Mitigation requires an understanding of the threats and hazards that, in turn, feed into the assessment of risk and disaster resilience in the community both now and in the future. The whole community, therefore, has a role in risk reduction, by recognizing, understanding, communicating, and planning for a community's future resilience. Mitigation links the long-term activities of the whole community to reduce or eliminate the risk of threats and hazards developing into disasters and the impacts of the disasters that occur.

Although Mitigation is the responsibility of the whole community, a great deal of mitigation activity occurs at the local level. Individual and community preparedness is fundamental to our success, as preparedness activities contribute to strengthening resilience and mitigate the impact of disasters through adaptability and capacity for rapid recovery. The assessment of risk and resilience must therefore begin at the community level and serve to inform our state, regional, and national planning. For risk information to result in specific risk reduction actions, leaders—whether elected in a jurisdiction, appointed in a given department, a nongovernmental director, a sector official, or in business or community—must have the ability to recognize, understand, communicate, and plan for a community's future resilience. The establishment of trusted relationships among leaders in a community prior to a disaster can greatly reduce the risks to life, property, the natural environment, and well-being. When these leaders are prepared, the whole community matures and becomes better prepared to reduce the risks over the long-term.

Table 4 defines and details the Mitigation core capabilities and the preliminary targets associated with each.

Table 4: Mitigation Mission Area Core Capabilities and Preliminary Targets

Mitigation Mission Area Core Capabilities and Preliminary Targets	
Planning	Conduct a systematic process engaging the whole community as appropriate in the development of executable strategic, operational, and/or tactical-level approaches to meet defined objectives.
1. Develop approved hazard mitigation plans that address relevant threats/hazards in accordance with the results of their risk assessment within all local, state, tribal, territorial, and Federal partners.	
Public Information and Warning	Deliver coordinated, prompt, reliable, and actionable information to the whole community through the use of clear, consistent, accessible, and culturally and linguistically appropriate methods to effectively relay information regarding any threat or hazard and, as appropriate, the actions being taken and the assistance being made available.
1. Communicate appropriate information, in an accessible manner, on the risks faced within a community after the conduct of a risk assessment.	
Operational Coordination	Establish and maintain a unified and coordinated operational structure and process that appropriately integrates all critical stakeholders and supports the execution of core capabilities.
1. Establish protocols to integrate mitigation data elements in support of operations with local, state, tribal, territorial, and insular area partners and in coordination with Federal agencies.	

Mitigation Mission Area Core Capabilities and Preliminary Targets	
Community Resilience	Enable the recognition, understanding, communication of, and planning for risk and empower individuals and communities to make informed risk management decisions necessary to adapt to, withstand, and quickly recover from future incidents.

1. Maximize the coverage of the U.S. population that has a localized, risk-informed mitigation plan developed through partnerships across the entire community.
2. Empower individuals and communities to make informed decisions to facilitate actions necessary to adapt to, withstand, and quickly recover from future incidents.

Long-term Vulnerability Reduction	Build and sustain resilient systems, communities, and critical infrastructure and key resources lifelines so as to reduce their vulnerability to natural, technological, and human-caused threats and hazards by lessening the likelihood, severity, and duration of the adverse consequences.

1. Achieve a measurable decrease in the long-term vulnerability of the Nation against current baselines amid a growing population base, changing climate conditions, increasing reliance upon information technology, and expanding infrastructure base.

Risk and Disaster Resilience Assessment	Assess risk and disaster resilience so that decision makers, responders, and community members can take informed action to reduce their entity's risk and increase their resilience.

1. Ensure that local, state, tribal, territorial, and insular area governments and the top 100 Metropolitan Statistical Areas (MSAs) complete a risk assessment that defines localized vulnerabilities and consequences associated with potential natural, technological, and human-caused threats and hazards to their natural, human, physical, cyber, and socioeconomic interests.

Threats and Hazards Identification	Identify the threats and hazards that occur in the geographic area; determine the frequency and magnitude; and incorporate this into analysis and planning processes so as to clearly understand the needs of a community or entity.

1. Identify the threats and hazards within and across local, state, tribal, territorial, and insular area governments, and the top 100 MSAs, in collaboration with the whole community, against a national standard based on sound science.

Mission Area: Response

Response includes those capabilities necessary to save lives, protect property and the environment, and meet basic human needs after an incident has occurred. It is focused on ensuring that the Nation is able to effectively respond to any threat or hazard, including those with cascading effects. Response emphasizes saving and sustaining lives, stabilizing the incident, rapidly meeting basic human needs, restoring basic services and technologies, restoring community functionality, providing universal accessibility, establishing a safe and secure environment, and supporting the transition to recovery.

Communities regularly deal with emergencies and disasters that have fewer impacts than those considered to be the greatest risk to the Nation. In addition, communities may have resident capacities to deal with the public's needs locally for many of these lesser incidents. Catastrophic

12

incidents require a much broader set of atypical partners to deliver equal access to the Response core capabilities other than those routinely addressed. Community involvement is a vital link to providing additional support to response personnel and may often be the primary source of manpower in the first hours and days after an incident. Because of this, community members should be encouraged to train, exercise, and partner with emergency management officials.

A catastrophic incident with cascading events may impact the execution of applicable laws and policies. Certain circumstances may trigger legal and policy exceptions that better aid delivery of core capabilities. Planners should identify applicable laws and policies with their respective counsel in the pre-planning phase.[8] These challenges should be identified during pre-incident planning to ensure they are accounted for during an incident.

Table 5 defines and details the Response core capabilities and the preliminary targets associated with each.

Table 5: Response Mission Area Core Capabilities and Preliminary Targets

Response Mission Area Core Capabilities and Preliminary Targets	
Planning	Conduct a systematic process engaging the whole community as appropriate in the development of executable strategic, operational, and/or tactical-level approaches to meet defined objectives.
1. Develop operational plans that adequately identify critical objectives based on the planning requirement, provide a complete and integrated picture of the sequence and scope of the tasks to achieve the objectives, and are implementable within the time frame contemplated in the plan using available resources.	
Public Information and Warning	Deliver coordinated, prompt, reliable, and actionable information to the whole community through the use of clear, consistent, accessible, and culturally and linguistically appropriate methods to effectively relay information regarding any threat or hazard and, as appropriate, the actions being taken and the assistance being made available.
1. Inform all affected segments of society by all means necessary, including accessible tools, of critical lifesaving and life-sustaining information to expedite the delivery of emergency services and aid the public to take protective actions. 2. Deliver credible and actionable messages to inform ongoing emergency services and the public about protective measures and other life-sustaining actions and facilitate the transition to recovery.	

---

[8] Given the scope and magnitude of a catastrophic incident, waivers, exceptions, and exemptions to policy, regulations, and laws may be available in order to save and sustain life, and to protect property and the environment. However, any such waivers, exceptions, and exemptions must be consistent with laws that preserve human and civil rights and protect individuals with disabilities and others with access and functional needs.

Response Mission Area Core Capabilities and Preliminary Targets

Operational Coordination	Establish and maintain a unified and coordinated operational structure and process that appropriately integrates all critical stakeholders and supports the execution of core capabilities.

1. Mobilize all critical resources and establish command, control, and coordination structures within the affected community and other coordinating bodies in surrounding communities and across the Nation and maintain as needed throughout the duration of an incident.
2. Enhance and maintain command, control, and coordination structures, consistent with the National Incident Management System (NIMS), to meet basic human needs, stabilize the incident, and transition to recovery.

Critical Transportation	Provide transportation (including infrastructure access and accessible transportation services) for response priority objectives, including the evacuation of people and animals, and the delivery of vital response personnel, equipment, and services into the affected areas.

1. Establish physical access through appropriate transportation corridors and deliver required resources to save lives and to meet the needs of disaster survivors.
2. Ensure basic human needs are met, stabilize the incident, transition into recovery for an affected area, and restore basic services and community functionality.
3. Clear debris from any route type (i.e., road, rail, airfield, port facility, waterway) to facilitate response operations.

Environmental Response/Health and Safety	Conduct appropriate measures to ensure the protection of the health and safety of the public and workers, as well as the environment, from all-hazards in support of responder operations and the affected communities.

1. Identify, assess, and mitigate worker health and safety hazards and disseminate health and safety guidance and resources to response and recovery workers.
2. Minimize public exposure to environmental hazards through assessment of the hazards and implementation of public protective actions.
3. Detect, assess, stabilize, and clean up releases of oil and hazardous materials into the environment, including buildings/structures, and properly manage waste.
4. Identify, evaluate, and implement measures to prevent and minimize impacts to the environment, natural and cultural resources, and historic properties from all-hazard emergencies and response operations.

Fatality Management Services	Provide fatality management services, including decedent remains recovery and victim identification, working with local, state, tribal, territorial, insular area, and Federal authorities to provide mortuary processes, temporary storage or permanent interment solutions, sharing information with mass care services for the purpose of reunifying family members and caregivers with missing persons/remains, and providing counseling to the bereaved.

1. Establish and maintain operations to recover a significant number of fatalities over a geographically dispersed area.

14

Response Mission Area Core Capabilities and Preliminary Targets

| **Fire Management and Suppression** | Provide structural, wildland, and specialized firefighting capabilities to manage and suppress fires of all types, kinds, and complexities while protecting the lives, property, and the environment in the affected area. |

1. Provide traditional first response or initial attack firefighting services.
2. Conduct expanded or extended attack firefighting and support operations through coordinated response of fire management and specialized fire suppression resources.
3. Ensure the coordinated deployment of appropriate local, regional, national, and international fire management and fire suppression resources to reinforce firefighting efforts and maintain an appropriate level of protection for subsequent fires.

| **Infrastructure Systems** | Stabilize critical infrastructure functions, minimize health and safety threats, and efficiently restore and revitalize systems and services to support a viable, resilient community. |

1. Decrease and stabilize immediate infrastructure threats to the affected population, to include survivors in the heavily-damaged zone, nearby communities that may be affected by cascading effects, and mass care support facilities and evacuation processing centers with a focus on life-sustainment and congregate care services.
2. Re-establish critical infrastructure within the affected areas to support ongoing emergency response operations, life sustainment, community functionality, and a transition to recovery.
3. Provide for the clearance, removal, and disposal of debris.
4. Formalize partnerships with governmental and private sector cyber incident or emergency response teams to accept, triage, and collaboratively respond to cascading impacts in an efficient manner.

| **Logistics and Supply Chain Management**[9] | Deliver essential commodities, equipment, and services in support of impacted communities and survivors, to include emergency power and fuel support, as well as the coordination of access to community staples. Synchronize logistics capabilities and enable the restoration of impacted supply chains. |

1. Mobilize and deliver governmental, nongovernmental, and private sector resources to save lives, sustain lives, meet basic human needs, stabilize the incident, and transition to recovery, to include moving and delivering resources and services to meet the needs of disaster survivors.
2. Enhance public and private resource and services support for an affected area.

| **Mass Care Services** | Provide life-sustaining and human services to the affected population, to include hydration, feeding, sheltering, temporary housing, evacuee support, reunification, and distribution of emergency supplies. |

1. Move and deliver resources and capabilities to meet the needs of disaster survivors, including individuals with access and functional needs.
2. Establish, staff, and equip emergency shelters and other temporary housing options (including accessible housing) for the affected population.
3. Move from congregate care to non-congregate care alternatives and provide relocation assistance or interim housing solutions for families unable to return to their pre-disaster homes.

[9] This replaces the previous "Public and Private Services and Resources" core capability.

15

Response Mission Area Core Capabilities and Preliminary Targets

Mass Search and Rescue Operations	Deliver traditional and atypical search and rescue capabilities, including personnel, services, animals, and assets to survivors in need, with the goal of saving the greatest number of endangered lives in the shortest time possible.

1. Conduct search and rescue operations to locate and rescue persons in distress.
2. Initiate community-based search and rescue support operations across a wide geographically dispersed area.
3. Ensure the synchronized deployment of local, regional, national, and international teams to reinforce ongoing search and rescue efforts and transition to recovery.

On-scene Security, Protection, and Law Enforcement	Ensure a safe and secure environment through law enforcement and related security and protection operations for people and communities located within affected areas and also for response personnel engaged in lifesaving and life-sustaining operations.

1. Establish a safe and secure environment in an affected area.
2. Provide and maintain on-scene security and meet the protection needs of the affected population over a geographically dispersed area while eliminating or mitigating the risk of further damage to persons, property, and the environment.

Operational Communications	Ensure the capacity for timely communications in support of security, situational awareness, and operations by any and all means available, among and between affected communities in the impact area and all response forces.

1. Ensure the capacity to communicate with both the emergency response community and the affected populations and establish interoperable voice and data communications between Federal, tribal, state, and local first responders.
2. Re-establish sufficient communications infrastructure within the affected areas to support ongoing life-sustaining activities, provide basic human needs, and transition to recovery.
3. Re-establish critical information networks, including cybersecurity information sharing networks, in order to inform situational awareness, enable incident response, and support the resiliency of key systems.

Public Health, Healthcare, and Emergency Medical Services	Provide lifesaving medical treatment via Emergency Medical Services and related operations and avoid additional disease and injury by providing targeted public health, medical, and behavioral health support, and products to all affected populations.

1. Deliver medical countermeasures to exposed populations.
2. Complete triage and initial stabilization of casualties and begin definitive care for those likely to survive their injuries and illness.
3. Return medical surge resources to pre-incident levels, complete health assessments, and identify recovery processes.

16

Response Mission Area Core Capabilities and Preliminary Targets	
Situational Assessment	Provide all decision makers with decision-relevant information regarding the nature and extent of the hazard, any cascading effects, and the status of the response.

1. Deliver information sufficient to inform decision making regarding immediate lifesaving and life-sustaining activities and engage governmental, private, and civic sector resources within and outside of the affected area to meet basic human needs and stabilize the incident.
2. Deliver enhanced information to reinforce ongoing lifesaving and life-sustaining activities, and engage governmental, private, and civic sector resources within and outside of the affected area to meet basic human needs, stabilize the incident, and transition to recovery.

Mission Area: Recovery

Recovery includes those capabilities necessary to assist communities affected by an incident to recover effectively. Support for recovery ensures a continuum of care for individuals to maintain and restore health, safety, independence and livelihoods, especially those who experience financial, emotional, and physical hardships. Successful recovery ensures that we emerge from any threat or hazard stronger and positioned to meet the needs of the future. Recovery capabilities support well-coordinated, transparent, and timely restoration, strengthening, and revitalization of infrastructure and housing; an economic base; health and social systems; and a revitalized cultural, historic, and environmental fabric.

The ability of a community to accelerate the recovery process begins with its efforts in pre-incident preparedness: increasing resilience; collaborative, inclusive planning; and developing capacity to manage disaster recovery effectively. Developing and maintaining Recovery core capabilities requires a multi-agency, interdisciplinary approach that engages the whole community, including a wide range of service and resource providers and stakeholders.

Whole community and government leaders have primary responsibility for planning and coordinating all aspects of their recovery and ensuring that organizations and individuals that play a key role in recovery are included and actively engaged. Following an incident, a well-coordinated management process allows recovery and community leaders to maintain open and transparent communication, share decision making, expand and engage traditional and non-traditional partners, identify needs and priorities more effectively, reallocate and share existing resources, and identify other potential resources and expertise from both inside and outside the community.

Following any incident, recovery efforts are an opportunity to leverage solutions that increase overall community resilience and capitalize on existing strengths, while addressing weaknesses that may have existed pre-incident. Lessons learned from the post-incident environment on establishing leadership, a coordinating structure, and developing whole community partnerships can help influence pre-incident planning and build capability for future incidents.

Table 6 defines and details the Recovery core capabilities and the preliminary targets associated with each.

Table 6: Recovery Mission Area Core Capabilities and Preliminary Targets

Recovery Mission Area Core Capabilities and Preliminary Targets	
Planning	Conduct a systematic process engaging the whole community as appropriate in the development of executable strategic, operational, and/or tactical-level approaches to meet defined objectives.

1. Convene the core of an inclusive planning team (identified pre-disaster), which will oversee disaster recovery planning.
2. Complete an initial recovery plan that provides an overall strategy and timeline, addresses all core capabilities, and integrates socioeconomic, demographic, accessibility, technology, and risk assessment considerations (including projected climate change impacts), which will be implemented in accordance with the timeline contained in the plan.

Public Information and Warning	Deliver coordinated, prompt, reliable, and actionable information to the whole community through the use of clear, consistent, accessible, and culturally and linguistically appropriate methods to effectively relay information regarding any threat or hazard and, as appropriate, the actions being taken and the assistance being made available.

1. Reach all populations within the community with effective actionable recovery-related public information messaging and communications that are accessible to people with disabilities and people with limited English proficiency, protect the health and safety of the affected population, help manage expectations, and ensure stakeholders have a clear understanding of available assistance and their roles and responsibilities.
2. Support affected populations and stakeholders with a system that provides appropriate, current information about any continued assistance, steady state resources for long-term impacts, and monitoring programs in an effective and accessible manner.

Operational Coordination	Establish and maintain a unified and coordinated operational structure and process that appropriately integrates all critical stakeholders and supports the execution of core capabilities.

1. Establish tiered, integrated leadership, and inclusive coordinating organizations that operate with a unity of effort and are supported by sufficient assessment and analysis to provide defined structure and decision-making processes for recovery activities.
2. Define the path and timeline for recovery leadership to achieve the jurisdiction's objectives that effectively coordinates and uses appropriate local, state, tribal, territorial, insular area, and Federal assistance, as well as nongovernmental and private sector resources. This plan is to be implemented within the established timeline.

Economic Recovery	Return economic and business activities (including food and agriculture) to a healthy state and develop new business and employment opportunities that result in an economically viable community.

1. Conduct a preliminary assessment of economic issues and identify potential inhibitors to fostering stabilization of the affected communities.
2. Ensure the community recovery and mitigation plan(s) incorporates economic revitalization and removes governmental inhibitors to post-disaster economic sustainability, while maintaining the civil rights of citizens.
3. Return affected area's economy within the specified time frame in the recovery plan.

Recovery Mission Area Core Capabilities and Preliminary Targets

Health and Social Services	Restore and improve health and social services capabilities and networks to promote the resilience, independence, health (including behavioral health), and well-being of the whole community.

1. Identify affected populations, groups and key partners in short-term, intermediate, and long-term recovery.
2. Complete an assessment of community health and social service needs, and prioritize these needs, including accessibility requirements, based on the whole community's input and participation in the recovery planning process, and develop a comprehensive recovery timeline.
3. Restore health care (including behavioral health), public health, and social services functions.
4. Restore and improve the resilience and sustainability of the health care system and social service capabilities and networks to promote the independence and well-being of community members in accordance with the specified recovery timeline.

Housing	Implement housing solutions that effectively support the needs of the whole community and contribute to its sustainability and resilience.

1. Assess preliminary housing impacts and needs, identify currently available options for temporary housing, and plan for permanent housing.
2. Ensure community housing recovery plans continue to address interim housing needs, assess options for permanent housing, and define a timeline for achieving a resilient, accessible, and sustainable housing market.
3. Establish a resilient and sustainable housing market that meets the needs of the community, including the need for accessible housing within the specified time frame in the recovery plan.

Infrastructure Systems	Stabilize critical infrastructure functions, minimize health and safety threats, and efficiently restore and revitalize systems and services to support a viable, resilient community.

1. Restore and sustain essential services (public and private) to maintain community functionality.
2. Develop a plan with a specified timeline for redeveloping community infrastructures to contribute to resiliency, accessibility, and sustainability.
3. Provide systems that meet the community needs while minimizing service disruption during restoration within the specified timeline in the recovery plan.

Recovery Mission Area Core Capabilities and Preliminary Targets

Natural and Cultural Resources	Protect natural and cultural resources and historic properties through appropriate planning, mitigation, response, and recovery actions to preserve, conserve, rehabilitate, and restore them consistent with post-disaster community priorities and best practices and in compliance with applicable environmental and historic preservation laws and executive orders.

1. Implement measures to protect and stabilize records and culturally significant documents, objects, and structures.

2. Mitigate the impacts to and stabilize the natural and cultural resources and conduct a preliminary assessment of the impacts that identifies protections that need to be in place during stabilization through recovery.

3. Complete an assessment of affected natural and cultural resources and develop a timeline for addressing these impacts in a sustainable and resilient manner.

4. Preserve natural and cultural resources as part of an overall community recovery that is achieved through the coordinated efforts of natural and cultural resource experts and the recovery team in accordance with the specified timeline in the recovery plan.

Conclusion and Next Steps

The Goal is designed to prepare our Nation for the risks that will severely stress our collective capabilities and resources. Each community contributes to the Goal by assessing and preparing for the risks that are most relevant and urgent for them individually, which in turn strengthens our collective security and resilience as a Nation. National preparedness is strengthened through collaboration and cooperation with international partners, including working closely with Canada and Mexico, with whom we share common borders.

The National Preparedness Goal is the cornerstone of implementing the National Preparedness System. Several National Preparedness System components contribute to building, sustaining, and delivering the core capabilities described in the National Preparedness Goal. These include:

- A National Planning System, which supports the integration of planning across all levels of government and the whole community to provide an agile, flexible, and accessible delivery of the core capabilities.

- A series of National Frameworks and Federal Interagency Operational Plans. The National Frameworks address the roles and responsibilities across the whole community to deliver the core capabilities. The Federal Interagency Operational Plans address the critical tasks, responsibilities, and resourcing, personnel, and sourcing requirements for the core capabilities.

- A National Preparedness Report, which provides a summary of the progress being made toward building, sustaining, and delivering the core capabilities described in the Goal. The annual National Preparedness Report provides an opportunity to measure the advancement the whole community has made in preparedness and to identify where challenges remain.

- A Campaign to Build and Sustain Preparedness, which provides an integrating structure for new and existing community-based, nonprofit, and private sector preparedness programs, research and development activities to include post-event evaluation of the use of science and technology tools, and preparedness assistance.

The results of these efforts inform current and future budget planning and decisions. Analysis of current performance against intended capabilities and associated performance measures can enable the whole community to individually and collectively determine necessary resource levels, inform resource allocation plans, and guide Federal preparedness assistance. This detailed information can augment assessments of budget implications across the preparedness enterprise. This approach allows for annual adjustments based on updated priorities and resource posture.

This National Preparedness Goal is a living document; regular reviews of the Goal will ensure consistency with existing and new policies, evolving conditions, and the National Incident Management System. These periodic reviews of the National Preparedness Goal will evaluate the Nation's progress toward building, sustaining, and delivering the core capabilities that are essential to a secure and resilient Nation.

21 |

Appendix A: Terms and Definitions

Access and Functional Needs: Persons who may have additional needs before, during and after an incident in functional areas, including but not limited to: maintaining health, independence, communication, transportation, support, services, self-determination, and medical care. Individuals in need of additional response assistance may include those who have disabilities; live in institutionalized settings; are older adults; are children; are from diverse cultures; have limited English proficiency or are non-English speaking; or are transportation disadvantaged.

All-of-Nation: See Whole Community.

All Hazard: A threat or an incident, natural or manmade, that warrants action to protect life, property, the environment, and public health or safety, and to minimize disruptions of government, social, or economic activities. It includes natural disasters, cyber incidents, industrial accidents, pandemics, acts of terrorism, sabotage, and destructive criminal activity targeting critical infrastructure. This also includes the effects climate change has on the threats and hazards.

Animal: Animals include household pets, service and assistance animals, working dogs, livestock, wildlife, exotic animals, zoo animals, research animals, and animals housed in shelters, rescue organizations, breeding facilities, and sanctuaries.

Community: Unified groups that share goals, values, or purposes; they may exist within geographic boundaries or unite geographically dispersed individuals. Communities bring people together in different ways for different reasons, but each provides opportunities for sharing information and promoting collective action.

Coordinating Structures: Groups composed of representatives from multiple departments or agencies, public and/or private sector organizations, or a combination of these for the purpose of facilitating the preparedness and delivery of capabilities. They share information and provide guidance, support, and integration to aid in the preparedness of the whole community and building resilience at the local, regional, and national levels.

Core Capabilities: Distinct critical elements necessary to achieve the National Preparedness Goal.

Critical Infrastructure: Systems and assets, whether physical or virtual, so vital to the United States that the incapacity or destruction of such systems and assets would have a debilitating impact on security, national economic security, national public health or safety, or any combination of those matters. The Nation's critical infrastructure is composed of 16 sectors: chemical; commercial facilities; communications; critical manufacturing; dams; defense industrial base; emergency services; energy; financial services; food and agriculture; government facilities; healthcare and public health; information technology; nuclear reactors, material, and waste; transportation systems; and water and wastewater systems.

Cultural Resources: Aspects of a cultural system that are valued by or significantly representative of a culture or that contain significant information about a culture.

Cybersecurity: The process of protecting information by preventing, detecting, and responding to attacks.

Disability: A physical or mental impairment that substantially limits one or more major life activities of such individual; a record of such an impairment; or being regarded as having such an impairment. This does not apply to impairments that are transitory and minor. A transitory impairment is an impairment with an actual or expected duration of six months or less.

Imminent Threat: Intelligence or operational information that warns of a credible, specific, and impending terrorist threat or ongoing attack against the United States.

Intelligence Cycle: The process of developing raw information into finished intelligence for policymakers, military commanders, law enforcement partners, and other consumers to use in making decisions. The cycle is highly dynamic and continuous. For the purposes of the National Prevention Framework, there are six steps in the intelligence cycle: planning and direction (establish the intelligence requirements of the consumer); collection (gather the raw data required to produce the desired finished product); processing and exploitation (convert the raw data into comprehensible form that is usable for producing the finished product); analysis and production (integrate, evaluate, analyze, and prepare the processed information for inclusion in the finished product); dissemination (deliver the finished product to the consumer who requested it and to others as applicable); and evaluation and feedback (acquire continual feedback during the cycle that aids in refining each individual stage and the cycle as a whole).

Mission Areas: Groups of core capabilities, including Prevention, Protection, Mitigation, Response, and Recovery.

Mitigation: The capabilities necessary to reduce loss of life and property by lessening the impact of disasters.

National Health Security: The Nation and its people are prepared for, protected from, and resilient in the face of health threats or hazards with potentially negative health consequences.

National Preparedness: The actions taken to plan, organize, equip, train, and exercise to build and sustain the capabilities necessary to prevent, protect against, mitigate the effects of, respond to, and recover from those threats that pose the greatest risk to the security of the Nation.

Prevention: The capabilities necessary to avoid, prevent, or stop a threatened or actual act of terrorism. For the purposes of the prevention framework, the term "prevention" refers to preventing imminent threats.

Protection: The capabilities necessary to secure the homeland against acts of terrorism and manmade or natural disasters.

Recovery: The capabilities necessary to assist communities affected by an incident to recover effectively.

Resilience: The ability to adapt to changing conditions and withstand and rapidly recover from disruption due to emergencies.

Response: The capabilities necessary to save lives, protect property and the environment, and meet basic human needs after an incident has occurred.

Risk Assessment: A product or process that collects information and assigns a value to risks for the purpose of informing priorities, developing or comparing courses of action, and informing decision making.

Security: The protection of the Nation and its people, vital interests, and way of life.

Stabilization: The process by which the immediate impacts of an incident on community systems are managed and contained.

Steady State: A condition where operations and procedures are normal and ongoing. Communities are considered to be at a steady state prior to disasters and after recovery is complete.

Terrorism: Any activity that involves an act that is dangerous to human life or potentially destructive of critical infrastructure or key resources and is a violation of the criminal laws of the United States or of any state or other subdivision of the United States; and, appears to be intended to intimidate or coerce a civilian population, or to influence the policy of a government by intimidation or coercion, or to affect the conduct of a government by mass destruction, assassination, or kidnapping. (Note that although the definition of terrorism includes both domestic and international acts of terrorism, the scope of the planning system is the prevention and protection against acts of terrorism in the homeland.)

Weapons of Mass Destruction: Materials, weapons, or devices that are intended or capable of causing death or serious bodily injury to a significant number of people through release, dissemination, or impact of toxic or poisonous chemicals or precursors, a disease organism, or radiation or radioactivity, to include, but not limited to, biological devices, chemical devices, improvised nuclear devices, radiological dispersion devices, and radiological exposure devices.

Whole Community: A focus on enabling the participation in national preparedness activities of a wider range of players from the private and nonprofit sectors, including nongovernmental organizations and the general public, in conjunction with the participation of all levels of government in order to foster better coordination and working relationships. Used interchangeably with "all-of-Nation."

GLOSSARY

Alien and Sedition Acts: Four laws passed during the administration of President John Adams granting Adams authority to suppress activism by immigrants. Known as the Alien and Sedition Acts, these statutes placed significant restrictions on the liberty of immigrants and political critics.

all-hazards umbrella: All-hazards preparation entails preparation for a wide range of natural and human-made disasters.

Animal Liberation Front (ALF): An American-based single-issue movement that protests animal abuse and is responsible for committing acts of violence such as arson and vandalism.

Anti-Terrorism and Effective Death Penalty Act of 1996: The United States passed its first comprehensive counterterrorism legislation, titled the Anti-Terrorism and Effective Death Penalty Act, in 1996. The purpose of the Anti-Terrorism Act was to regulate activity that could be used to mount a terrorist attack, provide resources for counterterrorist programs, and punish terrorism.

antiterrorism: Target hardening, enhanced security, and other defensive measures seeking to deter or prevent terrorist attacks.

Armed Forces for National Liberation (Fuerzas Armadas de Liberación Nacional, or FALN): A Puerto Rican independentista terrorist group active during the 1970s and 1980s and responsible for more bombings than any other single terrorist group in American history.

Army of God: A shadowy and violent Christian fundamentalist movement in the United States that has attacked moralistic targets, such as abortion providers.

Aryan Republican Army (ARA): A neo-Nazi terrorist group that operated in the midwestern United States from 1994 to 1996. Inspired by the example of the Irish Republican Army, the ARA robbed 22 banks in seven states before its members were captured. Their purpose had been to finance racial supremacist causes and to hasten the overthrow of the "Zionist Occupation Government."

Ásatrú: A mystical belief in the ancient Norse pantheon. Some Ásatrú believers are racial supremacists.

asymmetrical warfare: The use of unconventional, unexpected, and nearly unpredictable methods of political violence. Terrorists intentionally strike at unanticipated targets and apply unique and idiosyncratic tactics.

Aum Shinrikyō: A cult based in Japan and led by Shoko Asahara that was responsible for releasing nerve gas into the Tokyo subway system, injuring 5,000 people.

Aviation and Transportation Security Act of 2001 (ATSA): Passed in November 2001, the Aviation and Transportation Security Act required the following: (a) creation of the Transportation Security Administration, (b) replacement of the previous screening system of contractors with professional TSA screeners by November 2002, and (c) screening of all baggage through an explosives detection system by December 2002.

biological agents: "Living organisms . . . or infective material derived from them, which are intended to cause disease or death in man, animals, and plants, and which depend on their ability to multiply in the person, animal, or plant attacked."[1]

biometric technology: Technology that allows digital photographs of faces to be matched against those of wanted suspects; such technology is especially useful for antiterrorist screens at ports of entry, such as airports and border crossings.

Black Liberation Army (BLA): An African American terrorist group active during the 1970s. The BLA tended to target police officers and banks.

Black Panther Party for Self-Defense: An African American nationalist organization founded in 1966 in Oakland, California. The Black Panthers eventually became a national movement. It was not a terrorist movement, but some members eventually engaged in terrorist violence.

Black Power: An African American nationalist ideology developed during the 1960s that stressed self-help, political empowerment, cultural chauvinism, and self-defense.

Building and Infrastructure Protection Series (BIPS): A series of manuals released by the U.S. Department of Homeland Security's Science and Technology Directorate that report the latest recommendations on how to protect buildings and infrastructure in the United States.

Carnivore: An early software surveillance tool created to monitor e-mail.

CBRNE: Acronym for chemical agents, biological agents, radiological agents, nuclear weapons, and explosives.

Central Intelligence Agency (CIA): The principal intelligence agency in the United States and the theoretical coordinator of American foreign intelligence collection.

chemical agents: "Chemical substances, whether gaseous, liquid, or solid, which are used for hostile purposes to cause disease or death in humans, animals, or plants, and which depend on direct toxicity for their primary effect."[2]

Chemical Plot: A thwarted plot by Islamists to detonate liquid explosives aboard transatlantic flights in late 2006. The suspects intended to board flights from London to several destination cities in the United States and Canada. Liquid explosives were to be disguised as soft drinks in innocuous containers and detonated using household ingredients and an electronic source such as a camera.

Christian Identity: The American adaptation of Anglo-Israelism. A racial supremacist mystical belief that holds that Aryans are the chosen people of God, the United States is the Aryan "promised land," nonwhites are soulless beasts, and Jews are biologically descended from the devil.

Citizenship and Immigration Services (USCIS): An agency within the Department of Homeland Security responsible for safeguarding the procedures and integrity of the U.S. immigration system, particularly by supervising the conferral of citizenship and immigration status on legally eligible individuals.

Civil Liberties Act: During the 1980s, reparation payments of $20,000 were authorized to be disbursed to survivors of the internment of Japanese Americans during the Second World War, and in 1988, the U.S. government passed the Civil Liberties Act, which formally apologized for the internments and declared the internment program unjust.

clash of civilizations: Theoretical trends that occur as vestiges of the East–West ideological competition give way to patterns of religious extremism and seemingly interminable communal conflicts. This clash of civilizations scenario has been extensively debated since it was theorized by Professor Samuel Huntington.

contagion effect: Copycat terrorism in which terrorists imitate each other's behavior and tactics. This theory is still debated.

Container Security Initiative (CSI): "CSI is a program intended to help increase security for maritime containerized cargo shipped to the United States from around the world. CSI addresses the threat to border security and global trade posed by the potential for terrorist use of a maritime container to deliver a weapon."[3]

Convention to Prevent and Punish Acts of Terrorism Taking the Form of Crimes Against Persons and Related Extortion That Are of International Significance: A treaty among members of the Organization of American States that "sought to define attacks against internationally protected persons as common crimes, regardless of motives."[4]

core capabilities: Every document in the National Planning Frameworks contains specified core capabilities, which are defined as the elements, or *action items*, needed to achieve the National Preparedness Goal. The core capabilities of each framework are the central implementation mechanisms for achieving the National Preparedness Goal.

councils of government (COGs): Consortia of regional governing bodies that promote collaboration and cooperation among members. A large number of COGs (more than 500) have been established across the United States and are represented in every state.

counterterrorism: Proactive policies that specifically seek to eliminate terrorist environments and groups. The ultimate goal of counterterrorism is to save lives by proactively preventing or decreasing the number of terrorist attacks.

Country Reports on Terrorism: The U.S. Department of State regularly publishes a document titled *Country Reports on Terrorism*, which identifies and lists state sponsors of terrorism and foreign terrorist organizations (FTOs).

Creativity: A mystical belief practiced by the racial supremacist World Church of the Creator in the United States. Creativity is premised on a rejection of the white race's reliance on Christianity, which is held to have been created by the Jews as a conspiracy to enslave whites. According to Creativity, the white race itself should be worshipped.

criminal profiles: Descriptive composites that include the personal characteristics of suspects, such as their height, weight, race, gender, hair color, eye color, and clothing. Suspects who match these criminal profiles can be administratively detained for questioning.

critical infrastructure/key resources (CIKR): The assets, systems, and networks, whether physical or virtual, so vital to the United States that their incapacitation or destruction would have a debilitating effect on security, national economic security, national public health or safety, or any combination thereof.

critical tasks: The National Response Framework features 15 core capabilities that are designed as practical elements in achieving the National Preparedness Goal. Each of the 15 core capabilities includes identified *critical tasks* to be undertaken by response authorities tasked with implementing the framework.

Customs and Border Protection (CBP): An agency within the Department of Homeland Security that serves as the primary border security arm of the federal government.

cyberterrorism: The use of technology by terrorists to disrupt information systems.

cyberwar: As an offensive mode of warfare, terrorists use new technologies to destroy information and communications systems. As a counterterrorist option, cyberwar involves the targeting of terrorists' electronic activities by counterterrorist agencies. Bank accounts, personal records, and other data stored in digital databases can theoretically be intercepted and compromised.

cyclone: A hurricane in the Indian Ocean or South Pacific.

Days of Rage: Four days of rioting and vandalism committed by the Weathermen in Chicago in October 1969.

DCS-1000: The upgraded and renamed version of the Carnivore software surveillance tool.

Defense Intelligence Agency (DIA): The central agency for military intelligence of the U.S. armed forces.

DeFreeze, Donald: The leader of the Symbionese Liberation Army, a California-based American terrorist group active during the 1970s. DeFreeze adopted the nom de guerre Cinque.

Department of Agriculture (USDA): The primary mission of the Department of Agriculture is to ensure a "safe, sufficient and nutritious food supply for the American people."[5]

Department of Defense (DOD): The Department of Defense is tasked to manage the armed forces of the United States.

Department of Energy (DOE): The Department of Energy is tasked to manage the nation's energy security and promote scientific research to support this task. DOE is also tasked with environmental cleaning of the nuclear weapons complex.

Department of Health and Human Services (HHS): The Department of Health and Human Services is "the United States government's principal agency for protecting the health of all Americans and providing essential human services."[6]

Department of Homeland Security (DHS): The Homeland Security Act was enacted on November 25, 2002, and created the Department of Homeland Security. The new department was tasked with five main areas of responsibility, reflecting the underlying missions of the former emergency response and security agencies subsumed under the authority of DHS.

Department of Homeland Security Act of 2002: President George W. Bush, in June 2002, initiated a process that completely reorganized the American homeland security community. This process led to the enactment of the Department of Homeland Security Act of 2002, which was signed into law on November 25, 2002. The new law created a large, cabinet-level Department of Homeland Security.

Department of the Interior (DOI): The primary mission of the Department of the Interior "is to protect and provide access to our Nation's natural and cultural heritage and honor our trust responsibilities to Indian Tribes and our commitments to island communities."[7]

Department of the Treasury: The overarching mission of the Department of the Treasury is to "maintain a strong economy and create economic and job opportunities by promoting the conditions that enable economic growth and stability at home and abroad, strengthen national security by combating threats and protecting the integrity of the financial system, and manage the U.S. Government's finances and resources effectively."[8]

DHS Office of Intelligence and Analysis: The only Intelligence Community (IC) element statutorily charged with delivering intelligence to state, local, tribal, territorial, and private-sector partners, and developing intelligence from those partners for the Department of Homeland Security and the IC.

Diplomatic Security Service: A security bureau within the U.S. Department of State that protects diplomats and other officials.

Director of National Intelligence: Members of the IC are subsumed under the direction of the ODNI. President George W. Bush appointed John Negroponte, former U.S. ambassador to Iraq, as the United States' first director of national intelligence (DNI). The DNI is responsible for coordinating the various components of the IC.

dirty bomb: Non-weapons-grade radiological agents could theoretically be used to construct a toxic *dirty bomb* that would use conventional explosives to release a cloud of radioactive contaminants. Radioactive elements that could be used in a dirty bomb include plutonium, uranium, cobalt-60, strontium, and cesium-137.

disaster: Conditions rise to the level of a disaster when emergency response institutions are unable to contain or resolve one or more critical services, such as fire management, the restoration of order, attending to medical needs, or providing shelter. Official declarations of disaster are made by the president of the United States after he receives a request from a governor.

Earth Liberation Front (ELF): A single-issue movement that protests environmental degradation and pollution. A splinter group from the environmentalist group Earth First!, the ELF is potentially more radical than the Animal Liberation Front.

Echelon: A satellite surveillance network maintained by the U.S. National Security Agency. It is a kind of global wiretap that filters communications using antennae, satellites, and other technologies. Internet transfers, telephone conversations, and data transmissions are among the types of communications that can reportedly be intercepted.

emergency declarations: Presidential declarations that trigger aid that "protects property, public health, and safety and lessens or averts the threat of an incident becoming a catastrophic event."[9]

emergency event: An emergency event occurs when a hazard does, in fact, result in a condition of risk, necessitating intervention by emergency response institutions, such as law enforcement, medical personnel, or firefighting agencies.

Emergency Management Assistance Compact (EMAC): A national disaster relief network established by Congress in 1996. Membership in EMAC commits signatories to assist cosignatories with disaster relief assistance following disaster declarations.

emergency operations plans: Plans that explain the mission and duties of local response personnel and agencies.

emergency response operations: Coordinated interventions undertaken when emergencies occur.

emergency support functions: To ensure delivery of the 15 core capabilities, the National Response Framework identifies emergency support functions. The emergency support functions represent federal coordinating structures used to achieve the core capabilities, and like the core capabilities, they represent systematized collaboration among policy planners.

Enforcement and Removal Operations (ERO): A directorate within Immigration and Customs Enforcement that identifies and apprehends removable aliens, detains these individuals when necessary, and removes illegal aliens from the United States.

Enhanced Fujita–Pearson Scale: Reporting of tornado intensity is done in accordance with the Enhanced Fujita–Pearson Scale (Enhanced F-Scale) developed in 2007 by the National Oceanic and Atmospheric Administration (NOAA).

enhanced interrogation: Physical and psychological stress methods used during the questioning of suspects.

Environmental Protection Agency (EPA): The overarching mission of the EPA is to lead "the nation's environmental science, research, education and assessment efforts . . . [and] protect human health and the environment."[10]

Executive Order 13228: On October 8, 2001, President Bush issued Executive Order 13228, titled "Establishing the Office of Homeland Security and the Homeland Security Council."

extraordinary renditions: A method of covertly abducting and detaining suspected terrorists or affiliated operatives.

fallout: Explosions from nuclear bombs devastate the area within their blast zone, irradiate an area outside the blast zone, and are capable of sending dangerous radioactive debris into the atmosphere that descends to earth as toxic fallout.

Federal Air Marshal Service (FAMS): The federal "sky marshals" service. FAMS is a relatively clandestine law enforcement agency whose size, number of marshals, and presence on aircraft are classified. In 2005, FAMS was reassigned to the Transportation Security Administration from U.S. Immigration and Customs Enforcement.

Federal Aviation Administration (FAA): The Federal Aviation Administration administers civil aviation, issuing regulations and other requirements to control air traffic and promote safety.

Federal Bureau of Investigation (FBI): An investigative bureau within the U.S. Department of Justice. It is the largest federal law enforcement agency, and among its duties are domestic counterterrorism and intelligence collection.

Federal Response Plan: A procedural framework drafted in 1993 for federal response coordination should an incident occur. When activated, elements of the plan provided demonstrably effective guidance when actual terrorist incidents occurred.

Flood Insurance Rate Maps (FIRMs): Maps created, maintained, and regularly updated by the National Flood Insurance Program for "floodplain management and insurance purposes" that "generally show a community's base flood elevations, flood zones, and floodplain boundaries."[11]

flood zones: National land areas differentiated by the Federal Emergency Management Agency (FEMA) according to their risk of flooding.

foreign terrorist organizations (FTOs): *Country Reports on Terrorism* reports an updated list of organizations designated by the State Department as FTOs. This list is regularly revised in compliance with Section 219 of the Immigration and Nationality Act.

geospatial intelligence (GEOINT): The collection and assessment of topography and geographical features can provide actionable intelligence regarding locations, timeframes, and other information. GEOINT is the all-source analysis of imagery and geospatial information to describe, assess, and visually depict physical features and geographically referenced activities on earth.

Hague Convention of 1970: This treaty required signatories to extradite "hijackers to their country of origin or to prosecute them under the judicial code of the recipient state."[12]

hate crimes: Crimes motivated by hatred against protected groups of people. They are prosecuted as aggravated offenses rather than as acts of terrorism.

Haymarket Riot of 1886: During Chicago's Haymarket Riot of 1886, an anarchist threw a dynamite bomb at police officers, who then opened fire on protesters. Scores of officers and civilians were wounded.

hazard: A condition posing potential risks that can result in either an emergency event or a disaster.

Hazard Mitigation Assistance (HMA) grant programs: Grant programs that "provide funding for eligible mitigation activities that reduce disaster losses and protect life and property from future disaster damages."[13]

Homeland security enterprise: Homeland Security secretary Janet Napolitano explained that a new comprehensiveness, termed the *homeland security enterprise*, serves as the core concept for the *Quadrennial Homeland Security Review Report*'s strategic framework.

Homeland Security Investigations (HSI): A directorate within Immigration and Customs Enforcement responsible for investigating "immigration crime, human rights violations and human smuggling, smuggling of narcotics, weapons and other types of contraband, financial crimes, cybercrime and export enforcement issues."[14]

Homeland Security Presidential Directive 5 (HSPD-5): A 2003 directive by President George W. Bush that sought to institutionalize procedures for securing coordinated incident response protocols. The new protocols were intended to synchronize the Incident Command Systems of federal, state, and local authorities in the event of a terrorist incident.

Homeland Security Presidential Directive 7 (HSPD-7): A directive by President George W. Bush that specifically created a systematic approach to collaborative efforts among different levels of government. HSPD-7 classified 17 sectors as critical infrastructure sectors and assigned responsibility for these sectors to specified federal sector-specific agencies. It was superseded in 2013 by Presidential Policy Directive 21.

Homeland Security Presidential Directive 21 (HSPD-21): A 2007 directive by President George W. Bush that established a National Strategy for Public Health and Preparedness. Written as an all-hazards approach to homeland security, the purpose of HSPD-21 was to "transform our national approach to protecting the health of the American people against all disasters."[15]

Homeland Security Presidential Directives (HSPDs): On October 29, 2001, President Bush released the first of many HSPDs, which implement policies and procedures constituting the homeland security enterprise.

Homestead Steel Strike of 1892: During the Homestead Steel Strike of 1892 on the Monongahela River near Pittsburgh, a strike by steelworkers resulted in a pitched gun battle between striking workers and hundreds of Pinkerton agents (in which the strikers prevailed). The strike was eventually suppressed following intervention by the Pennsylvania state militia.

human intelligence (HUMINT): Intelligence that has been collected by human operatives rather than through technological resources.

hurricane: A tropical cyclone with maximum sustained winds of 74 miles per hour or higher.

Hurricane Andrew: In 1992, Hurricane Andrew struck the United States, causing an estimated $26 billion in damage. The federal response was widely criticized as unnecessarily slow and uncoordinated.

Hurricane Katrina: In 2005, Hurricane Katrina struck the Gulf Coast of the United States, causing nearly 2,000 deaths and billions of dollars in destruction. The federal response was strongly criticized as unacceptably inadequate and ponderous.

Hurricane Sandy: At its peak, Sandy was a Category 3 hurricane, but its eventual merger with a frontal weather system made it an unusually large storm, eventually becoming the geographically largest Atlantic hurricane ever recorded. For this reason, Hurricane Sandy was given the popular nickname "Superstorm Sandy."

imagery intelligence (IMINT): Images are regularly collected to provide actionable intelligence. Collection technologies range from relatively routine hand-held equipment to very sophisticated means. IMINT includes intelligence information derived from the collection by visual photography, infrared sensors, lasers, electro-optics, and radar sensors.

Immigration and Customs Enforcement (ICE): An agency within the Department of Homeland Security that serves as the investigative and enforcement arm for federal criminal and civil laws overseeing immigration, border control, customs, and trade. ICE is the second largest federal law enforcement investigative agency.

Information Sharing and Analysis Centers (ISACs): Information Sharing and Analysis Centers are conceptually designed to assist in planning for the 16 identified critical infrastructure sectors. Each ISAC is assigned to a specific critical infrastructure sector.

Intelligence Community: The greater network of intelligence agencies. In the United States, the Central Intelligence Agency is the theoretical coordinator of intelligence collection.

International Association of Emergency Managers (IAEM): "[A] non-profit educational organization dedicated to promoting the 'Principles of Emergency Management' and representing those professionals whose goals are saving lives and protecting property and the environment during emergencies and disasters."[16]

international conventions: Multinational agreements between partner countries that state that specified protocols and procedures will be respected among signatories. The underlying purposes of international conventions enacted during specific historical periods reflect the quality of terrorist threats existing at the time.

International Court of Justice: The principal judicial arm of the United Nations. The court hears disputes between nations and gives advisory opinions to recognized international organizations.

International Criminal Court: A court established to prosecute crimes against humanity, such as genocide. Its motivating principle is to promote human rights and justice.

International Criminal Tribunal for Rwanda (ICTR): The ICTR has investigated allegations of war crimes and genocide that resulted from the breakdown of order in Rwanda during the 1990s.

International Criminal Tribunal for the Former Yugoslavia (ICTY): The ICTY has investigated allegations of war crimes and genocide arising out of the wars that broke out after the fragmentation of Yugoslavia during the 1990s.

Joint Field Office: "A temporary Federal facility that provides a central location for coordination of response efforts by the private sector, NGOs, and all levels of government."[17]

Joint Terrorism Task Forces (JTTFs): "Small cells of highly trained, locally based, passionately committed investigators, analysts, linguists, SWAT experts, and other specialists from dozens of U.S. law enforcement and intelligence agencies. [This] is a multi-agency effort led by the Justice Department and FBI designed to combine the resources of Federal, state, and local law enforcement."[18]

Ker-Frisbie Rule: A doctrine that permits coercive abductions and appearances before U.S. judicial authorities, named after two cases permitting such abductions and appearances.

Ku Klux Klan (KKK): A racial supremacist organization founded in 1866 in Pulaski, Tennessee. During its five eras, the KKK was responsible for thousands of acts of terrorism.

kuklos: Literally "circle" in Greek; the insignia of the Ku Klux Klan, consisting of a cross and a teardrop-like symbol enclosed by a circle.

leaderless resistance: A cell-based strategy of the Patriot and neo-Nazi movements in the United States, requiring the formation of "phantom cells" to wage war against the government and enemy interests. Dedicated Patriots and neo-Nazis believe that this strategy will prevent infiltration from federal agencies.

Local Emergency Planning Committees (LEPCs): Community-based associations that allow local planners to prepare for emergency scenarios. LEPCs are tasked pursuant to the Emergency Planning and Community Right-to-Know Act of 1986 to prepare for emergencies involving hazardous materials.

Macheteros: A Puerto Rican independentista terrorist group active during the 1970s and 1980s.

May 19 Communist Organization (M19CO): An American Marxist terrorist group that was active in the late 1970s and early 1980s. It was composed of remnants of the Republic of New Africa, the Black Liberation Army, the Weather Underground, and the Black Panthers. M19CO derived its name from the birthdays of Malcolm X and Vietnamese leader Ho Chi Minh.

McCarthyism: During the 1950s, Republican senator Joseph McCarthy of Wisconsin sought to expose communist infiltration and conspiracies in government, private industry, and the entertainment industry. The manner in which McCarthy promoted his cause was to publicly interrogate people from these sectors in a way that had never been done before: on television. Hundreds of careers were ruined, and many people were blacklisted—that is, nationally barred from employment. McCarthy was later criticized for overstepping the bounds of propriety, and the pejorative term *McCarthyism* has come to mean a political and ideological witch hunt.

McVeigh, Timothy: A member of the Patriot movement in the United States and probably a racial supremacist. McVeigh was responsible for constructing and detonating an ANFO bomb that destroyed the Alfred P. Murrah Federal Building in Oklahoma City, Oklahoma, on April 19, 1995. One hundred sixty-eight people were killed.

measurements and signatures intelligence (MASINT): The use of a broad array of technical and scientific disciplines to measure the characteristics of specified subjects; for example, tracking communications signatures or measuring water and soil samples. MASINT is intelligence information obtained by quantitative and qualitative analysis of data derived from specific technical sensors for the purpose of identifying any distinctive features associated with the source, emitter, or sender.

medical surge: A management methodology for responding to large-scale medical incidents.

Medical Surge Capacity and Capability Handbook: A Management System for Integrating Medical and Health Resources During Large-Scale Emergencies **(MSCC):** A management handbook published by the U.S. Department of Health and Human Services that promoted a system for integrating medical and health resources when responding to large-scale emergencies.

militias: Organized groups of armed citizens who commonly exhibit antigovernment tendencies and subscribe to conspiracy theories. The armed manifestation of the Patriot movement.

mission areas: The Federal Emergency Management Agency published five National Planning Frameworks, each dedicated to promoting efficient preparedness within the context of one of five mission areas. The five mission areas are prevention, protection, mitigation, response, and recovery.

mitigation of risk: "The effort to reduce loss of life and property by lessening the impact of disasters."[19] Mitigation mechanisms are pre-emergency initiatives that theoretically reduce the potential costs and destructiveness of disasters when they occur.

Modified Mercalli Intensity Scale: A measurement scale for earthquakes.

Montreal Convention of 1971: This treaty extended international law to cover "sabotage and attacks on airports and grounded aircraft, and laid down the principle that all such offenses must be subject to severe penalties."[20]

National Association of Counties (NACo): An organization representing county governments. Its services include research initiatives, surveys, technical assistance, newsworthy publications, policy reports, and advocacy to the federal government.

National Association of Regional Councils (NARC): An organization that promotes regionalism among governmental bodies within and across states. NARC "serves as the national voice for regionalism by advocating for regional cooperation as the most effective way to address a variety of community planning and development opportunities and issues."[21]

National Counterterrorism Center (NCTC): A center established to integrate the counterterrorism efforts of the intelligence community in the wake of the September 11, 2001, attacks.

National Emergency Management Association (NEMA): A professional association, essentially an umbrella organization, "*of* and *for* emergency management directors from all 50 states, eight territories, and the District of Columbia" (emphasis in original).[22]

National Flood Insurance Program: A program initiated by FEMA that regularly disburses funds for flood-related insurance claims.

National Homeland Security Consortium: The National Emergency Management Association established the National Homeland Security Consortium as a public forum for government personnel and private practitioners to "coalesce efforts and perspectives about how best to protect America in the 21st century" in the aftermath of the September 11 attacks.[23]

National Incident Management System (NIMS): Implemented in 2008, the National Incident Management System was designed to serve as "the template for the management of incidents"[24] on a national scale. NIMS was established as a "systematic, proactive approach"[25] for homeland security authorities at all levels of government. Significantly, NIMS is an integrated system for implementing the National Response Framework.

National Infrastructure Protection Plan (NIPP): The National Infrastructure Protection Plan identifies 16 critical infrastructure sectors, within which collaborative initiatives will be undertaken between the public and private sectors. This will be done in concert with sector-specific plans assigned to each critical infrastructure sector. The NIPP's purpose is to ensure efficient private sector participation in homeland security planning and preparedness initiatives.

National Intelligence Priorities Framework: Intelligence policy priorities are governed by the National Intelligence Priorities Framework (NIPF), which promulgates policy and establishes responsibilities for setting national intelligence priorities and translating them into action.

National Joint Terrorism Task Force: The Federal Bureau of Investigation has been tasked with coordinating JTTFs nationally through the National Joint Terrorism Task Force, an interagency office operating from FBI headquarters.

National League of Cities (NLC): An organization providing advocacy support on behalf of over 19,000 cities, towns, and villages.

National Planning Frameworks: In synchrony with NIMS, the Federal Emergency Management Agency published five National Planning Frameworks, each dedicated to promoting efficient preparedness within the context of one of five mission areas. The purpose of the National Planning Frameworks and the identified mission areas is to achieve the National Preparedness Goal. The National Planning Frameworks comprise the National Prevention Framework, the National Protection Framework, the National Mitigation Framework, the National Response Framework (second edition), and the National Disaster Recovery Framework.

National Preparedness Goal: The National Preparedness Goal, approved in September 2011, seeks to achieve "a secure and resilient nation with the capabilities required across the whole community to prevent, protect against, mitigate, respond to, and recover from the threats and hazards that pose the greatest risk."[26] The National Preparedness Goal introduces and

explains the five mission areas, which have been identified as necessary elements for successful management of homeland security incidents.

National Reconnaissance Office: responsible for designing, building, launching, and maintaining America's intelligence satellites. NRO provides satellite reconnaissance support to the IC and Department of Defense

National Response Framework: A procedural roadmap for federal coordination of a national effort to institute an efficient disaster and emergency response system. The document "is built on scalable, flexible, and adaptable concepts identified in the National Incident Management System to align key roles and responsibilities across the Nation."[27]

National Security Agency (NSA): An American intelligence agency charged with signals intelligence collection, code making, and code breaking.

National Security Strategy: The process of clarifying the new homeland security enterprise is ongoing, and publications regularly clarify the homeland security enterprise. One of those publications is the 2010 *National Security Strategy*.

National Strategy for Counterterrorism: The process of clarifying the new homeland security enterprise is ongoing, and publications regularly clarify the homeland security enterprise. One of those publications is the 2011 *National Strategy for Counterterrorism*.

National Strategy for Homeland Security: Published in 2002, this was the first clarification of the newly emerging homeland security culture. The document explicated the concept of homeland security, identified essential homeland security missions, and established priorities for coordinating the protection of critical domestic infrastructures.

National Strategy for Public Health and Preparedness: The National Strategy for Public Health and Preparedness was established in October 2007 by Homeland Security Presidential Directive 21 as part of an all-hazards approach to homeland security.

nativism: American cultural nationalism.

natural hazards: Natural hazards and disasters are emergency incidents arising from nonhuman causes. Such incidents are the consequence of phenomena arising from natural environmental conditions. Events such as earthquakes, hurricanes, tornadoes, and floods are examples of natural hazards.

netwar: An important concept in the new terrorist environment is the netwar theory, which refers to "an emerging mode of conflict and crime . . . in which the protagonists use network forms of organization and related doctrines, strategies, and technologies attuned to the information age. These protagonists are likely to consist of dispersed small groups who communicate, coordinate, and conduct their campaigns in an internetted manner, without a precise central command."[28]

New Left: A movement of young leftists during the 1960s who rejected orthodox Marxism and took on the revolutionary theories of Frantz Fanon, Herbert Marcuse, Carlos Marighella, and other new theorists.

New Terrorism: A typology of terrorism characterized by a loose, cell-based organizational structure; asymmetrical tactics; the threatened use of weapons of mass destruction; potentially high casualty rates; and usually a religious or mystical motivation.

New World Liberation Front: An American terrorist group active during the mid-1970s, organized as a "reborn" manifestation of the Symbionese Liberation Army by former SLA members and new recruits.

Non-Special Flood Hazard Areas: Areas that have a moderate to low risk of flooding.

non-wildland fires: Fires that occur in populated areas and affect human-constructed infrastructure.

nuclear weapons: High-explosive military weapons using weapons-grade plutonium and uranium. Nuclear explosions devastate the area within their blast zone, irradiate an area outside the blast zone, and are capable of sending dangerous radioactive debris into the atmosphere that descends to earth as toxic fallout.

Office of Nuclear Security and Incident Response: The U.S. Nuclear Regulatory Commission office that has primary responsibility for emergency planning and preparedness.

Office of the Director of National Intelligence (ODNI): In December 2004, the intelligence community was reorganized with the passage of the Intelligence Reform and Terrorism Prevention Act. Members of the community were subsumed under the direction of a new Office of the Director of National Intelligence, responsible for coordinating the various components of the intelligence community.

"Old Terrorism": Prior to the September 11, 2001, terrorist attacks, the global terrorist environment was characterized by a relatively uncomplicated and predictable terrorist operational profile.

Omega 7: An anticommunist Cuban American terrorist group that targeted Cuban interests.

open source intelligence (OSINT): Information collected from publicly available electronic and print outlets. It is information that is readily available to the public, but used for intelligence analysis. Examples of open sources include newspapers, the Internet, journals, radio, videos, television, and commercial outlets.

Operation Iraqi Freedom: The U.S.-led invasion of Iraq involving the commitment of large conventional military forces.

Order, The: An American neo-Nazi terrorist group founded by Robert Jay Mathews in 1983 and centered in the Pacific Northwest. The Order's methods for fighting its war against what it termed the Zionist Occupation Government were counterfeiting, bank robberies, armored car robberies, and murder. The Order had been suppressed by December 1985.

Palmer Raids: During a domestic bombing campaign allegedly conducted by communists and anarchists, President Woodrow Wilson authorized U.S. attorney general A. Mitchell Palmer to conduct a series of raids—the so-called Palmer Raids—against labor activists, including American labor unions, socialists, communists, anarchists, and leftist labor groups.

Phineas actions: Acts of violence committed by individuals who are "called" to become Phineas priests. Adherents believe that Phineas actions will hasten the ascendancy of the Aryan race.

Phineas Priesthood: A shadowy movement of Christian Identity fundamentalists in the United States who believe that they are called by God to purify their race and Christianity. They are opposed to abortion, homosexuality, interracial mixing, and whites who "degrade" white racial supremacy. This is a calling for men only, so no women can become Phineas priests. The name is taken from the Bible at Chapter 25, Verse 6 of the Book of Numbers, which tells the story of a Hebrew man named Phineas who killed an Israelite man and his Midianite wife in the temple.

preparedness planning: The design and adoption of emergency management contingencies prior to the occurrence of an emergency situation. Preparations are made to increase the likelihood that an initial response effort and the subsequent recovery period will ultimately be efficient and successful.

Presidential Decision Directive/NSC-63: A 1998 directive released by President Bill Clinton following terrorist attacks upon U.S. installations abroad. It was a comprehensive policy directive designed to consolidate efforts to protect critical infrastructure.

presidential declarations: Significant exercises of the authority of the executive branch of the federal government that bring into effect an array of options for federal assistance to state and local authorities.

presidential major disaster declarations: Presidential declarations, issued after catastrophes occur, that give broad authority to federal agencies "to provide supplemental assistance to help state and local governments, families and individuals, and certain nonprofit organizations recover from the incident."[29]

Presidential Policy Directive 21 (PPD-21): A directive released in 2013 by President Barack Obama, superseding HSPD-7. Titled *Critical Infrastructure Security and Resilience*, PPD-21 identified 16 critical infrastructure sectors, designated specified federal agencies as sector-specific agencies, and tasked those agencies with implementing policies and programs for the protection of their assigned sectors.

Prevention and Punishment of Crimes Against Internationally Protected Persons, Including Diplomatic Agents: A multilateral treaty adopted by the United Nations in 1973 that sought to establish a common international framework for suppressing extremist attacks against those protected by internationally recognized status.

PRISM: A covert surveillance operation coordinated by the U.S. National Security Agency.

***Quadrennial Homeland Security Review Report: A Strategic Framework for a Secure Homeland* (QHSR):** A document published by the Department of Homeland Security intending to consolidate the definition of homeland security by presenting the concept as encompassing a broader and more comprehensive mission than previously envisioned. The QHSR is a documentary acknowledgment that homeland security is evolving conceptually.

racial holy war (RaHoWa): A term given by racial supremacists to a future race war that they believe will inevitably occur in the United States.

racial profiling: The unconstitutional detention of people because of their ethnonational or racial heritage.

radiological agents: Materials that emit radiation that can harm living organisms when inhaled or otherwise ingested. Non-weapons-grade radiological agents could theoretically be used to construct a toxic *dirty bomb*.

recovery systems: Recovery systems are implemented in the aftermath of emergency events and attempt to return affected regions to predisaster baselines. This requires, at a minimum, rebuilding damaged infrastructure and restoring affected populations to their pre-emergency norms of living. Recovery operations are often quite expensive and long term.

Red Scares: Periods in U.S. history when a generalized climate of political anxiety occurred in response to perceived threats from communist, anarchist, and other leftist subversion.

***Reference Manual to Mitigate Potential Terrorist Attacks Against Buildings, Edition 2* (BIPS 06/FEMA 426):** A widely distributed guide for government and private agencies and organizations that reports a considerable number of infrastructure security and target-hardening recommendations.

Rewards for Justice Program: An international bounty program managed by the U.S. Diplomatic Security Service. The program offers cash rewards for information leading to the arrest of wanted terrorists.

Richter Scale: A measurement scale for earthquakes.

Robert T. Stafford Disaster Relief and Emergency Assistance Act: Legislation that "authorizes the President to issue 'major disaster' or 'emergency' declarations before or after catastrophes occur."[30]

sector-specific agencies: Sector-specific homeland security missions have been identified for federal agencies in addition to establishing the Department of Homeland Security. These agencies are known as sector-specific agencies.

shadow wars: Covert campaigns to suppress terrorism.

Shakur, Assata: The symbolic leader of the Black Liberation Army (BLA) in the United States, formerly known as JoAnne Chesimard. A former Black Panther, she was described by admirers as the "heart and soul" of the BLA.

signals intelligence (SIGINT): Intelligence that has been collected by technological resources.

single-issue terrorism: Terrorism that is motivated by a single grievance.

sleeper agents: A tactic used by international terrorist movements in which operatives theoretically establish residence in another country to await a time when they will be activated by orders to carry out a terrorist attack.

Snowden, Edward: A former Central Intelligence Agency employee and National Security Agency contractor who leaked details of covert surveillance operations to the media prior to becoming an international fugitive. The information was apparently delivered to *The Guardian*, the *Washington Post*, and a documentary filmmaker.

Special Flood Hazard Areas: Areas that "are at high risk for flooding" and have "a 26 percent chance of suffering flood damage during the term of a 30-year mortgage."[31]

state sponsors of terrorism: Governments that support terrorist activity.

storm surge: A wall of water pushed forward by an approaching hurricane.

Symbionese Liberation Army (SLA): A violent terrorist cell that gained notoriety for several high-profile incidents in the mid-1970s.

target hardening: An antiterrorist measure that makes potential targets more difficult to attack. Target hardening includes increased airport security, the visible deployment of security personnel, and the erection of crash barriers at entrances to parking garages beneath important buildings.

Tempora: A British covert surveillance operation conducted in cooperation with the U.S. National Security Agency.

terrorist profile: A descriptive composite similar to standard criminal profiles. The modern terrorist profile includes the following characteristics: Middle Eastern heritage, temporary visa status, Muslim faith, male gender, and young-adult age.

3-1-1 rule: The Transportation Security Administration requires passengers to comply with the 3-1-1 rule for carrying liquids on airliners—no more than three ounces of liquid placed in one clear quart-sized plastic bag in one carry-on bag. The reason for this requirement is a thwarted plot by Islamists to detonate liquid explosives aboard transatlantic flights in late 2006.

Tokyo Convention on Offences and Certain Other Acts Committed on Board Aircraft: Enacted in 1963 as the first international airline crimes treaty, the Tokyo Convention required all signatories to "make every effort to restore control of the aircraft to its lawful commander and to ensure the prompt onward passage or return of the hijacked aircraft together with its passengers, crew, and cargo."[32]

Transportation Security Administration (TSA): Responsibility for civil aviation security was transferred from the Federal Aviation Administration to the Transportation Security Administration in 2002. The agency promulgates and manages the implementation of transportation security regulations to support its aviation security mission.

tropical cyclone: A "rotating, organized system of clouds and thunderstorms that originates over tropical or subtropical waters and has a closed low-level circulation."[33]

tropical depression: A tropical cyclone with maximum sustained winds of up to 38 miles per hour.

tropical storm: A tropical cyclone with maximum sustained winds of 39 to 73 miles per hour.

tsunami: A series of massive waves created when seawater is displaced by earthquakes or other disturbances on the ocean floor.

typhoon: A hurricane in the western North Pacific.

U.S. Nuclear Regulatory Commission (NRC): The U.S. Nuclear Regulatory Commission is authorized to regulate safety and security protocols at nuclear plants.

unified coordination: Conceptually defined as "the term used to describe the primary state/Federal incident management activities conducted at the incident level."[34] Its underlying purpose is to provide support to personnel and agencies directly involved with operations at the incident site.

Unified Coordination Group: Because Joint Field Offices are charged with delivering unified coordination, the National Response Framework recommends that each Joint Field Office be managed under the direction of a Unified Coordination Group. The Unified Coordination Group comprises senior leaders representing federal and state interests and, in certain circumstances, tribal governments, local jurisdictions, and the private sector.

United Freedom Front (UFF): A leftist terrorist group in the United States that was active from the mid-1970s through the mid-1980s.

United States Conference of Mayors (USCM): An organization representing urban areas on issues affecting major cities and other urban governments. USCM engages in advocacy and publicizing of issues on behalf of governments representing a large proportion of the population of the United States.

USA PATRIOT Act of 2001: On October 26, 2001, President George W. Bush signed the "Uniting and Strengthening America by Providing Appropriate Tools Required to Intercept and Obstruct Terrorism Act of 2001," commonly known as the USA PATRIOT Act, into law. It was an omnibus law whose stated purpose was, in part, to "deter and punish terrorist acts in the United States and around the world" by expanding the investigative and surveillance authority of law enforcement agencies.

USA PATRIOT Improvement and Reauthorization Act of 2005: Congressional oversight and proactive management of renewal processes for the USA PATRIOT Act were enacted in sunset provisions to counterbalance enhanced authority granted to the executive branch under the USA PATRIOT Act. Thus, the USA PATRIOT Act was first renewed by Congress in March 2006 after passage of the USA PATRIOT Improvement and Reauthorization Act of 2005.

wars of national liberation: A series of wars fought in the developing world in the postwar era. These conflicts frequently pitted indigenous guerrilla fighters against European colonial powers or governments perceived to be pro-Western. Insurgents were frequently supported by the Soviet bloc or China.

Weather Bureau: The designation adopted by the leaders of the Weatherman faction of Students for a Democratic Society.

Weather Underground Organization: The adopted name of the Weathermen after they moved underground.

Weathermen: A militant faction of Students for a Democratic Society that advocated and engaged in violent confrontation with the authorities. Some Weathermen engaged in terrorist violence.

Whiskey Rebellion: One of the early post-independence disturbances was popularly known as the Whiskey Rebellion, an anti-tax uprising in western Pennsylvania (1791–1794).

wildland fires: Fires that occur in woodland areas.

XKeyscore: A covert surveillance operation, coordinated by the U.S. National Security Agency, capable of collecting real-time data on chat rooms, browsing history, social networking media, and e-mail.

NOTES

CHAPTER 1

1. Executive Order 13228 was amended by Executive Order 12656, which clarified that policies enacted in reply to terrorist events outside of the United States would remain within the authority of the National Security Council.
2. U.S. Department of Homeland Security, *The 2014 Quadrennial Homeland Security Review* (Washington, DC: U.S. Department of Homeland Security, 2014), 6–8.
3. U.S. Department of Homeland Security, "Preventing Terrorism Overview," accessed February 22, 2019, www.dhs.gov/topic/preventing-terrorism-overview.
4. U.S. Department of Homeland Security, *Quadrennial Homeland Security Review Report: A Strategic Framework for a Secure Homeland* (Washington, DC: U.S. Department of Homeland Security, 2010), iii.
5. Ibid., iii.
6. U.S. Department of Homeland Security, *The 2014 Quadrennial Homeland Security Review* (Washington, DC: U.S. Department of Homeland Security, 2014), 13.
7. Ibid., 76–79.
8. Ibid., 28.
9. Ibid., 28.

CHAPTER 2

1. U.S. Department of Homeland Security, *Quadrennial Homeland Security Review Report: A Strategic Framework for a Secure Homeland* (Washington, DC: Author, 2010), viii.
2. For further information, see the National Science and Technology Council, Committee on Homeland and National Security, *A National Strategy for CBRNE Standards* (Washington, DC: National Science and Technology Council, 2011).
3. Federal Emergency Management Agency, "What Is Mitigation?" accessed November 21, 2015, www.fema.gov/what-mitigation.
4. FEMA, "Whole Community," accessed November 21, 2015, www.fema.gov/whole-community.
5. Ibid.
6. Ibid.
7. Erin G. Richardson and David Hemenway, "Homicide, Suicide, and Unintentional Firearm Fatality: Comparing the United States With Other High-Income Countries, 2003," *Journal of Trauma* 70, no. 1 (2011): 238–43.
8. U.S. Department of Homeland Security, *Active Shooter: How to Respond* (Washington, DC: Author, 2008), http://www.dhs.gov/xlibrary/assets/active_shooter_booklet.pdf.
9. National Weather Service, *Tropical Cyclones: A Preparedness Guide* (Washington, DC: U.S. Department of Commerce, 2012), 2.
10. Ibid.
11. See Eric S. Blake, Christopher Landsea, and Ethan J. Gibney, *The Deadliest, Costliest and Most Intense United States Tropical Cyclones From 1851 to 2010* (Technical Memorandum NWS NHC-6) (Washington, DC: U.S. National Oceanic and Atmospheric Administration, 2011).
12. See National Oceanic and Atmospheric Association, *Hurricane/Post-Tropical Cyclone Sandy, October 22–29, 2012* (Service Assessment) (Washington, DC: National Oceanic and Atmospheric Association, 2013).
13. See National Hurricane Center, "Tropical Cyclone Naming History and Retired Names," accessed November 21, 2015, www.nhc.noaa.gov/aboutnames_history.shtml. See also World Meteorological Organization, "Tropical Cyclone Naming," accessed November 21, 2015, www.wmo.int/pages/prog/www/tcp/Storm-naming.html.
14. U.S. Geological Survey, "Earthquake Glossary: Earthquake," accessed November 21, 2015, http://earthquake.usgs.gov/learn/glossary/?term=earthquake.
15. U.S. Geological Survey, "Earthquake Glossary: Richter Scale," accessed November 21, 2015, http://earthquake.usgs.gov/learn/glossary/?term=Richter scale.
16. U.S. Geological Survey, "The Modified Mercalli Intensity Scale," accessed November 21, 2015, http://earthquake.usgs.gov/learn/topics/mercalli.php.
17. National Weather Service, Storm Prediction Center, "Enhanced F Scale for Tornado Damage," accessed November 21, 2015, www.spc.noaa.gov/faq/tornado/ef-scale.html.
18. National Flood Insurance Program, "What Are Flood Zones?" accessed November 21, 2015, www.floodsmart.gov/floodsmart/pages/faqs/what-are-flood-zones.jsp.
19. National Flood Insurance Program, "What Is a Special Flood Hazard Area (SFHA)?" accessed November 21, 2015,

www.floodsmart.gov/floodsmart/pages/faqs/what-is-a-special-flood-hazard-area.jsp.

20. National Flood Insurance Program, "What Is a Flood Insurance Rate Map (FIRM) and How Do I Use It?" accessed November 21, 2015, www.floodsmart.gov/floodsmart/pages/faqs/what-is-a-flood-insurance-rate-map-and-how-do-i-use-it.jsp.

21. See U.S. Fire Administration, "Total Wildland Fires and Acres," accessed November 21, 2015, www.usfa.fema.gov/statistics/estimates/wildfire.shtm.

22. Committee on Enhancing the Robustness and Resilience of Future Electrical Transmission and Distribution in the United States to Terrorist Attack; Board on Energy and Environmental Systems; Division on Engineering and Physical Sciences; and National Research Council, *Terrorism and the Electric Power Delivery System* (Washington, DC: National Academies of Sciences, Engineering, Medicine, 2012).

23. See National Fire Protection Association, "Fire Loss in the United States," accessed November 21, 2015, www.nfpa.org/itemDetail.asp?categoryID=2453&itemID=55585&URL=Research/Statistical%20reports/Overall%20fire%20statistics.

CHAPTER 3

1. U.S. Department of State, Office of the Coordinator of Counterterrorism, *Country Reports on Terrorism 2014* (Washington, DC: U.S. Department of State, 2015), 283.

2. Ibid., 328.

3. Ibid., 328.

4. Paul Wilkinson, "Fighting the Hydra: Terrorism and the Rule of Law," in *International Terrorism: Characteristics,* *Causes, Controls,* ed. Charles W. Kegley (New York: St. Martin's, 1990), 255.

5. Ibid.

6. Ibid.

7. Ibid.

CHAPTER 4

1. See Glenn Greenwald, "NSA PRISM Program Taps in to User Data of Apple, Google and Others," *The Guardian,* June 6, 2013.

2. See Glenn Greenwald, "XKeyscore: NSA Tool Collects 'Nearly Everything a User Does on the Internet,'" *The Guardian,* July 31, 2013.

3. Richard Lardner, "NSA Leak Details Citizen Records," *Boston Globe,* July 21, 2013.

4. See *Frisbie v. Collins,* 342 U.S. 519, 522 (1954). See also *Ker v. Illinois,* 119 U.S. 436, 444 (1886).

5. See Craig Whitlock, "From CIA Jails, Inmates Fade Into Obscurity," *Washington Post,* October 27, 2007.

6. See Tracy Wilkinson, "Italy Orders Arrest of 13 CIA Operatives," *Los Angeles Times,* June 25, 2005.

CHAPTER 5

1. Office of Homeland Security, *National Strategy for Homeland Security* (Washington, DC: White House, 2002).

2. These five areas are periodically updated. See U.S. Department of Homeland Security, "Our Mission," accessed November 16, 2018, www.dhs.gov/our-mission.

3. U.S. White House, *National Security Strategy* (Washington, DC: Author, 2010).

4. U.S. White House, *National Strategy for Counterterrorism* (Washington, DC: White House, 2011).

5. For more information, see Appendix B.

6. For more information, see Appendix A.

7. U.S. Department of Homeland Security, "Our Mission."

8. Ibid.

9. Office functions are reported as described at U.S. Department of Homeland Security, "Office of the Secretary," accessed November 16, 2018, www.dhs.gov/office-secretary.

10. U.S. Department of Homeland Security, "Office of the Secretary."

11. Ibid.

12. Ibid.

13. Ibid.

14. Ibid.

15. Ibid.

16. Ibid.

17. Ibid.

18. Ibid.

19. Ibid.

20. Ibid.

21. U.S. Department of Homeland Security, "Operational and Support Components," accessed November 16, 2018, www.dhs.gov/department-components.

22. Ibid.

23. Ibid.
24. Ibid.
25. Ibid.
26. Ibid.
27. Ibid.
28. Ibid.
29. Ibid.
30. Ibid.
31. Ibid.
32. Ibid.
33. Ibid.
34. Ibid.
35. Ibid.
36. U.S. Department of Agriculture, "About the U.S. Department of Agriculture," accessed November 16, 2018, https://www.usda.gov/our-agency/about-usda.
37. U.S. Department of Defense, "Our Story," accessed November 16, 2018, https://www.defense.gov/Our-Story/.
38. U.S. Department of Energy, "About Us," accessed November 16, 2018, www.energy.gov/about/index.htm.
39. U.S. Department of Homeland Security, "Energy Sector," accessed November 21, 2015, http://www.dhs.gov/energy-sector.
40. U.S. Department of Health and Human Services, "About HHS," accessed November 16, 2018, www.hhs.gov/about.
41. U.S. Department of Health and Human Services, "National Health Security Strategy," accessed November 16, 2018, http://www.phe.gov/Preparedness/planning/authority/nhss/Pages/default.aspx.
42. U.S. Department of the Interior, "About Interior," accessed November 16, 2018, https://www.doi.gov/whoweare.
43. U.S. Department of the Interior, "Departments of the Interior, Homeland Security Announce $6.8 Million in Conservation Projects," October 13, 2010, https://www.doi.gov/news/pressreleases/Departments-of-the-Interior-and-Homeland-Security-Announce-6-point-8-Million-in-Conservation-Projects.
44. U.S. Department of the Treasury, "Role of the Treasury," accessed November 16, 2018, https://home.treasury.gov/about/general-information/role-of-the-treasury.
45. U.S. Department of Treasury, "Resource Center," accessed November 16, 2018, https://www.treasury.gov/resource-center/terrorist-illicit-finance/Pages/default.aspx.
46. U.S. Environmental Protection Agency, "About EPA," accessed November 16, 2018, http://www2.epa.gov/aboutepa.
47. Steve Bowman, *Homeland Security: The Department of Defense's Role* (Washington, DC: Congressional Research Service, 2003), 1.
48. Ibid.
49. David E. Sanger, "In Speech, Bush Focuses on Conflicts Beyond Iraq," *New York Times*, May 1, 2003.

CHAPTER 6

1. U.S. Intelligence Community Mission Statement. https://www.intelligence.gov/mission.
2. U.S. Department of Homeland Security, "Homeland Security Information Network (HSIN)," accessed November 21, 2015, http://www.dhs.gov/homeland-security-information-network-hsin# and http://www.dhs.gov/law-enforcement.
3. U.S. Department of Justice, *The National Criminal Intelligence Sharing Plan* (Washington, DC: U.S. Department of Justice, 2003), https://it.ojp.gov/documents/National_Criminal_Intelligence_Sharing_Plan.pdf.
4. Regional Information Sharing Systems Website, http://www.riss.net, accessed November 21, 2015.
5. *The 9/11 Commission Report: Final Report of the National Commission on Terrorist Attacks Upon the United States*. New York: Norton, 2004.
6. Ibid., 408–410.
7. The Six Steps in the Intelligence Cycle. https://www.intelligence.gov/how-the-ic-works#start.
8. Ibid.
9. Ibid.
10. Ibid.
11. Ibid.
12. Ibid.
13. Craig Whitlock, "After a Decade at War With the West, Al-Qaeda Still Impervious to Spies," *Washington Post*, March 20, 2008.
14. Robert M. Clark, *Intelligence Collection* (Thousand Oaks, CA: CQ Press, 2014), 490.
15. Ibid., 492.
16. Ibid., 489.
17. Intelligence Community Directive 204, "National Intelligence Priorities Framework," Office of the Director of National Intelligence (January 2015).
18. Ibid.
19. Office of the Director of National Intelligence. https://www.dni.gov/index.php/who-we-are/organizations. Accessed November 2, 2018.
20. Ibid.
21. Ibid.
22. Ibid.
23. Ibid.
24. National Geospatial-Intelligence Agency. https://www.intelligence.gov/how-the-ic-works/our-organizations/414-nga.
25. National Reconnaissance Office. https://www.intelligence.gov/how-the-ic-works/our-organizations/415-nro.

26. DHS Office of Intelligence and Analysis. https://www.intelligence.gov/how-the-ic-works/our-organizations/420-dhs-office-of-intelligence-and-analysis.

27. Paul R. Pillar, *Terrorism and U.S. Foreign Policy* (Washington, DC: Brookings Institution Press, 2001), 115.

28. U.S. Select Committee on Intelligence, *Report on the U.S. Intelligence Community's Prewar Intelligence Assessments on Iraq* (Washington, DC: U.S. Senate, 2004).

29. Ibid., 14.

30. U.S. Department of Homeland Security, "Office of the Secretary."

31. Ibid.

32. Dafna Linzer, "Search for Banned Arms in Iraq Ended Last Month," *Washington Post*, January 12, 2005.

CHAPTER 7

1. 32 USC § 905.

2. 10 USC §§ 10001–18506.

3. National Emergency Management Association, "Welcome—The National Emergency Management Association," accessed November 21, 2015, www.nemaweb.org.

4. NEMA, "What Is NEMA," accessed November 18, 2018, http://www.nemaweb.org/index.php/about/what-is-nema.

5. NEMA, "NEMA Strategic Plan," accessed November 18, 2018, www.nemaweb.org/index.php?option=com_content&view=article&id=78&Itemid=244.

6. NEMA, "National Homeland Security Consortium," accessed November 18, 2018, www.nemaweb.org/index.php?option=com_content&view=article&id=122&Itemid=215.

7. Ibid.

8. National Association of Regional Councils (NARC), "About the Association," accessed November 18, 2018, http://narc.org/about-narc/about-the-association.

9. NARC, "Vision and Mission," accessed November 18, 2018, http://narc.org/about-narc/about-the-association/vision-and-mission.

10. NARC, "Policy Positions," accessed November 18, 2018, http://narc.org/issueareas/homeland-security/policy-positions.

11. See Public Law 104-3.

12. Emergency Management Assistance Compact, "What Is EMAC?" accessed November 18, 2018, https://www.emacweb.org/index.php/learn-about-emac/what-is-emac.

13. U.S. Census Bureau, "Census Bureau Reports There Are 89,000 Local Governments in the United States," accessed November 18, 2018, https://www.census.gov/newsroom/releases/archives/governments/cb12-161.html.

14. U.S. Environmental Protection Agency, "Local Emergency Planning Committees," accessed March 1, 2019, https://www.epa.gov/epcra/local-emergency-planning-committees.

15. National Association of Counties, *Counties and Homeland Security: Policy Agenda to Secure the People of America's Counties* (Washington, DC: National Association of Counties, 2002).

16. Ibid.

17. National Association of Counties, "NACo Fights Massive Cuts to Homeland Security," May 13, 2011.

18. National Association of Counties, "Homeland Security Assistance Faces Severe Reductions," *NACo County News: The Voice of America's Counties* 43, no. 10 (2011).

19. Mitchel Herckis, "Congressional Appropriators Side With NLC, First Responders on Homeland Security Grants," May 21, 2012, http://www.nlc.org/media-center/news-search/congressional-appropriators-side-with-nlc-first-responders-on-homeland-security-grants.

20. United States Conference of Mayors, "About the U.S. Conference of Mayors," accessed November 18, 2018, https://www.usmayors.org/the-conference/about/.

21. International Association of Emergency Managers, "About IAEM," accessed March 1, 2019, https://www.iaem.com/page.cfm?p=about/intro.

22. Council of State Governments and Eastern Kentucky University, *The Impact of Terrorism on State Law Enforcement: Adjusting to New Roles and Changing Conditions* (2005), www.csg.org/knowledgecenter/docs/Misc0504Terrorism.pdf.

23. Ibid., 8–13.

24. U.S. Department of Justice, Office of the Inspector General, *The Department of Justice's Terrorism Task Forces* (Washington, DC: Author, 2005), https://oig.justice.gov/reports/plus/e0507/final.pdf.

25. Ibid.

26. Ibid.

27. Federal Bureau of Investigation, "Joint Terrorism Task Forces," accessed November 18, 2018, https://www.fbi.gov/investigate/terrorism/joint-terrorism-task-forces.

CHAPTER 8

1. Gus Martin, "Globalization and International Terrorism," in *The Blackwell Companion to Globalization*, ed. George Ritzer (Malden, MA: Blackwell, 2006), 695.

2. Ibid., 645.

3. Ibid., 1.

4. Office for the Protection of the Constitution. See David J. Whittaker, ed., *The Terrorism Reader*, 4th ed. (London: Routledge, 2012).

5. Ibid.

6. U.S. Department of Defense, *Department of Defense Dictionary of Military and Associated Terms* (Washington, DC: Author, 2010/2014), 266.

7. 18 U.S.C. 3077.

8. Terrorist Research and Analytical Center, National Security Division, Federal Bureau of Investigation, *Terrorism in the United States 1995* (Washington, DC: U.S. Department of Justice, 1996), ii.

9. Bureau of Counterterrorism, *Country Reports on Terrorism 2013* (Washington, DC: U.S. Department of State, 2014).

10. For a discussion of the ambiguities about defining combatants and noncombatants, see Peter C. Sederberg, *Terrorist Myths: Illusion, Rhetoric, and Reality* (Englewood Cliffs, NJ: Prentice Hall, 1989), 37–39.

11. For a discussion of the ambiguities of defining indiscriminate force, see Sederberg, 39–40.

12. From Paul R. Pillar, *Terrorism and U.S. Foreign Policy* (Washington, DC: Brookings Institution, 2001), 44–45.

13. Martin, "Globalization and International Terrorism."

14. The term was used in a study conducted by the National Defense Panel titled "Transforming Defense: National Security in the 21st Century." The 1997 report warned that "unanticipated asymmetries" in the international security environment would likely result in an attack on the American homeland. Richard Leiby, "Rueful Prophets of the Unimaginable: High-Level Studies Warned of Threat," *Washington Post*, September 22, 2001.

15. Holly Yan, "Vehicles as Weapons: Muenster Part of a Deadly Trend," *CNN*, April 7, 2018.

16. See Jessica Stern, *The Ultimate Terrorists* (Cambridge, MA: Harvard University Press, 1999), 70ff.

17. David Ronfeldt Arquilla and Michele Zanini, "Networks, Netwar, and Information-Age Terrorism," in *Countering the New Terrorism*, ed. Ian O. Lesser, Bruce Hoffman, John Arquilla, David Ronfeldt, and Michele Zanini (Santa Monica, CA: RAND, 1999), 47.

18. Ibid., 49.

19. Michael Moss, "An Internet Jihad Aims at U.S. Viewers," *New York Times*, October 15, 2007. See also Mary Beth Sheridan, "Terrorism Probe Points to Reach of Web Networks," *Washington Post*, January 24, 2008.

20. See Geisal G. Mohamed, "The Globe of Villages: Digital Media and the Rise of Homegrown Terrorism," *Dissent* (2007): 61–64. See also John Gray, "A Violent Episode in the Virtual World," *New Statesman*, July 18, 2005.

21. Ariana Eunjung Cha, "From a Virtual Shadow, Messages of Terror," *Washington Post*, October 2, 2004.

22. Paul R. Pillar, *Terrorism and U.S. Foreign Policy* (Washington, DC: Brookings Institution, 2001), 123.

CHAPTER 9

1. For a good history of the American labor movement, see Thomas R. Brooks, *Toil and Trouble: A History of American Labor* (New York: Dell, 1971).

2. Some of the lifestyle issues, such as drug legalization, have been endorsed by libertarian conservatives.

3. Robert J. Kelly, "The Ku Klux Klan: Recurring Hate in America," in *Hate Crime: The Global Politics of Polarization*, ed. Robert J. Kelly and Jess Maghan (Carbondale: Southern Illinois University Press, 1998), 54.

4. Bruce Hoffman, *Inside Terrorism* (New York: Columbia University Press, 1998), 107. These numbers declined during the late 1990s and then rebounded after the September 11, 2001, attacks on the U.S. homeland and the 2008 election of President Barack Obama.

5. The film was directed by Italian filmmaker Gillo Pontecorvo, who died in October 2006. See Times Staff and Wire Services, "Gillo Pontecorvo, Movie Director Best Known for 'The Battle of Algiers,' Dies at 86," *Los Angeles Times*, October 14, 2006.

6. Southern Poverty Law Center, *Intelligence Report*, Spring 2019.

7. National Abortion Federation. https://prochoice.org/escalation-picketing-trespassing-obstruction-clinic-blockades-invasions-threats-harm-anti-abortion-extremists/, accessed March 6, 2019.

8. Richard Kelly Hoskins, *Vigilantes of Christendom: The History of the Phineas Priesthood* (Lynchburg, VA: Virginia Pub Co., 1995).

9. Mark Hamm, *In Bad Company: America's Terrorist Underground* (Boston: Northeastern University Press, 2002), 147.

10. Ibid.

11. Ramon Spaaij, *Understanding Lone Wolf Terrorism: Global Patterns, Motivations, and Prevention* (New York: Springer, 2012).

12. *United States of America v. James Wenneker von Brunn*, Criminal Complaint, United States District Court for the District of Columbia.

13. Associated Press, "Did Shoe Bombing Suspect Act Alone?" *Christian Science Monitor*, December 27, 2001.

CHAPTER 10

1. U.S. Department of Homeland Security, "What Is Critical Infrastructure?" accessed September 17, 2013, www.dhs.gov/what-critical-infrastructure.
2. U.S. Departnent of Homeland Security, *National Strategy for Aviation Security* (Washington, DC: Author, 2007), 6.
3. President's Commission on Critical Infrastructure, *The National Strategy for the Physical Protection of Critical Infrastructures and Key Assets* (Washington, DC: White House, 2003), 7.
4. Ibid., 8.
5. Ibid., 55.
6. Ibid., 55.
7. Jens Manuel Krogstad, Jeffrey S. Passel, and D'Vera Cohn, *5 Facts About Illegal Immigration in the U.S.* (Washington, DC: Pew Research Center, November 28, 2018).
8. See Janice L. Kephart, "Immigration and Terrorism: Moving Beyond the 9/11 Staff Report on Terrorist Travel," Center for Immigration Studies, Center Paper 24 (2005).
9. U.S. Drug Enforcement Administration, "Drugs and Terrorism: A New Perspective," *Drug Intelligence Brief*, September 2002. Data derived from estimates by the Office of National Drug Control Policy.
10. June S. Beittel, *Mexico: Organized Crime and Drug Trafficking Organizations* (Washington, DC: Congressional Research Service, July 3, 2018), 2.
11. Ibid., 1.
12. National Atlas of the United States, "Profile of the Land and the People of the United States," accessed April 22, 2013, http://nationalatlas.gov/articles/mapping/a_general.html.
13. Ibid.
14. U.S. Immigration and Customs Enforcement (ICE), "Overview," accessed November 21, 2015, www.ice.gov/about/overview.
15. Ibid.
16. ICE, "Enforcement and Removal Operations," accessed November 21, 2015, www.ice.gov/about/offices/enforcement-removal-operations.
17. ICE, "Homeland Security Investigations," accessed November 21, 2015, www.ice.gov/about/offices/homeland-security-investigations.
18. ICE, "Border Enforcement Security Task Force (BEST)," accessed November 21, 2015, www.ice.gov/best.
19. ICE, "National Security Investigations Division," accessed November 21, 2015, www.ice.gov/national-security-investigations-division.
20. ICE, "Counter-Proliferation Investigations Unit," accessed November 21, 2015, www.ice.gov/counter-proliferation-investigations.
21. ICE, "National Security Unit," accessed November 21, 2015, www.ice.gov/national-security-unit.
22. ICE, "Counterterrorism and Criminal Exploitation Unit," accessed November 21, 2015, www.ice.gov/counterterrorism-criminal-exploitation.
23. ICE, "Joint Terrorism Task Force," accessed November 21, 2015, www.ice.gov/jttf.
24. U.S. Customs and Border Protection (CBP), "About CBP," accessed November 21, 2015, http://www.cbp.gov/about.
25. CBP, "Border Security," accessed November 21, 2015, http://www.cbp.gov/border-security#.
26. Ibid.
27. Ibid., 60.
28. U.S. Customs and Border Protection, "Obtaining Information About the Container Security Initiative (CSI)," accessed November 21, 2015, https://help.cbp.gov/app/answers/detail/a_id/406/~/obtaining-information-about-the-container-security-initiative-%28csi%29. Interestingly, the website directed readers to a hotline, stating, "If you have information concerning possible terrorist use of shipping containers to smuggle weapons or other items, you may be entitled to a reward for the information provided."
29. President's Commission on Critical Infrastructure, 61.
30. Ibid., 61.

CHAPTER 11

1. Analysis is derived from Michael Kenney, "Cyber-Terrorism in a Post-Stuxnet World," *Orbis* 59, no. 1 (2015): 111–28.
2. Ibid., 113.
3. Ibid., 114.
4. Ibid., 117.
5. Ibid., 121; Barry Collin, "The Future of Cyberterrorism," *Crime and Justice International*, March 1997, 15–18.
6. Ian O. Lesser, "Countering the New Terrorism: Implications for Strategy," in *Countering the New Terrorism*, ed. Ian O. Lesser, Bruce Hoffman, John Arquilla, David Ronfeldt, and Michele Zanini (Santa Monica, CA: RAND, 1999), 95.
7. U.S. Department of Homeland Security, "What Is Critical Infrastructure?" accessed November 21, 2015, www.dhs.gov/what-critical-infrastructure.
8. U.S. Department of Homeland Security, "Critical Infrastructure Sectors," accessed November 21, 2015, www.dhs.gov/critical-infrastructure-sectors.

9. U.S. Department of Homeland Security, "National Infrastructure Protection Plan Overview," accessed November 21, 2015, www.dhs.gov/sites/default/files/publications/nipp_consolidated_snapshot.pdf.

10. U.S. Department of Homeland Security, "Building and Infrastructure Protection Series (BIPS)," accessed November 21, 2015, www.dhs.gov/xlibrary/assets/st-building-infrastructure-protection-series.pdf.

11. U.S. Department of Homeland Security, *Preventing Structures From Collapsing to Limit Damage to Adjacent Structures and Additional Loss of Life When Explosives Devices Impact Highly Populated Urban Centers* (Washington, DC: Author, 2011), www.dhs.gov/xlibrary/assets/preventing-structures-from-collapsing-062011.pdf.

12. Federal Emergency Management Agency (FEMA), *Reference Manual to Mitigate Potential Terrorist Attacks Against Buildings*, 2nd ed. (Washington, DC: FEMA, 2011), www.dhs.gov/xlibrary/assets/st/st-bips-06.pdf.

13. FEMA, *Primer to Design Safe School Projects in Case of Terrorist Attacks and School Shootings*, 2nd ed. (Washington, DC: Author, 2012), www.dhs.gov/xlibrary/assets/st/bips07_428_schools.pdf.

14. U.S. Department of Homeland Security, "Aviation Security Policy, National Security Presidential Directive 47/Homeland Security Presidential Directive 16," accessed March 11, 2019, http://www.dhs.gov/hspd-16-aviation-security-policy.

CHAPTER 12

1. U.S. Department of Homeland Security, "Resilience," accessed December 1, 2018, https://www.dhs.gov/topic/resilience.

2. Office of the President, *National Security Strategy of the United States of America* (Washington, DC: U.S. White House, November 2017), 14.

3. U.S. Department of Homeland Security, "Resilience," accessed December 1, 2018, https://www.dhs.gov/topic/resilience.

4. U.S. White House, "Homeland Security Presidential Directive/HSPD-5: Management of Domestic Incidents," February 28, 2003.

5. Ibid.

6. Ibid.

7. U.S. Department of Homeland Security, *National Incident Management System* (Washington, DC: Author, 2013).

8. U.S. Department of Homeland Security, "National Incident Management System," accessed November 21, 2015, http://www.fema.gov/national-incident-management-system.

9. Ibid.

10. Ibid.

11. U.S. Federal Emergency Management Agency, *A Whole Community Approach to Emergency Management: Principles, Themes, and Pathways for Action* (Washington, DC: FEMA, December 2011).

12. Ibid., 3.

13. Ibid., 4.

14. U.S. Department of Homeland Security, "About the Private Sector Office," accessed November 21, 2015, www.dhs.gov/about-private-sector-office.

15. U.S. Federal Emergency Management Agency (FEMA), "Private Sector," accessed November 21, 2015, http://www.fema.gov/private-sector.

16. Ibid.

17. U.S. Department of Homeland Security, "National Infrastructure Protection Plan," accessed November 21, 2015, www.dhs.gov/national-infrastructure-protection-plan.

18. FEMA, "National Domestic Preparedness Consortium," accessed March 12, 2019, https://www.ndpc.us/.

19. Ibid., at https://www.ndpc.us/about.aspx.

20. U.S. White House, "Homeland Security Presidential Directive/HSPD-21: Public Health and Medical Preparedness," October 18, 2007, www.fas.org/irp/offdocs/nspd/hspd-21.htm.

21. Ibid.

22. U.S. Department of Health and Human Resources, *Medical Surge Capacity and Capability: A Management System for Integrating Medical and Health Resources During Large-Scale Emergencies*, 2nd ed. (Washington, DC: U.S. Department of Health and Human Resources, 2007), www.phe.gov/Preparedness/planning/mscc/handbook/Pages/default.aspx.

23. Ibid.

24. Ibid.

25. James E. Moore II, Harry W. Richardson, and Peter Gordon, "The Economic Impacts of a Terrorist Attack on Downtown Los Angeles Financial District," *Non-published Research Reports*, Paper 22 (2009).

26. U.S. Department of Homeland Security, *Buildings and Infrastructure Protection Series: Reference Manual to Mitigate Potential Terrorist Attacks Against Buildings*, 2nd ed. (BIPS-06/FEMA-426) (Washington, DC: U.S. Department of Homeland Security, 2011), www.dhs.gov/xlibrary/assets/st/st-bips-06.pdf.

27. FEMA, "Hazard Mitigation Assistance," accessed November 21, 2015, www.fema.gov/hazard-mitigation-assistance.

28. Ibid.

29. Jessica Eve Stern, "The Covenant, the Sword, and the Arm of the Lord," in *Toxic Terror: Assessing Terrorist Use of Chemical and Biological Weapons*, ed. Jonathan B. Tucker (Cambridge, MA: MIT Press, 2000), 139–157.

30. The threat scenarios are indeed very plausible. In early 2007, Iraqi insurgents detonated several chorine bombs, killing a number of people and injuring hundreds. See Damien Cave and Ahmad Fadam, "Iraq Insurgents Employ

Chlorine in Bomb Attacks," *New York Times*, February 22, 2007. See also Karin Brulliard, "Chlorine Bombs Kill 10, Injure at Least 350 in Iraq," *Washington Post*, March 17, 2007; and Garrett Therolf and Alexandra Zavis, "Bomb Releases Chlorine in Iraq's Diyala Province," *Los Angeles Times*, June 3, 2007.

31. See Joby Warrick, "An Easier, but Less Deadly, Recipe for Terror," *Washington Post*, December 31, 2004.

32. Stern.

33. See John Mintz, "Technical Hurdles Separate Terrorists From Biowarfare," *Washington Post*, December 30, 2004.

34. Ali S. Khan, Alexandra M. Levitt, and Michael J. Sage, with CDC Strategic Planning Workgroup, "Biological and Chemical Terrorism: Strategic Plan for Preparedness and Response," *Morbidity and Mortality Weekly Report* 49, no. 4 (2000): 1–14.

35. Ibid.

36. Centers for Disease Control and Prevention, "First Hours: Terrorism Emergencies," accessed November 21, 2015, http://emergency.cdc.gov/cerc/index.asp.

37. Stern.

38. For a discussion of the nuclear threat, see Walter Laqueur, *The New Terrorism: Fanaticism and the Arms of Mass Destruction* (New York: Oxford University Press, 1999), 70.

39. See Dafna Linzer, "Nuclear Capabilities May Elude Terrorists, Experts Say," *Washington Post*, December 29, 2004.

40. U.S. Nuclear Regulatory Commission, "Nuclear Security and Safeguards," accessed November 21, 2015, www.nrc.gov/security.html.

41. Ibid.

42. Ibid.

43. Ibid.

44. U.S. Nuclear Regulatory Commission, "Emergency Preparedness and Response," accessed November 21, 2015, www.nrc.gov/about-nrc/emerg-prepared ness.html.

45. U.S. Nuclear Regulatory Commission, "Nuclear Security and Incident Response," accessed November 21, 2015, www.nrc.gov/about-nrc/organization/nsirfuncdesc.html.

46. Ibid.

CHAPTER 13

1. U.S. Federal Emergency Management Agency, "National Preparedness Goal," accessed November 21, 2015, www.fema.gov/national-preparedness-goal.

2. U.S. Department of Homeland Security, *National Preparedness Goal* (Washington, DC: Author, 2001), iii.

3. Occupational Safety and Health Administration, "Incident Command System," accessed November 21, 2015, https://www.osha.gov/SLTC/etools/ics/what_is_ics.html.

4. Federal Emergency Management Agency, "Introduction to the Incident Command System (ICS 100)," accessed November 19, 2015, https://emilms.fema.gov/IS100b/index.htm.

5. Francis X. McCarthy, *FEMA's Disaster Declaration Process: A Primer* (Washington, DC: Congressional Research Service, 2011).

6. U.S. Federal Emergency Management Agency, *National Response Framework* (Washington, DC: U.S. Department of Homeland Security, 2016), 43. https://www.fema.gov/media-library-data/1466014682982-9bcf8245ba4c60c120aa915abe74e15d/National_Response_Framework3rd.pdf, i.

7. Ibid.

8. Ibid.

9. U.S. Federal Emergency Management Agency, *National Response Framework* (Washington, DC: U.S. Department of Homeland Security, 2016), 43. https://www.fema.gov/media-library-data/1466014682982-9bcf8245ba4c-60c120aa915abe74e15d/National_Response_Framework3rd.pdf, 33.

10. Ibid., 42.

11. Ibid.

12. Ibid.

13. Ibid.

14. For a discussion of several myths about terrorism, including the notion that it is highly effective, see Walter Laqueur, "The Futility of Terrorism," in *International Terrorism: Characteristics, Causes, Controls*, ed. Charles W. Kegley Jr. (New York: St. Martin's, 1990), 69ff. Another good discussion of effectiveness is found in Peter C. Sederberg, *Terrorist Myths: Illusion, Rhetoric, and Reality* (Englewood Cliffs, NJ: Prentice Hall, 1989), 96ff. See also Philip B. Heymann, *Terrorism and America: A Commonsense Strategy for a Democratic Society* (Cambridge, MA: MIT Press, 1998), 12ff.

CHAPTER 14

1. Andrew MacDonald [William Pierce], *The Turner Diaries* (New York: Barricade Books, 1996).

2. E. L. Anderson [David L. Hoggan], *The Myth of the Six Million* (Newport Beach, CA: Noontide, 1969).

3. Samuel P. Huntington, *The Clash of Civilizations and the Remaking of World Order* (New York: Touchstone, 1996).

GLOSSARY

1. Jessica Eve Stern, "The Covenant, the Sword, and the Arm of the Lord," in *Toxic Terror: Assessing Terrorist Use of Chemical and Biological Weapons*, ed. Jonathan B. Tucker (Cambridge, MA: MIT Press, 2000), 139–57.

2. Ibid.

3. U.S. Customs and Border Protection, "Obtaining Information About the Container Security Initiative (CSI)," accessed December 9, 2013, https://help.cbp.gov/app/answers/detail/a_id/406/~/obtaining-information-about-the-container-security-initiative-%28csi%29.

4. Paul Wilkinson, "Fighting the Hydra: Terrorism and the Rule of Law," in *International Terrorism: Characteristics, Causes, Controls*, ed. Charles W. Kegley (New York: St. Martin's, 1990), 255.

5. U.S. Department of Agriculture, "USDA Biographies: Secretary of Agriculture – Tom Vilsack," accessed March 19, 2013, www.usda.gov/wps/portal/!ut/p/_s.7_0_A/7_0_10B?contentidonly=true&contentid=bios_vilsack.xml.

6. U.S. Department of Health and Human Services, "About HHS," accessed March 19, 2013, www.hhs.gov/about.

7. U.S. Department of the Interior, "About Interior," accessed March 19, 2013, www.doi.gov/whoweare/interior.cfm.

8. U.S. Department of the Treasury, "About," accessed March 19, 2013, www.treasury.gov/about/role-of-treasury/Pages/default.aspx.

9. Francis X. McCarthy, *FEMA's Disaster Declaration Process: A Primer* (Washington, DC: Congressional Research Service, 2011).

10. U.S. Environmental Protection Agency, "About EPA," accessed March 19, 2013, www.epa.gov/epahome/aboutepa.htm.

11. National Flood Insurance Program, "What Is a Flood Insurance Rate Map (FIRM) and How Do I Use It?" accessed March 10, 2013, www.floodsmart.gov/floodsmart/pages/faqs/what-is-a-flood-insurance-rate-map-and-how-do-i-use-it.jsp.

12. Ibid.

13. FEMA, "Hazard Mitigation Assistance," accessed May 21, 2013, www.fema.gov/hazard-mitigation-assistance.

14. U.S. Immigration and Customs Enforcement, "Homeland Security Investigations," accessed December 10, 2013, www.ice.gov/about/offices/homeland-security-investigations.

15. U.S. White House, "Homeland Security Presidential Directive/HSPD-21," October 18, 2007, www.fas.org/irp/offdocs/nspd/hspd-21.htm.

16. International Association of Emergency Managers, "About IAEM," accessed March 1, 2019, https://www.iaem.com/page.cfm?p=about/intro.

17. FEMA, *National Response Framework*, 2nd ed. (Washington, DC: U.S. Department of Homeland Security, 2013).

18. U.S. Department of Justice, "Joint Terrorism Task Force," accessed April 16, 2013, www.justice.gov/jttf.

19. U.S. Federal Emergency Management Agency (FEMA), "What Is Mitigation?" accessed March 9, 2013, www.fema.gov/what-mitigation.

20. Ibid.

21. National Association of Regional Councils (NARC), "About the Association," accessed December 7, 2013, http://narc.org/about-narc/about-the-association.

22. National Emergency Management Association (NEMA), "Welcome – The National Emergency Management Association," accessed December 7, 2013, www.nemaweb.org.

23. NEMA, "National Homeland Security Consortium," accessed December 7, 2013, www.nemaweb.org/index.php?option=com_content&view=article&id=122&Itemid=215.

24. U.S. Department of Homeland Security, "About National Incident Management System," accessed July 27, 2013, www.fema.gov/about-national-incident-management-system.

25. Ibid.

26. FEMA, "National Preparedness Goal," accessed May 29, 2013, www.fema.gov/national-preparedness-goal.

27. U.S. Federal Emergency Management Agency, *National Response Framework* (Washington, DC: U.S. Department of Homeland Security, 2016), 43. https://www.fema.gov/media-library-data/1466014682982-9bcf8245ba4c-60c120aa915abe74e15d/National_Response_Framework3rd.pdf, i.

28. David Ronfeldt Arquilla and Michele Zanini, "Networks, Netwar, and Information-Age Terrorism," in *Countering the New Terrorism*, ed. Ian O. Lesser, Bruce Hoffman, John Arquilla, David Ronfeldt, and Michele Zanini (Santa Monica, CA: RAND, 1999), 47.

29. McCarthy.

30. Ibid.

31. National Flood Insurance Program, "What Is a Special Flood Hazard Area (SFHA)?" accessed March 10, 2013, www.floodsmart.gov/floodsmart/pages/faqs/what-is-a-special-flood-hazard-area.jsp.

32. Ibid.

33. National Weather Service, *Tropical Cyclones: A Preparedness Guide* (Washington, DC: U.S. Department of Commerce, 2012), Section 2.

34. FEMA, *National Response Framework*, 40.

INDEX

Immigration. *See* Border security; Customs and Border Protection (CBP)

Immigration Act of 1918, 6

Immigration and Customs Enforcement (ICE), 97, 98t, 99t, 218t, 227, 228, 230f, 231

Immigration and Nationality Act (INA), 49

Immigration and Naturalization Service (INS), 77
  creation of, 227
  role after 9/11, 92t, 227
  role before 9/11, 91t, 227

Imperial wizard, 190t, 308p

Improvised explosive device (IED), 28, 202

INA (Immigration and Nationality Act), 49

Incarceration, 11, 46, 57, 188, 209, 306

Incident Accident Plans (IAPs), 284

Incident Command System (ICS), 263, 282–285, 286f

Inclusion, 76–77

Independentistas, 185–187

Indiscriminate use of force, 150

Information security, target hardening and, 239–245. *See also* Technology

Information Sharing and Analysis Centers (ISACs), 262

Infrastructure. *See* Nodes

Infrastructure Protection and Disaster, 98t, 99t, 100t

Inhalation anthrax, 268

Inland Regional Center attack, 165, 202–203

INS. *See* Immigration and Naturalization Service (INS)

Integration, operational, 11–12

Intelligence, 107–122
  defined, 18t, 107
  interagency disconnect and, 119–120
  international collaboration, 107, 306
  National Intelligence Strategy, 313–342
  problems with collection/analysis of, 108
  risk identification, 264
  sources of, 114–115
  state and local collection of, 109
  See also Surveillance

Intelligence Community (IC), 108–113
  agencies, 116–118
  challenges, 110–111, 118–122
  definition of, 108
  evolution of modern, 109–112
  post 9/11, 112–113, 120–122
  reorganization of, 111

Intelligence cycle, 113–115
  collection types, 114
  phases of, 113–114

Intelligence oversight, 115–116

Intelligence Reform and Terrorism Prevention Act of 2004 (IRTA), 111

Internal security, 137

International Association of Emergency Managers (IAEM), 136

International conventions, 52–53t, 55. *See also specific conventions by name*

International Court of Justice, 56

International courts/tribunals, 56–57

International Criminal Court (ICC), 56

International Criminal Police Organization (INTERPOL), 253

International Criminal Tribunal for Rwanda (ICTR), 57

International Criminal Tribunal for the Former Yugoslavia (ICTY), 56–57p

International law
  counterterrorist cooperation, 19, 20t
  overview, 47
  policy challenges and, 48
  See also specific conventions by name

International terrorism. *See* Terrorism

Internet
  cybersecurity, 98t, 239–245
  data mining, 74–76, 75p, 244
  terrorist use of, 161–162, 169
  See also Media

Internetted movements, 161

Internment camps, 68t, 71, 72p

Interoperability, 10

INTERPOL (International Criminal Police Organization), 253

Interrogations, enhanced, 78t, 80–81

The Interview, 241

Inversion process, 184

Invisible Empire, 190t, 191

IRA (Irish Republican Army), 27, 82, 179

Iran, 50, 157

Iraq
  as state sponsor of terrorism, 50
  WMD and, 96, 119, 121–122, 262, 266, 267

Iraqi Kurds, 121, 303

Iraq invasion
  intelligence miscalculation and, 121–122
  media access during, 66

Iraq's Continuing Programs for Weapons of Mass Destruction, 119

Irish Republican Army (IRA), 27, 82, 179

IRTA (Intelligence Reform and Terrorism Prevention Act of 2004), 111

ISACs (Information Sharing and Analysis Centers), 262

Islamic Jihad, 179

Islamic State of Iraq and al-Sham (ISIS), 164–165, 204

Islamic State of Iraq and the Levant (ISIL), 4, 166–167. *See also* New Terrorism

Islam/Islamists
  extremists, 17
  media/Internet use by, 161–162, 169, 216–217
  symbolic targets of, 17–18
  See also al-Qaeda; Jihad

Israel, 122, 140, 154, 159, 192, 303

Italy, 17, 80

HOMELAND SECURITY TIMELINE

1797–1801 ◆ Restrictive laws known as the Alien and Sedition Acts were passed during the administration of President John Adams.

1861–1865 ◆ During the American Civil War, the administration of President Abraham Lincoln suspended the right to the writ of habeas corpus. Approximately 38,000 civilians were detained by the military.

1865 or 1866 ◆ The Ku Klux Klan was founded in the United States in Pulaski, Tennessee, by former Confederate general Nathan Bedford Forrest and other upper-class southerners.

1870s ◆ The Molly Maguires, a secret organization of Irish coal miners in Pennsylvania, committed acts of sabotage and terrorism against mining companies.

1882–1968 ◆ Nearly 5,000 African Americans (mostly men) died when they were lynched by mobs or smaller groups of white Americans.

1919 ◆ The First Red Scare occurred when a series of bombs were detonated and letter bombs were intercepted. Anarchists and communists were accused of the bombing campaign.

September 1920 ◆ A bomb was detonated on Wall Street in New York City. Thirty-five people were killed, and hundreds were injured.

1930s ◆ The Second Red Scare occurred.

1938 ◆ The House Un-American Activities Committee was formed to investigate allegations of subversion.

1940 ◆ The Smith Act was passed, making advocacy of the violent overthrow of the government a federal crime.

1941–1945 ◆ The administration of President Franklin Delano Roosevelt established a War Relocation Authority, and the U.S. Army was tasked with moving ethnic Japanese to internment facilities on the West Coast and elsewhere. More than 100,000 ethnic Japanese, approximately two-thirds of whom were American citizens, were forced to relocate to the internment facilities.

1950s ◆ The Third Red Scare occurred when Republican senator Joseph McCarthy of Wisconsin held a series of hearings to counter fears of spying by communist regimes, primarily China and the Soviet Union. The hearings reflected and encouraged a general fear that communists were poised to overthrow the government and otherwise subvert the "American way of life."

November 1, 1950 ◆ Two Puerto Rican nationalists attempted to assassinate President Harry Truman at Blair House in Washington, D.C.

March 1965 ◆ Members of the United Klans of America shot to death Viola Liuzzo in Alabama and wounded a traveling companion.

October 1969 ◆ A bomb in Chicago destroyed a monument dedicated to the Chicago police. The leftist Weatherman group was responsible for the attack.

1970s ◆ The Weathermen—renamed the Weather Underground Organization—committed at least 40 bombings.

Mid-1970s ◆ The leftist Symbionese Liberation Army (SLA), a violent terrorist cell, gained notoriety for several high-profile incidents.

1970s ◆ The Black Liberation Army is suspected to have committed a number of attacks in New York and California. They are thought to have been responsible for numerous bombings, ambushes of police officers, and bank robberies.

1975–1983 ◆ The Puerto Rican independentista group Armed Forces for National Liberation (Fuerzas Armadas de Liberación Nacional, or FALN) was responsible for approximately 130 bombings.

1978–1995	During a 17-year FBI manhunt, Theodore Kaczynski, also known as the Unabomber, killed three people and injured 22 in a series of bombings.
May 1981	A Brink's armored car was robbed in Nyack, New York, by former members of the Weather Underground Organization, Students for a Democratic Society, the Black Panther Party, the Republic of New Africa, and the Black Liberation Army.
April 1983	A neo-Nazi group calling itself The Order initiated a campaign of violence, hoping to foment a race war in the United States.
August 1992	Hurricane Andrew struck the United States, causing an estimated $26 billion in damage. The federal response was widely criticized as unnecessarily slow and uncoordinated.
February 1993	In the first terrorist attack on the World Trade Center, a large vehicular bomb exploded in a basement parking garage; it was a failed attempt to topple one tower into the other.
1994–1996	The Aryan Republican Army (ARA) operated in the Midwest and robbed 22 banks in seven states.
April 19, 1995	The Alfred P. Murrah Federal Building was bombed in Oklahoma City, Oklahoma, by rightist extremists.
April 24, 1996	During the administration of President Bill Clinton, the United States passed its first comprehensive counterterrorism legislation, titled the Anti-Terrorism and Effective Death Penalty Act. The purpose of the Anti-Terrorism Act was to regulate activity that could be used to mount a terrorist attack, provide resources for counterterrorist programs, and punish terrorism.
January 1997	A bomb was detonated in Atlanta, Georgia, at a family health clinic that provided abortion services. A second bomb was detonated soon thereafter.
January 16, 1997	Two bombs exploded at an abortion clinic in Sandy Springs, Georgia. The Army of God was suspected.
August 7, 1998	Two car bombs exploded at the U.S. embassies in Nairobi, Kenya, and Dar es Salaam, Tanzania, killing more than 250 and wounding about 5,000.
August 19, 1999	A mass shooting by two teenagers at Columbine High School in Littleton, Colorado, resulted in 13 dead and 21 wounded.
September 11, 2001	Terrorists hijacked four airliners. Two of the planes were crashed into the Twin Towers of the World Trade Center in New York City, causing them to collapse. One plane was crashed into the army section of the Pentagon building. The final plane crashed into rural Pennsylvania.
October 26, 2001	President George W. Bush signed new antiterrorist legislation into law. It was labeled the "Uniting and Strengthening America by Providing Appropriate Tools Required to Intercept and Obstruct Terrorism Act of 2001," commonly known as the USA PATRIOT Act.
Late 2001	Letters containing anthrax were sent through the U.S. postal system in the New York and Washington, D.C., areas.
May 2002	The U.S. Department of Justice expanded the FBI's surveillance authority. New guidelines were promulgated that permitted field offices to conduct surveillance of religious institutions, websites, libraries, and organizations without an a priori (before the fact) finding of criminal suspicion. A broad investigative net was cast, using the rationale that verifiable threats to homeland security must be detected and preempted.
August 2005	Hurricane Katrina struck the Gulf Coast of the United States, causing nearly 2,000 deaths and billions of dollars in destruction. The federal response was strongly criticized as unacceptably inadequate and ponderous. State emergency response authorities in Louisiana were also widely criticized as images of thousands of refugees were broadcast globally.
May 2008	The U.S. Department of Justice's inspector general released an extensive report revealing that FBI agents had complained repeatedly since 2002 about harsh interrogations conducted by military and CIA interrogators.
June 1, 2009	In Little Rock, Arkansas, Abdulhakim Mujahid Muhammad opened fire at an armed forces recruiting office, killing Private William Long and wounding Private Quinton Ezeagwula.
November 5, 2009	At the Fort Hood military base in Killeen, Texas, Major Nidal Malik Hasan shot and killed 13 people and wounded at least 30. He was an army psychiatrist.

December 25, 2009 ◆	Umar Farouk Abdulmutallab, a Nigerian national, attempted to detonate an explosive compound hidden in his underwear aboard an aircraft flying from Amsterdam to Detroit.
May 1, 2010 ◆	Times Square in New York City was evacuated after the discovery of a car bomb.
April 25–28, 2011 ◆	More than 350 tornadoes struck in the United States, the largest recorded tornado outbreak in history. During the outbreak, on April 27, an EF4 tornado struck the Tuscaloosa–Birmingham, Alabama, area. Its path of destruction stretched more than 80 miles, and more than 60 people were killed.
May 2, 2011 ◆	Al-Qaeda founder Osama bin Laden was killed during a raid by U.S. Navy SEAL commandos in Abbottabad, Pakistan.
October 2012 ◆	"Superstorm" Sandy struck the United States' eastern seaboard. The hurricane came ashore in the densely populated urban northeastern region of the United States, causing widespread damage.
December 14, 2012 ◆	At Sandy Hook Elementary School in Newtown, Connecticut, 20 children and six adults were killed by gunman Adam Lanza.
April 15, 2013 ◆	Two bombs were detonated at the crowded finish line of the Boston Marathon. Three people were killed, and more than 260 were wounded, many severely.
May 20, 2013 ◆	An EF5 tornado with wind velocity exceeding 200 miles per hour touched down near Moore, Oklahoma. Hundreds of residents were injured, and 23 were killed.
June–July 2013 ◆	The British newspaper *The Guardian* published a series of articles reporting covert surveillance operations coordinated by the U.S. National Security Agency (NSA). The operations involved the acquisition of European and U.S. telephone metadata and Internet surveillance.
September 16, 2013 ◆	A mass shooting occurred at the Washington Navy Yard in Washington, D.C., in which lone-gunman Aaron Alexis shot and killed 12 people before being shot and killed by police.
May 23, 2014 ◆	Six people were killed and seven wounded by a gunman in Isla Vista, California. The gunman committed suicide after the shootings.
June 18, 2015 ◆	Racial supremacist Dylann Roof shot and killed nine at the Emanuel African Methodist Episcopal Church. Roof attended a prayer meeting with the victims before the assault, and all victims were African Americans.
October 1, 2015 ◆	Eight students and an instructor were shot and killed at Umpqua Community College in Roseburg, Oregon.
November 27, 2015 ◆	Robert Lewis Dear attacked a Planned Parenthood clinic in Colorado Springs, Colorado. Three people were shot to death, including a University of Colorado police officer.
December 2, 2015 ◆	Armed assailants killed and wounded dozens of victims during an assault on a regional center in San Bernardino, California.
January 2, 2016 ◆	Armed activists and militia adherents occupied the headquarters of the Malheur National Wildlife Refuge in Oregon. Calling themselves Citizens for Constitutional Freedom, the activists demanded that the federal refuge be "returned" to private ownership. They initially occupied the site to protest the convictions and imprisonment of two ranchers convicted of arson.